Mathematics for Digital Science 2

SCIENCES

Computer Science,
Field Director – Jean-Charles Pomerol

Operational Research and Decision,
Subject Head – Patrick Siarry

Mathematics for Digital Science 2

Digital Information

Gérard-Michel Cochard
Mhand Hifi

iSTE

WILEY

First published 2025 in Great Britain and the United States by ISTE Ltd and John Wiley & Sons, Inc.

ISTE Ltd
27-37 St George's Road
London SW19 4EU
UK

www.iste.co.uk

John Wiley & Sons, Inc.
111 River Street
Hoboken, NJ 07030
USA

www.wiley.com

Library of Congress Control Number: 2025930227

British Library Cataloguing-in-Publication Data
A CIP record for this book is available from the British Library
ISBN 978-1-78945-195-5

ERC code:
PE1 Mathematics
PE1_17 Mathematical aspects of computer science
PE1_20 Control theory, optimisation and operational research

Contents

Preface

The term “digital” relates to information and communication sciences and technologies, covering areas such as computing, digital electronics and telecommunications. Over the last 50 years, advances in these fields, through discoveries, developments and applications, have grown exponentially. Such progress has profoundly transformed human activities, making the transition to an “all digital” world a significant economic and political issue. Concurrently, professions related to digital technology are constantly evolving. It is noteworthy that these technological advances are grounded in a substantial mathematical foundation. Thus, any engineer or researcher aiming to drive innovation must possess a strong knowledge of mathematics.

Many students opt for computer science-related courses early on without first acquiring the essential mathematical foundations they might need in the future. While they may excel as technicians, they might not be as well prepared to become effective engineers or researchers.

It should be recognized that computer science cannot be fully understood without a solid mathematical foundation. Advances in digital technologies have been closely linked to mathematical support. The pioneers of computing were primarily mathematicians: Alan Turing, Claude Shannon, John von Neumann and Charles Babbage, often referred to as the “grandfather of computing”. They could never have designed computers without their mathematical expertise. Other historical figures such as Euclid, who formulated the first algorithm, George Boole, the father of binary algebra, Ada Lovelace, a pioneer in creation the first computer programs, Grace Hopper, who developed the first language compiler, and Margaret Hamilton, a key figure in software engineering, were also mathematicians. Today, areas such as artificial intelligence, managing large datasets, and information security are

Mathematics for Digital Science 2,
by Gérard-Michel COCHARD and Mhand HIFI.

at the center of computer science research. Once again, it is evident that without a robust mathematical foundation, innovation in these areas would be challenging.

The digital mathematics courses offered are designed to provide the fundamental mathematical knowledge essential for mastering and advancing digital technologies. While the book primarily targets university and engineering students, it also serves as a resource for IT professionals seeking to enhance their mathematical skills as part of their professional development in the field.

The three-volume work compiles the lessons taught to numerous generations of students in the first two university cycles, specifically in bachelor's/master's degrees in Computer Science or in Computer Methods Applied to Business Management. The first volume covers the essential mathematical foundations for approaching digital technologies. The second – and present – volume focuses on digital information, while the third volume is devoted to data analysis and optimization.

Chapters 1 and 2 of this second volume address the representation of information (such as numbers, texts, images, sounds, videos, labels) in the form of combinations of 0s and 1s. This binary representation facilitates the processing, analysis and transmission of information using computer resources.

Chapter 3 addresses the classification of communication signals, both analog and digital, and introduces Fourier series analysis for periodic signals.

Chapter 4 presents mathematical tools for signal analysis, including z-transforms, Fourier transforms and Laplace transforms.

Chapter 5 is dedicated to the digitalization of analog signals, while Chapters 6 and 7 focus on signal processing, specifically modulation and filtering.

The following four chapters are devoted to digital images. Chapter 8 presents the characteristics of digital images. Chapter 9 describes the main techniques of 2D computer graphics. Chapter 10 covers the fundamental elements of image processing and analysis, while Chapter 11 explains the essential concepts of image compression.

Finally, Chapter 12 describes the main algorithms of numerical analysis used to find approximate solutions to problems that are difficult to solve.

To achieve the goal of mastering the application of mathematical results, demonstrations are presented whenever they are accessible to the reader, who is assumed to be familiar with basic knowledge. In some cases, the validity of the results is accepted without detailed proof, with references to more advanced works where such demonstrations can be found.

Each chapter includes numerous examples to illustrate the concepts presented. These examples are generally elaborated in detail to ensure a better understanding for the reader.

Finally, each topic is presented exhaustively, starting with the initial definitions and hypotheses. While many readers may already be familiar with these basic concepts, this thorough approach is intended to assist those with potential knowledge gaps, reducing the need to consult additional sources.

January 2025

1

Representation of Numbers

CONCEPTS COVERED IN THIS CHAPTER.–

This first part, devoted to the digital representation of information, primarily addresses numbers. Starting with numbers is appropriate because computers were originally designed for calculation purposes.

Since computers operate using only 0s and 1s, it is essential to understand the binary system and the conversions between the decimal and the binary systems, as well as vice versa. In addition, two other systems commonly used in computing, the octal system and the hexadecimal system, are discussed.

The approach to the machine representation of numbers involves working within a fixed range, as computers cannot recognize infinity. This necessitates methods for representing negative numbers and rational numbers. The common solutions used for these cases are described.

References: [STA 07, VEL 19, LIP 83].

Using a computer assumes that the information transmitted to it is in a format or language understandable by its components. Transforming data from a human-readable format into a format understandable by the computing machine is called "information coding".

Modern computers work exclusively with the binary system, which uses a reduced alphabet of $\{0, 1\}$. This raises the question of how to represent of common types of information (such as numbers, texts, images, sounds and videos), in binary. The ellipsis suggests that in the future, it might also be possible to represent other sensory experiences, such as smells and tastes, in binary form.

Mathematics for Digital Science 2,
by Gérard-Michel COCHARD and Mhand HIFI.

The study will begin with the coding of numbers, as computers were originally designed for computation. The following chapter will then cover the coding of other types of information, including texts, images, sounds and videos.

1.1. Representation of numbers

The decimal system is commonly used for manipulating numbers today, but historically, some civilizations used sexagesimal numbering systems. This approach is still evident in the measurement of time and angles: 1 hour = 60 minutes; 1 minute = 60 seconds, and angles are measured such that 1 flat angle = 180° = 3×60°.

Roman numerals were not very suitable for calculation, serving primarily for counting. The decimal system, which is quite ancient (originating in Egypt in the 3rd millennium BCE and later refined with the invention of 0), is based on the 10 symbols: 0, 1, 2, 3, 4, 5, 6, 7, 8, 9, and uses positional numbering. As a result, the number 1234 is distinct from the number 3124, despite using the same digits, because the position of each digit is important for determining the number's value.

1.1.1. Position numbering

EXAMPLE 1.1.– Consider the number 1234. This number can be understood as follows: 1×1,000 + 2×100 + 3×10 + 4, or 1234 = 1×103 + 2×102 + 3×101 + 4×100.

In the decimal system, which as a base of 10 (with 10 digits), the position of each digit in a number corresponds to a power of 10.

In the binary system, which uses only the digits 0 and 1 and has a base of 2, a decimal number like 1234 is written as: 10011010010. This binary representation is understood as follows:

$$10011010010 = 1\times2^{10} + 0\times2^{9} + 0\times2^{8} + 1\times2^{7} + 1\times2^{6} + 0\times2^{5} + 1\times2^{4} + 0\times2^{3} + 0\times2^{2} + 1\times2^{1} + 0\times2^{0}.$$

Just like in the decimal system, the position numbering is used in the binary system. In the binary system, the base is 2, and the digits are 0 and 1, which are referred to as bits in computing. The binary can be verified to correspond to the number 1234 (from Example 1.1) by using powers of 2 (see Figure 1.1).

It should be noted that $2^{10} = 1{,}024$ is referred to as a kilobyte (KB) in computing, commonly abbreviated as "K". In this context, 1,024 bytes are equivalent to 1 KB.

n	0	1	2	3	4	5	6	7	8	9	10
2^n	1	2	4	8	16	32	64	128	256	512	1024

Figure 1.1. *Powers of 2*

Unlike human thinking, computers cannot process infinite numbers and are limited in the range of numbers they can manipulate. To represent an n-bit number, the largest integer that can be represented is $2^n - 1$. To represent the largest integer with m decimal digits in the binary system, which is $10^m - 1$, how many n bits are needed to represent this number in a computer?

To represent the largest integer with m decimal digits, the $10^m - 1 = 2^n - 1$ must be satisfied (assuming this is possible). Taking logarithms gives:

$$m \times \log_{10}(10) = n \times \log_{10}(2).$$

Since $\log_{10}(10) = 1$, this simplifies to

$$m = n \times \log_{10}(2)$$

Solving for n:

$$\mathrm{n} = \frac{m}{\log_{10}(2)} = \mathrm{m} \times \frac{\ln(10)}{\ln(2)} \approx 3.32 \times \mathrm{m}$$

It therefore takes approximately $(3.32 \times m)$ bits to represent an integer of m decimal digits in the binary system.

1.1.2. *The binary system*

As noted previously, the binary system is based on two symbols (bits): 0 and 1. Unlike the decimal system, operations in binary are extremely simple. Some may recall learning decimal multiplication tables in elementary school!

This is because, in the binary system, basic operations such as addition and multiplication are much more direct and less complex compared to the decimal system. Binary addition is performed bit by bit (see column 1 of Figure 1.2), where

$0 + 0 = 0, 0 + 1 = 1, 1 + 0 = 1$ (without carry) and $1 + 1 = 10$ (with carry). Binary multiplication follows simple rules (see column 2 of Figure 1.2), where $0 \times 0 = 0$, $0 \times 1 = 0$, $1 \times 0 = 0$ and $1 \times 1 = 1$.

addition	multiplication
0 + 0 = 0	0 x 0 = 0
0 + 1 = 1	0 x 1 = 0
1 + 0 = 1	1 x 0 = 0
1 + 1 = 10	1 x 1 = 1

Figure 1.2. *Addition and multiplication in binary*

This simplicity in binary operations is one of the reasons why computers use this system for internal calculations. This allows for efficient data manipulation and rapid calculations at the electronic circuit level.

The primary problem to solve is converting between decimal and binary systems. To convert a decimal number into a binary number, a straightforward method based on powers of 2 is used.

Consider the example of covering the decimal number 1234 to binary. Start by finding the largest power of 2 less than or equal to 1234, which is $1{,}024 = 2^{10}$. Subtract 1,024 from 1234 to get 210. Next, find the largest power of 2 closest to 210, which is $128 = 2^7$. Subtract 128 from 210 to get 82. Continue this process by finding the closest power of 2 to each resulting value and performing the corresponding subtractions. The sequence of operations is shown in Figure 1.3.

1234	
1024	2^{10}
210	
128	2^7
82	
64	2^6
18	
16	2^4
2	
2	2^1
0	

Figure 1.3. *Decimal-to-binary conversion*

The powers of 2 present in this decomposition are 2^1, 2^4, 2^6, 2^7 and 2^{10} and the missing powers of 2 are 2^0, 2^2, 2^3, 2^5, 2^8 and 2^9. To write the binary number, use the position numbering rule: place a 1 for each present power of 2 and a 0 for each absent power of 2. Thus, the binary representation is:

$$(1\times2^{10}) + (0\times2^9) + (0\times2^8) + (1\times2^7) + (1\times2^6) + (0\times2^5) + (1\times2^4) + (0\times2^3) + (0\times2^2) + (1\times2^1) + (0\times2^0) = (10011010010)_2$$

When necessary, the subscript 2 indicates a number in the binary system, and the subscript 10 indicates a number in the decimal system.

Now let us move on to the reverse problem: converting a binary number into a decimal number. It is actually simpler because you just write it down with the position numbering rule and do the math. Therefore, to convert this binary number to decimal, each bit of the binary number is multiplied by 2 raised to the power of its position, starting from zero at the far right.

$$(1001010010)_2 = (1\times 2^{10}) + (0\times 2^9) + (0\times 2^8) + (1\times 2^7) + (1\times 2^6) + (0\times 2^5) + (1\times 2^4) + (0\times 2^3) + (0\times 2^2) + (1\times 2^1) + (0\times 2^0) = 1024 + 128 + 64 + 16 + 2 = (1234)_{10}$$

Now, consider how to represent a rational number in base 2. Take the example of the decimal number $(14.80078125)_{10}$. To express this in base 2 follow this rule for a binary number of the form $(0.xyzt\ \ldots)_2$ where $x, y, z, t, \ldots$ are bits:

$$0 + x\times 2^{-1} + y\times 2^{-2} + z\times 2^{-3} + t\times 2^{-4} + \cdots$$

This method can be used to convert a decimal number with fractions into base 2. Applying this method will illustrate the conversion process for the example provided.

EXAMPLE 1.2.– Consider the decimal number 14.80078125 in base 10. First, convert the integer part $(14)_{10}$ into binary, which is $(1110)_2$. For the decimal part $(0.80078125)_{10}$, perform successive multiplications by 2,while recording the integer part of the result at each step:

$2\times 0.80078125 = 1.6015625$ →1 (integer part: 1)
$2\times 0.6015625 = 1.203125$ →1 (integer part: 1)
$2\times 0.203125 = 0.40625$ →0 (integer part: 0)
$2\times 0.40625 = 0.8125$ →0 (integer part: 0)
$2\times 0.8125 = 1.625$ →1 (integer part: 1)

$2 \times 0.625 = 1.25$ →1 (integer part: 1)
$2 \times 0.25 = 0.5$ →0 (integer part: 0)
$2 \times 0.5 = 1$ →1 (integer part: 1)

The decimal part can be expressed as a sum of fractions:

$$(0.80078125)_{10} = \frac{1}{2} + \frac{1}{4} + \frac{1}{32} + \frac{1}{64} + \frac{1}{256}$$

This corresponds to the binary representation $(0.11001101)_2$.

Combining this with the integer part $(14)_{10} = (1110)_2$, the full binary representation of $(14.80078125)_{10}$ is:

$$(14.80078125)_{10} = (1110.11001101)_2$$

The calculation performed in the previous example will be useful later. By converting the decimal number $(14.80078125)_{10}$ to binary, the complete binary representation is $(1110.11001101)_2$. This conversion illustrates the general method for converting decimal numbers to binary numbers using the position numbering rule. This method will be explored in more detail later.

1.1.3. *Octal system and hexadecimal system*

In the octal system , the base is 8, so there are eight digits available: 0, 1, 2, 3, 4, 5, 6 and 7. Converting a decimal number to octal follows the same principle as converting to binary. Using powers of 8, the number $(1234)_{10}$ can be represented in octal base. Given that

$$8^0 = 1, \quad 8^1 = 8, \quad 8^2 = 64, \quad 8^3 = 512, \quad 8^4 = 4096,$$

Break down $(1234)_{10}$ as follows:

$$1234 = 2 \times 512 + 3 \times 64 + 2 \times 8 + 2 \times 1$$

Converting to octal:

$$1234 = 2 \times 8^3 + 3 \times 8^2 + 2 \times 8^1 + 2 \times 8^0 = (2322)_8$$

The conversion process involves starting with the binary representation of the number, grouping the into packets of 3 from the right, and then using the table in Figure 1.4 to convert each group to octal. For example:

$$(1234)_{10} = (10011010010)_2$$

Group the binary digits:

$$(10011010010)_2 = (010\,011\,010\,010)_2$$

Convert each group to octal:

$$010_2 = 2_8, \quad 011_2 = 3_8, \quad 010_2 = 2_8, \quad 010_2 = 2_8$$

Thus,

$$(1234)_{10} = (2322)_8$$

binary	000	001	010	011	100	101	110	111
octal	0	1	2	3	4	5	6	7

Figure 1.4. *Binary-to-octal conversion*

This conversion method can also be applied for other bases, such as the hexadecimal system (base 16), using the corresponding powers.

In the hexadecimal system, the base is 16 and the 16 digits used are 0, 1, 2, 3, 4, 5, 6, 7, 8, 9, A, B, C, D, E and F. The letters A to F represent the values 10 to 15, respectively, and are considered as digits.

To convert a decimal number to hexadecimal base, use powers of 16. By applying the corresponding powers of 16 $(16^0, 16^1, 16^2, 16^3, \ldots)$, the decimal number $(1234)_{10}$ can be decomposed into its hexadecimal representation:

– Find the largest power of 16 less than 1234:

$$16^2 = 256, \quad 16^1 = 16, \quad 16^0 = 1$$

– Decompose 1234 (and the successive values) using these powers:

$$1234 = 4 \times 256 + 210; \quad 210 = 13 \times 16 + 2$$

Here, 13 in hexadecimal is represented as D.

– Combine these results:

$$1234 = 4 \times 16^2 + 13 \times 16^1 + 2 \times 16^0$$

Thus,

$$(1234)_{10} = (4D2)_{16}$$

The conversion process can also be done using the table in Figure 1.5. First, convert the number to binary. By grouping the binary digits into groups of four from the right, the representation of $(1234)_{10}$ in binary is:

$$(1234)_{10} = (10011010010)_2$$

To ensure each group has four digits, add leading zeros if necessary:

$$(1234)_{10} = (10011010010)_2 = (\mathbf{0}100\ 1101\ 0010)_2 = (4D2)_{16}$$

binary	0000	0001	0010	0011	0100	0101	0110	0111
hexadecimal	0	1	2	3	4	5	6	7

binary	1000	1001	1010	1011	1100	1101	1110	1111
hexadecimal	8	9	A	B	C	D	E	F

Figure 1.5. *Binary-to-hexadecimal conversion*

1.2. Representation of numbers in a machine

So far, the discussion has focused mainly on positive integers in different number systems. However, it is also important to consider negative numbers and rational numbers (fractions) in this context.

1.2.1. *Machine representation of negative numbers*

Several approaches can be considered:

1) Sign-magnitude representation: a sign bit (0 or 1) is used to indicate whether a number is positive (0 in front) or negative (1 in front). For example,

– $(-1234)_{10}$ can be represented as $(\mathbf{1}10011010010)_2$ in sign-magnitude notation, where "**1**" denotes the negative sign and "10011010010" represents the magnitude.

– In this representation, a number of *n* bits has *n*-1 bits dedicated to the magnitude.

A drawback of this approach is the representation of “0”, which can be ambiguous: $(0)_{10}$ →(**0** 0) and $(0)_{10}$ →(**1** 0). It is important to distinguish between the binary representation of numbers and their machine representation, which includes bit (→).

2) One’s complement representation: in one’s *complement* notation, positive numbers are represented in their usual binary form (1 or 0) with a leading sign bit of “0”. For negative numbers, the bitwise complement (inversion of all bits) of the positive number’s binary representation is used, with a leading sign bit of “1”. For example,

– $(+1234)_{10}$ in binary is $(+10011010010)_2$. In one’s complement notation, this becomes $(\mathbf{0}10011010010)_2$ (with a sign bit).

– $(-1234)_{10}$ is represented by inverting the bits: $(101100101101)_2$

This method involves representing positive numbers in their normal binary form and converting negative numbers by flipping all bits of their positive counterparts.

The same difficulty with “0” arises, as it has two representations. Another issue is the imbalance between positive and negative numbers. For example, if a machine can represent $(1234)_{10}$, it is expected that it should also be able to represent $-(1234)_{10}$. In other words, the system should have the possibility to represent an equal number of positive and negative numbers.

In a machine, numbers are typically coded using n bits (often 32 or 64). For simplicity, consider a machine with 8-bit coding (7 bits for the absolute value). The largest positive number that can be represented is $2^7 - 1 = 127$. Its representation is therefore $(\mathbf{0}1111111)_2$.

In the one’s complement method, -127 is represented by $(10000000)_2$ i.e. zero. This representation effectively indicates that there are not as many positive numbers as negative numbers in one’s complement notation: positive numbers range from 0 to 127, while negative numbers range from -1 to -126. In Figure 1.6, negative numbers are obtained by taking the one’s complement of positive numbers, which involves flipping all the bits. This results in a representation, where 1 denotes the sign bit for negative values.

One solution could be to use a representation where 0 is denoted by a number containing only zeros. However, this approach presents challenges for simple operations like addition. For example, consider $(127)_{10}-(127)_{10}$ in one’s complement:

Representation: $(127)_{10}$ is represented as 01111111_2, and $-(127)_{10}$ is represented as 10000000_2 in one's complement.

Performing the addition:

$$01111111_2 \text{ (for } 127_{10}) + 10000000_2 \text{ (for } -127_{10}) = 11111111_2$$

In one's complement, 11111111_2 could be interpreted as -0. This result is problematic because it introduces an additional representation for zero and does not easily fit into a straightforward arithmetic interpretation.

decimal	representation 1
127	01111111
126	01111110
125	01111101
124	01111100
....	....
4	00000100
3	00000011
2	00000010
1	00000001
0	00000000
-1	11111110
-2	11111101
-3	11111100
-4	11111011
....	....
-124	10000011
-125	10000010
-126	10000001
-127	10000000

Figure 1.6. *Representation 1*

3) Two's *complement representation:* the two's *complement*, which involves adding 1 to the one's complement of a number. Let us illustrate this with the previous example and complete the table (see representation 2 in Figure 1.7).

For two's complement:

Representation: 127_{10} in binary is $\rightarrow 01111111_2$. To find -127_{10} in two's complement: first, get the one's complement of 127_{10}: 10000000_2; second, add 1 to this result: $10000000_2 + 1 = 10000001_2$.

Performing the subtraction:

$$127_{10} - 127_{10} = 01111111_2 - 10000001_2 = 00000000_2 = 0_{10}$$

The last carry on the left is ignored. With the two's complement method, it is possible to encode the number -128_{10} but not $+128_{10}$. In two's complement representation on n bits, the numbers can be encoded in the range from 0 to $2^{n-1} - 1$ and from –1 to -2^{n-1}

The two's complement representation allows additions to be performed "normally" by following the rules of binary addition.

Therefore, on 8 bits (where the left carry is always ignored), consider the following example.

decimal	representation 1	representation 2
127	01111111	01111111
126	01111110	01111110
125	01111101	01111101
124	01111100	01111100
----	----	----
4	00000100	00000100
3	00000011	00000011
2	00000010	00000010
1	00000001	00000001
0	00000000	00000000
-1	11111110	11111111
-2	11111101	11111110
-3	11111100	11111101
-4	11111011	11111100
----	----	----
-124	10000011	10000100
-125	10000010	10000011
-126	10000001	10000010
-127	10000000	10000001
-128		10000000

Figure 1.7. *Representation 2*

EXAMPLE 1.3.– Representation of the numbers $(60)_{10}$ and $(8)_{10}$:

$$(60)_{10} = (34)_{10} + (26)_{10} \rightarrow (00100010)_2 + (00011010)_2 = (00111100)_2 \rightarrow (00111100)_2 \rightarrow (60)_{10}$$

$$(8)_{10} = (34)_{10} - (26)_{10} \rightarrow (00100010)_2 + (11100110)_2 = (00001000)_2 \rightarrow (00001000)_2 \rightarrow (8)_{10}$$

Of course, the result of an operation on binary numbers must fit within the n bits used by the computer. When an arithmetic operation provides a result that exceeds the capacity of n bits, an overflow occurs.

To illustrate this, consider the following example:

EXAMPLE 1.4.– Let the number be $(280)_{10}$.

$$(280)_{10} = (120)_{10} + (160)_{10} \rightarrow (01111000)_2 + (10100000)_2 = (00011000)_2 \rightarrow (24)_{10}$$

The *result is incorrect* because $(280)_{10}$ cannot indeed be represented in 8 bits as $(00011000)_2$.

In general, if a number exceeds the representational range of a given number of bits, systems may produce an error or overflow message.

1.2.2. Representation of rational numbers

Rational numbers , also known as fractions, are numbers expressed as the ratio of two integers. They can be represented in several ways: as fractions, as decimal numbers with an integer part and a decimal part, or using scientific notation, which uses powers of 10. The focus here is on the latter representation, also known as *floating point* notation.

In scientific notation , a number is written as $s.m \times 10^p$, where s is the sign (+ or −), m is the mantissa (which is always positive), and p is the exponent (which can be positive or negative). For example, consider the rational number $x = \frac{22}{7}$, whose approximate decimal value is 3.1428 ... (an approximation of π). The scientific notation for this number could be:

$$x = 3.1428 \approx +0.31428 \times 10^{1}$$

where + is the sign, 0.31428 is the mantissa, and 1 is the exponent. Alternatively, this number can be represented as:

$$x = 3.1428 \approx +31{,}428 \times 10^{-6}$$

where + is the sign, 31,428 is the mantissa, and -6 is the exponent.

As demonstrated, the mantissa can vary depending on the value of the exponent. To standardize the representation, it is common to impose the following condition for the mantissa: $0.1 \leq m < 1$. This means that the mantissa must be between 0.1 (inclusive) and 1 (exclusive). This rule ensures a standardized representation of numbers in scientific notation and facilitates calculations and comparisons between numbers. With this mantissa rule, the digit before the decimal point is the first significant digit in floating-point notation. This notation is useful for expressing very small numbers or very large numbers. For example, the mass of an electron is approximately $0.9 \times 10\text{-}^{-30}$ kg, and the mass of the Sun is approximately 0.2×10^{31} kg.

Using the mantissa rule , express the decimal part of the mantissa that is after the decimal point, which always starts with 0. For clarity, this part of mantissa is denoted as m', where $m' = 10m$. The range of values for m' is therefore $[1, 10[$.

Consider a representation where the mantissa is expressed using four digits and the exponent uses three digits. In this configuration, the range of positive numbers that can be represented is from 0.1000×10^{-999} to 0.9999×10^{999}. This range covers an extremely large number of values, yet each number can be represented with just four digits for the mantissa and three digits for the exponent, totaling seven digits including the exponent sign.

However, an exponent can be negative. In such cases, if the usual representation is maintained, the first digit will determine the sign of the exponent. If t bits are allocated to store the exponent, then $t - 1$ bits will be used for the absolute value of the exponent.

To handle negative exponent, a fixed value e, called the *excess* or bias , can be added to the actual exponent to ensure that the encoded exponent is always positive. For t bits reserved for the exponent, *the excess e* is given by 2^{t-1}. For instance, with $t = 8$ bits for the exponent, the excess $e = 2^{7} - 1$. This allows the exponent to be represented in the form 0 to 254. Therefore, an exponent range of $[-127, +127]$ is mapped to the encoded range $[0, 254]$. This method enables the representation of both negative and positive exponents using t bits.

To recap, a number in scientific notation (or floating-point representation) is written as:

$$s.m' \times 2^E$$

where s represents the sign ($+$ or $-$), m' is the mantissa satisfying $1 \leq m' < 10$ in *decimal*, e is the fixed excess or bias, and E is the encoded exponent, which is the true exponent plus the excess (E = real exponent + e).

This scientific notation can be translated into binary representation as follows (see Figure 1.8):

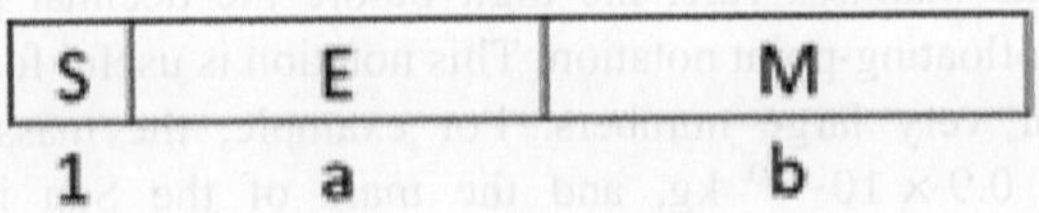

Figure 1.8. *Structure of a rational number*

– S on 1 bit represents the sign, with 0 indicating a positive number and 1 indicating a negative number (i.e. 0 for + and 1 for –).

– E on a bits represents the exponent. It is calculated using the rule $E = p + e$, where p is the real exponent of the number and e is a fixed excess (bias) chosen to ensure that E is always positive. The number of bits allocated to the exponent is a.

– M on b bits represents the mantissa, which contains the significant part of the number. The number of bits allocated to the mantissa is b.

By combining these elements, the complete binary representation of a rational number in a machine is obtained. The IEEE 754 standard specifies the following rules for the binary representation:

a) Formats: there are two formats, single precision and double precision.

b) Representation: a number is coded in the form $S.1, M' \times 2^c$. The bit 1 preceding the decimal point of the mantissa is not explicitly encoded (it is called the hidden bit); only the part $M'$of the mantissa is encoded. The exponent is encoded as $E = c + e$, where e is a fixed excess.

c) The single precision format: it has the following structure (see Figure 1.9). The exponent is encoded on 8 bits, with an excess of $e = (127)_{10} = (01111111)_2$.

The mantissa (excluding the hidden bit) M' is encoded on 23 bits. The total number of bits for encoding is 32 bits.

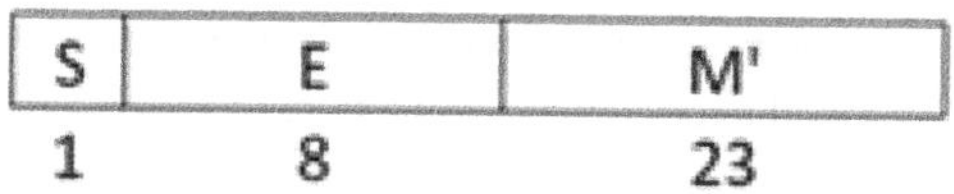

Figure 1.9. *IEEE 754 single precision*

d) The double precision format: it has the following structure (see Figure 1.10).

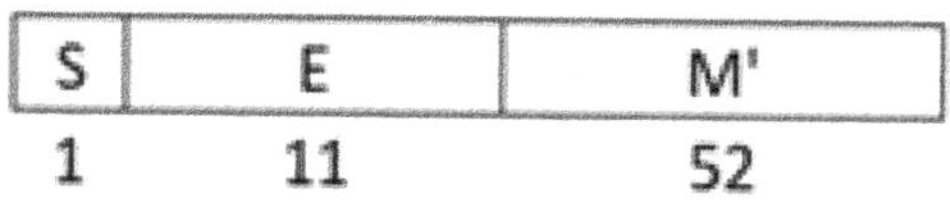

Figure 1.10. *IEEE 754 double precision*

The exponent is coded using 11 bits, with an excess of $e = (1023)_{10} = (01111111111)_2$; the mantissa (excluding the hidden bit) M' is encoded using 52 bits. The enter number is encoded using 64 bits.

Let us consider an example using the single precision format.

EXAMPLE 1.5.– To encode the number $(14.80078125)_{10}$ in single precision format, follow these steps:

– Determine the sign bit: since the number is positive, the sign bit S is 0.

– Convert the number to binary: the integer part $(14)_{10}$ converts to $(1110)_2$, and the decimal part $(0.80078125)_{10}$ converts to $(0.11001101)_2$ through binary fraction conversion.

– Combining these two parts, the binary representation is as follows:

$(14.80078125)_{10} = (1110.11001101)_2$

– Normalize the binary number: normalize (1110.11001101)2 to scientific notation in binary provides the following result:

$(1.11011001101)_2 \times 2^3$.

– Determine the exponent: to find the binary value of the exponent, first calculate the exponent E by adding the true exponent to the fixed excess e. For this example, the true exponent p is 3, the fixed excess e is 127. Hence, the encoded E is:

$$E = p + e = (00000011)_2 + (01111111)_2 = (10000010)_2 = 130.$$

– Determine the mantissa: the reduced mantissa M', omitting the hidden bit, is:

$M' = 11011001101000000000000_2$ (padded with zeros to fit 23 bits)

– Combine the components: the sign bit $S = 0$, the exponent $E = (10000010)_2$, and the mantissa $M' = 11011001101000000000000_2$. Therefore, the 32-bit single precision representation of $(14.80078125)_{10}$ is:

0 10000010 11011001101000000000000

This representation is the binary encoding of the number $(14.80078125)_{10}$ in single precision format according to the IEEE 754 standard.

2

Media Representation

CONCEPTS COVERED IN THIS CHAPTER.–

Following the discussion on digital representation in Chapter 1, the digital representation of text characters is explored using various standard encodings. The chapter then covers the principles of image encoding, as well as the digitalization of sound and video.

It concludes with an examination of tagging codes, particularly barcodes and QR codes.

References: [HOL 94, LIP 83, MAC 87, POL 76, STA 07, THO 14].

2.1. Character coding

Texts contain a significant amount of information and are composed of characters from various alphabets. To manipulate them on a machine, they must be converted into sequences of 0s and 1s. The characters in texts are highly diverse, including letters, numbers, as well as a wide array of signs and symbols.

Figure 2.1. *Morse code*

Mathematics for Digital Science 2,
by Gérard-Michel COCHARD and Mhand HIFI.

Morse code: a significant example of coding with only two symbols is Morse code (see Figure 2.1), which predates the computer y a significant margin. Morse code was invented in 1832 for use with the telegraph. Its invention is often attributed to Samuel Morse, although this is not definitively established. In Morse code, the dot (•) corresponds to 0 and the dash (–) to 1. Although theoretically applicable to computing, Morse code has not been used in modern computing due to limitations such as the restricted character set (e.g. lack of distinction between upper and lower case letters and absence of special symbols) and the inconsistent coding (variable number of dots and dashes).

BCD coding: another code specifically designed for representing numbers, treated as characters, is BCD (Binary-Coded Decimal) coding. In this system, each digit is represented by a sequence of 4 bits, as detailed in the correspondence table presented in Figure 2.2. BCD coding provides a standardized method for representing digits in binary form.

digit	0	1	2	3	4	5	6	7	8	9
BCD	0000	0001	0010	0011	0100	0101	0110	0111	1000	1001

Figure 2.2. *BCD code*

To represent the text "1942" (emphasizing that it is treated as text rather than a number for calculations) in BCD coding, the bit sequence 0001 1001 0100 0010 is used. It is important to note that this representation is specific to BCD coding and does not support numerical calculations. It also differs from the binary representation of the number 1942, which is $(11110010110)_2$ in binary. BCD coding is a 4-bit code, allowing for the encoding of $2^4 = 16$ different characters. However, it only uses 10 of these combinations to represent decimal digits. This leaves six unused combinations, which is insufficient to represent all letters of the Latin alphabet in addition to the numbers. Given that there are 26 letters and 10 digits, at least 36 different characters are needed. Therefore, a code with at least 6 bits is required, as $2^6 = 64$, which is greater than 36.

ITA code: another coding system is the ITA code (International Telegraph Alphabet), which uses 5-bit sequences. This code can represent 32 different characters. Despite its limited character set, it can transmit both letters and numbers through special sequences. For example, the sequence "11111" signals a series of letters, while "11011" indicates a series of numbers. However, this method is complex and not very practical.

ISO6 code: another code to consider is the ISO6 code, which uses 6-bit sequences. Although standardized, it is rarely used today. The code is defined by a grid where the binary positions are categorized into high-order (the first two most significant bits) and low-order (the remaining four bits). This coding scheme serves as a basis for subsequent coding systems.

ASCII code: the ASCII (American Standard Code for Information Interchange) code is a widely used encoding system that uses 7-bit sequences. It remains in use today, as shown in Figure 2.3.

most significants bits

least significant bits	0 000	1 001	2 010	3 011	4 100	5 101	6 110	7 111
0 0000	NUL	DLE	SP	0	@	P	\	p
1 0001	SOH	DC1	!	1	A	Q	a	q
2 0010	STX	DC2	//	2	B	R	b	r
3 0011	ETX	DC3	#	3	C	S	c	s
4 0100	EOT	DC4	$	4	D	T	d	t
5 0101	ENQ	NA K	%	5	E	U	e	u
6 0110	ACK	SYN	&	6	F	V	f	v
7 0111	BEL	ETB	,	7	G	W	g	w
8 1000	BS	CAN	(	8	H	X	h	x
9 1001	HT	EM	)	9	I	Y	i	y
A 1010	LF	SUB	•	:	J	Z	j	z
B 1011	VT	ESC	+	;	K	[	k	{
C 1100	FF	FS	'	<	L	/	l	|
D 1101	CR	GS	-	=	M	]	m	}
E 1110	SO	RS	.	>	N	^	n	~
F 1111	SI	US	/	?	O	_	o	DEL

Figure 2.3. *ASCII code*

As with the ISO6 code, the ASCII code divides bits into high-order (the most significant 3 bits) and low-order (the least significant 4 bits) segments. However, since computers typically operate with, at least, 8-bit words (bytes), an additional

8-th bit is included. This extra bit can be used for various purposes, such as a parity bit to detect errors. The parity bit is set to 1 if there is an odd number of 1s among the first seven bits, and 0 otherwise. With $2^7 = 128$ possible combinations, ASCII code offers a broader range of characters. It encodes not only letters (both uppercase and lowercase) and numbers but also many special characters. In addition, the first two columns of the ASCII table include control characters used to provide instructions to display or printing devices. Thus, ASCII code provides significant flexibility for representing and communicating various types of characters and commands. With the ASCII code (including the parity bit), the phrase "*Vive les vacances*!"1 would be encoded as follows:

01010110 01101001 11110110 01100101 10100000 01101100 01100101 11110011 10100000 11110110 11100001 01100011 11100001 111011 10 01100011 01100101 11110011 10100000 00100001

With (sp: space)

V: 01010110 i: 01101001 v: 01110110 e: 01100101

sp: 00100000 l: 01101100 e: 01100101 s: 01110011

sp: 00100000 v: 01110110 a: 01100001 c: 01100011

a: 01100001 n: 01101110 c: 01100011 e: 01100101

s: 01110011 sp: 00100000 !: 00100001

However, standard ASCII code is limited in its character set, lacking support for accented letters and offering a fixed number of special characters.

ISO-8859 code: by extending the ASCII code to 8 bits, ISO-8859 code, also known as Latin code, broadens the character range. The 8-th bit allows for the inclusion of additional characters, enabling the encoding of texts in most European languages. Figure 2.4 shows the ISO-8859-15 version. In this system, a byte starting with 0 represents a standard ASCII character, while a byte starting with 1 represents an extended character. This extension accommodates a greater variety of special characters and accented letters.

Unicode: to encode *all* the characters used across the world, more than the 256 possibilities provided by an extended 8-bit encoding are needed. To address this, the Unicode standard was developed. Unicode supports all global alphabets and provides an extensive character set. It includes 144,697 characters, each encoded in

1. Long live the holidays!

the form U+H, where H represents a value expressed in hexadecimal system. This coding system is referred to as a "code point". Unicode is organized into layers, called planes. The first plane, known as Plane 0, includes characters coded between U+0000 and U+FFFF, while the last plane, Plane 16, contains characters coded between U+100000 and U+10FFFF. Each plane allows a specific set of characters. For example, the character U+178D represents a symbol from the Khmer alphabet and is part of Unicode's repertoire. Similarly, the Euro character, encoded as U+20AC, is included in Unicode and can be represented and interpreted consistently by systems that support this universal encoding. Figure 2.5 provides an overview of these two characters.

most significant bits

least significant bits	0 0000	1 0001	2 0010	3 0011	4 0100	5 0101	6 0110	7 0111	8 1000	9 1001	A 1010	B 1011	C 1100	D 1101	E 1110	F 1111
0 0000	NUL	DLE	SP	0	@	P	\	p	PAD	DCS	NBSP	°	À	Ð	à	ð
1 0001	SOH	DC1	!	1	A	Q	a	q	HOP	PU1	¡	±	Á	Ñ	á	ñ
2 0010	STX	DC2	//	2	B	R	b	r	BPH	PU2	¢	²	Â	Ò	â	ò
3 0011	ETX	DC3	#	3	C	S	c	s	NHB	STS	£	³	Ã	Ó	ã	ó
4 0100	EOT	DC4	$	4	D	T	d	t	IND	CCH	€	Ž	Ä	Ô	a	ô
5 0101	ENQ	NA K	%	5	E	U	e	u	NEL	MW	¥	µ	Å	Õ	å	o
6 0110	ACK	SYN	&	6	F	V	f	v	SSA	SPA	|	¶	Æ	Ö	æ	ö
7 0111	BEL	ETB	,	7	G	W	g	w	ESA	EPA	§	·	Ç	×	ç	÷
8 1000	BS	CAN	(	8	H	X	h	x	HTS	SOS	š	ž	È	Ø	è	ø
9 1001	HT	EM	)	9	I	Y	i	y	HTJ	SGCI	©	¹	É	Ù	é	ù
A 1010	LF	SUB	*	:	J	Z	j	z	VTS	SCI	ª	º	Ê	Ú	ê	ú
B 1011	VT	ESC	+	;	K	[	k	{	PLD	CSI	«	»	Ë	Û	ë	û
C 1100	FF	FS	'	<	L	/	l	|	PLU	STS	¬	Œ	Ì	Ü	ì	ü
D 1101	CR	GS	-	=	M	]	m	}	RI	OSC		œ	Í	Ý	í	ý
E 1110	SO	RS	.	>	N	^	n	~	SS2	PM	®	Ÿ	Î	Þ	î	þ
F 1111	SI	US	/	?	O	_	o	DEL	SS3	APC	¯	¿	Ï	β	ï	ÿ

Figure 2.4. *ISO 8859 code*

Figure 2.5. *Some Unicode characters*

To facilitate the practical use of Unicode, various encoding schemes have been developed, such as UTF-8, UTF-16 and UTF-32. Among these, UTF-8 is the most widely adopted and used. UTF-8 enables the encoding of Unicode characters in a flexible manner. For characters in the range U+0000 to U+007F (i.e. 128 characters), UTF-8 uses a single byte. This byte follows the format 0xxxxxxx, which is fully compatible with standard ASCII code. The 7 bits following the initial zero bit directly correspond to ASCII characters.

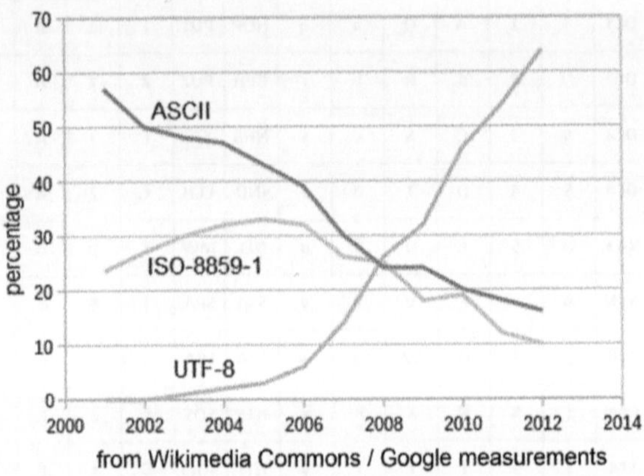

Figure 2.6. *Evolution of codes on the Web. For a color version of this figure, see www.iste.co.uk/cochard/mathematics2.zip*

For characters encoded from U+0800 to U+FFFF, UTF-8 uses a 3-byte encoding scheme. The first byte begins with 1110xxxx, where the first four bits are reserved for the format and the remaining four bits contain data. The following two bytes start with 10xxxxxx, each providing 6 bits of data.

Finally, for characters encoded from U+10000 to U+10FFFF, UTF-8 uses a 4-byte encoding scheme. The first byte starts with 11110xxx, where the first five bits are reserved for the format and the remaining three bits contain data. The following three bytes begin with 10xxxxxx, each providing 6 bits of data. Over time,

UTF-8 has gained widespread popularity due to its compatibility with standard ASCII code and its efficiency in encoding a wide range of characters. Figure 2.6 shows the increasing adoption of UTF-8 in Unicode usage.

2.2. Image coding

Coding images are essential for displaying them on a screen. Unlike a continuous surface, a digital screen is composed of pixels, which are discrete points arranged in a grid. Each pixel has a specific position and a key attribute: its color. To understand color coding, it is important to comprehend some fundamental concepts related to color.

2.2.1. *Representation of digital colors*

To represent colors digitally, various coding systems are used. The most commonly used system is the RGB (Red, Green, Blue) model, which uses the three primary colors of light – red, green and blue – to provide a wide range of colors. Each component is usually represented by an integer value ranging from 0 to 255, where 0 indicates no presence of that component and 255 signifies full intensity. By combining intensities of these three components, a specific color can be achieved.

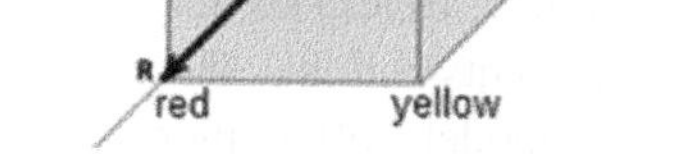

Figure 2.7. *RGB model*

Thus, using the vectors **R** , **V** and **B** , which correspond to the axes of the color cube shown in Figure 2.7 and have a range of 256, a specific color can be represented as a linear combination of these vectors: $\mathbf{C} = x\mathbf{R} + y\mathbf{V} + z\mathbf{B}$, where x, **V** and z are integers between 0 and 255, representing the respective intensities of red, green and blue required to achieve the desired color. This system allows for the representation of a total of $256^3 = 16{,}777{,}216$ different colors, which is more than sufficient to cover the range perceptible by the human eye.

2.2.2. *Scanning an image*

Digitalizing an image involves converting it into a digital format. This process requires overlaying a grid of vertical and horizontal lines spaced at a certain distance d (see Figure 2.8). The intersections of these lines define the pixels, which are the fundamental units captured during scanning. A smaller distance d between the lines results in higher image quality, as it increases the number of pixels and thus the level of detail in the digital representation. This distance d determines the resolution of the image, which refer to the amount of details and the clarity with which the image is represented digitally.

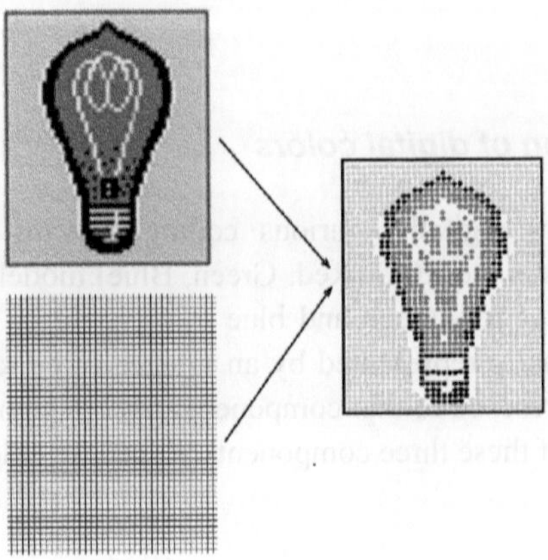

Figure 2.8. *Rasterizing an image*

In addition to the RGB model, other color coding systems are used. For example, the CMYK (Cyan, Magenta, Yellow, Black) model is primarily used in printing and represents colors in terms of pigments and ink. More complex color spaces, such as the Lab model or the HSV model, offer more accurate and flexible color representations by considering aspects such as human perception of color.

2.2.3. *Image quality*

The quality of the resulting image depends on the density of the scanning grid. Generally, a smaller distance d between lines yields better image quality. A reduced distance d enables the capture of more details, making the image more precise. The distance d between pixels defines the image's *resolution*. Higher resolution indicates more pixels per unit area, allowing finer details and a sharper image.

Image resolution can be expressed in various ways, such as the total number of pixels or the number of pixels per inch (dpi, dot per inch). Higher resolution is generally preferred for images intended for high-quality media, such as computer monitors or high-resolution printers, as it allows for better detail and color reproduction. However, higher resolutions can also require more storage space and greater processing power for manipulation and display.

To create an image file, the pixels are scanned from left to right and top to bottom, with each pixel assigned a color. Using the RGB system as described previously, each pixel is represented by 3 bytes: the first byte for the red component (ranging from 0 to 255), the second byte for the green component (0 to 255), and the third byte for the blue component (0 to 255). Consequently, each pixel is encoded using 24 bits. This can result in large file sizes, making it necessary to apply compression algorithms.

Several compression algorithms define image formats. Common formats for the web include JPEG format, which performs lossy compression (it is suitable for complex images such as photographs), GIF or PNG formats , which use lossless compression (often used for simpler images such as diagrams or illustrations). Compression techniques help reduce file size while preserving acceptable quality for image display.

For binary image, which consists of only two colors (such as black and white), pixel encoding is simpler: 0 represents black and 1 represents white. In this case, each pixel is encoded using just 1 bit.

In grayscale images, pixels are represented on a scale from 0 (black) to 255 (white). In this case, each pixel is encoded using 8 bits, which correspond to the *dose (intensity)* of gray.

The size of an uncompressed image (i.e. the file size) depends on the number of pixels and the image type (binary, grayscale, color). For example:

– A binary image with a resolution of 1366×768 pixels has a size of 1,049,088 bits, or approximately 128 KB.

– A grayscale image of the same resolution is eight times larger, totaling approximately 1,024 KB.

– A color image with a resolution of 1366×768 pixels, using the RGB system with 24-bit color depth (8 bits per color) has a size of $1366 \times 768 \times 24 =$ 25,178,112 bits, or approximately 3 MB.

2.3. Sound coding

What is sound? It is a physical phenomenon caused by sound vibrations that propagate through the surrounding environment.

These vibrations create pressure variations that are perceived by the human ear as sounds. Sound pressure is a continuous analogue quantity.

To digitalize sound, this analogue quantity must be covered into a digital form. Measuring sound pressure over time results in a continuous curve on a coordinate system, with time in seconds on the x-axis and pressure in Pascals on the y-axis (see Figure 2.9).

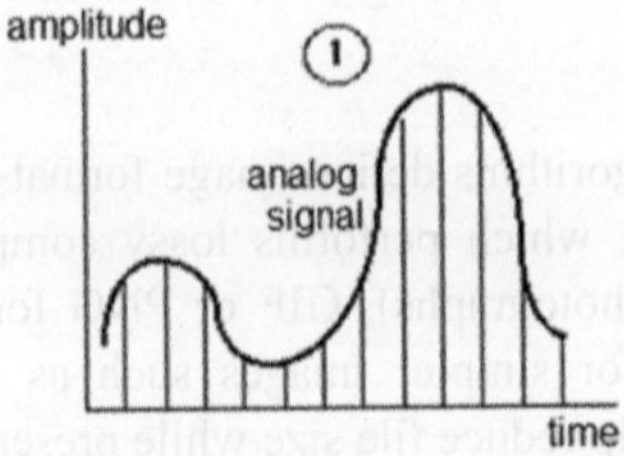

Figure 2.9. *Analog signal*

Similar to image representation, the curve can be simplified into a set of points (or vertical bars) by measuring the pressure at regular intervals (see Figure 2.10). This process is known as *sampling*. Sampling occurs at a specific frequency, denoted as f, which represents the number of measurements taken per second.

Thus, $f = \frac{1}{D}$, where D is the time interval between each measurement. A higher frequency results in points that are very close together, while a lower frequency results in points that are spaced further apart. Frequency is measured in Hertz (Hz). For example, a frequency of 11 kHz (11,000 Hz) means that 11,000 measurements of the sound signal are made per second.

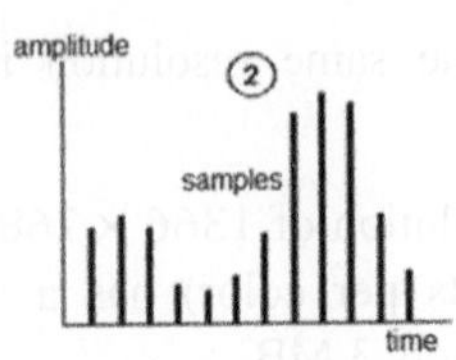

Figure 2.10. *Analog signal sampling*

The next step involves quantifying the sampled sound to modify the scale of the amplitudes for digital encoding. Typically, a maximum height is chosen based on a power of 2. For simplicity, consider using a scale from 0 to 7, which corresponds to $2^3 = 8$ possible values.

The scale in Pascals is therefore replaced with this new graduation (see Figure 2.11). However, it is important to note that the points (or vertical bars) do not correspond exactly to any of the eight tick marks. The ordinates are then adjusted to best match one of the eight available graduations. This adjustment may introduce a slight distortion known as *quantization noise*. In this example, the graduation is coarse, but with a finer scale, the distortion would likely be negligible and might go unnoticed by the human ear.

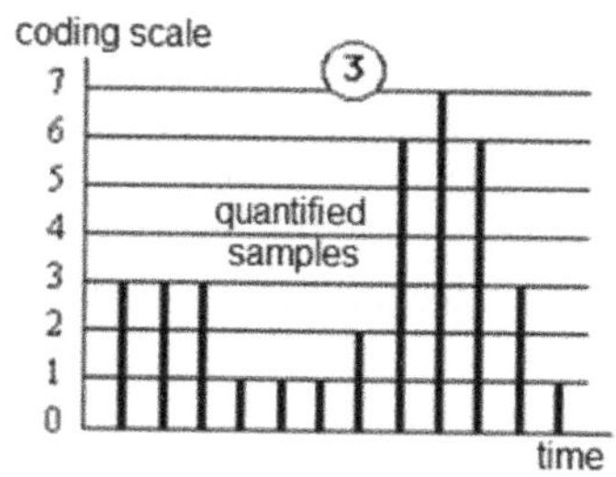

Figure 2.11. *Scale adjustment*

To finalize the process, coding is performed by assigning each point (or each bar) its corresponding value. In this example, with only eight possible values, the sound will be encoded on 3 bits. Consequently, the resulting sound file will consist of a sequence of bit triples (see Figure 2.12).

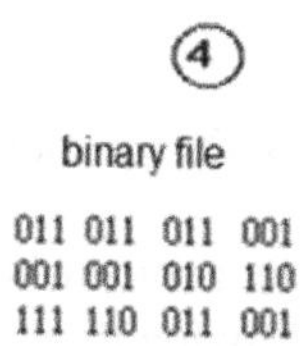

Figure 2.12. *Coding*

The human ear can detect frequencies between 20 Hz and 20,000 Hz. According to the Nyquist–Shannon theorem in signal theory, the sampling frequency should be at least twice the maximum audible frequency. Therefore, a sampling rate of at least 40,000 Hz is recommended to ensure good quality in digital audio.

The frequency range of the human voice is between 100 Hz and 3,400 Hz. Consequently, when digitalizing the human voice, a sampling frequency higher than 7,800 Hz is recommended. For telephone lines, the sampling frequency is generally set to 8,000 Hz. "Telephone" quality corresponds to a sampling frequency of 11,000 Hz. "Radio" quality corresponds to a sampling frequency of 22,000 Hz, "HiFi" or "audio CD" quality requires a sampling frequency of 44,000 Hz.

Returning to coding and the digital measurement scale, reducing quantization noise requires a sufficiently wide range of values. In general, sound sample are encoded using 8-bit or 16-bit coding, which is significantly higher than in the previous example. "Telephone" quality uses 8-bit coding, while "radio" quality also uses 8-bit coding but can include two signals in stereo. "Audio CD" quality, on the contrary, uses 16-bit coding with two signals in stereo.

Thus, one minute of "radio" quality sound (mono) results in a file size calculated as follows: with a sampling frequency of 22,000 Hz, 8-bit coding, and 60 seconds, the size is $22{,}000 \times 60 \times 8 = 10{,}560{,}000$ bits, or approximately 1.26 MB. This illustrates the need compression techniques for audio, similar to those used for images.

2.4. Video coding

A video is a sequence of images synchronized with sound.

The techniques for digitalizing images and sound are the same as those previously mentioned. However, there are two additional difficulties:

1) There are many images in a video, making it impractical to represent them all in full detail due to the enormous file size, even with efficient compression techniques.

2) The images must be synchronized with sound.

Excessive technical details will be avoided, focusing instead on the principle of one method defined by the MPEG (Moving Pictures Expert Group) standard.

This method classifies images into three types (see Figure 2.13): *intra-images*, *predicted images* and *interpolated images*.

Intra-images are lightly compressed and serve as reference points (usually one intra-image for 10 successive images). These are the only ones directly digitized from the original.

Predicted images are created by encoding and compressing the differences with the intra-images or the previous predicted images (usually one predicted image every three images).

Interpolated images are generated as intermediate frames based on the preceding images.

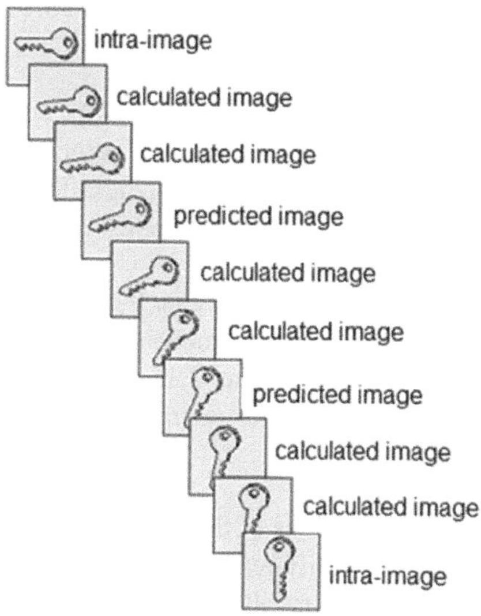

Figure 2.13. *MPEG image types*

2.5. Tagging codes

Tagging codes are 1D or 2D image representations designed for reading by optical readers. They are commonly used to store information related to objects, such as identifiers, serial numbers, traceability data, web links and prices. These codes facilitate quick information retrieved through optical readers, including barcode scanner and mobile application.

2.5.1. *Figures*

If we have a check book, we will notice that check numbers are marked with special numbers represented by vertical bars.

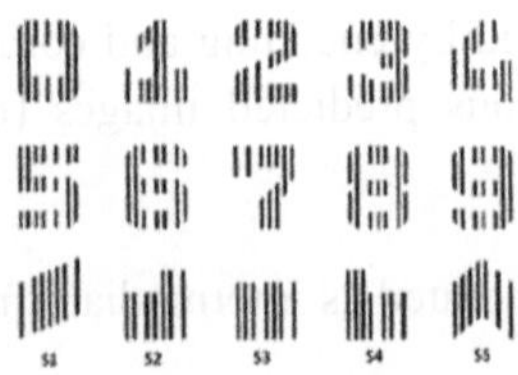

Figure 2.14. *Numbers defined by bars*

These numbers use the CMC7 code (see Figure 2.14), which was invented in the 1960s. On closer inspection, we will see that the vertical bars are arranged with either long or spacings, totaling six different spacings. These varying spacings are used for coding that can be read by magnetic or optical readers.

In CMC7 code, in addition to numeric digits, there are five symbols: S1 to S5. Automatic reading of these numbers and symbols involves detecting long spacings based on the coding specified in Figure 2.15. This coding allows automatic readers – whether magnetic or optical – to interpret the information encoded in the CMC7 numbers and symbols.

Figure 2.15 shows the long spacings of the CMC7 code, represented by "x". The letters a, b, …, f denote the six different spaces between the seven bars. These longer spacings in the vertical bars allow automatic readers to recognize and interpret the numbers and symbols encoded in the CMC7 code.

	a	b	c	d	e	f
0			x	x		
1	x				x	
2		x	x			
3	x		x			
4	x			x		

	a	b	c	d	e	f
5				x	x	
6			x		x	
7	x	x				
8		x			x	
9		x		x		

	a	b	c	d	e	f
S1	x					x
S2		x				x
S3			x			x
S4				x		x
S5					x	x

Figure 2.15. *CMC7*

2.5.2. *Bank cards*

Bank cards are now the most popular means of payment, especially for online transactions. Equipped with a chip (invented by Roland Moreno and Michel Ugon in 1974), the card contains information readable in plain text, encrypted data and a secret code (which is not readable). Figure 2.16 shows the information that can be read from a card (note that this is an expired card!).

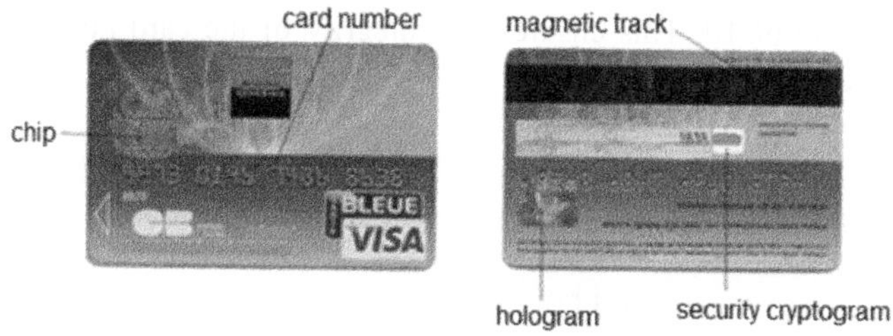

Figure 2.16. *Credit card*

Now, let us examine a card number coding, which is divided into three distinct parts (see Figure 2.17).

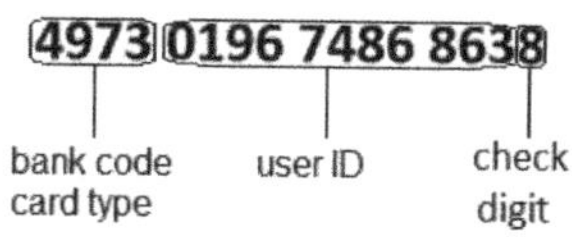

Figure 2.17. *Coding of a bank card*

The check digit is crucial for validating the card number. The validation process involves the following steps: digits in odd positions are multiplied by 2, and the sum of the digits of these results is calculated. Simultaneously, the sum of the digits in even positions (without multiplication) is also calculated (see Figure 2.18).

	odd rank	even rank
	4 x 2 = 8	9
	7 x 2 = 14 --> 1 + 4 = 5	3
	0 x 2 = 0	1
	9 x 2 = 18 --> 1 + 8 = 9	6
	7 x 2 = 14 --> 1 + 4 = 5	4
	8 x 2 = 16 --> 1 + 6 = 7	6
	8 x 2 = 16 --> 1 + 6 = 7	6
	3 x 2 = 6	8
sum	47	43
total sum	90	

Figure 2.18. *Card number validation*

For the card number to be considered valid, the total of the sum of the digits in odd positions (after multiplying by 2) plus the sum of the digits in even positions must be a multiple of 10. This ensures the integrity of the card number and helps detect possible entry errors or fraud.

Figure 2.19. *Entering the code received constitutes confirmation*

When paying by bank card online, a secure connection to the merchant's site is required. Generally, the merchant site requests for the card number, expiration date and security code. It then connects to the bank, which requires transaction confirmation. This confirmation can be performed in several ways, with the most commonly used methods being the following:

1) SMS code: the bank sends a specific code via SMS to an individual's cell phone, which they must enter on the merchant's website (see Figure 2.19). This code must be entered to confirm the transaction.

2) Online account confirmation: the bank sends an SMS asking the buyer to log in to their online account management application. This requires a connection code previously established by the buyer. After logging in, the buyer can review the transaction and either confirm or reject it.

These additional confirmation methods are implemented to enhance the security of online bank card transactions. They are designed to prevent fraud by requiring extra validation from the cardholder.

2.5.3. *Barcodes*

Barcodes are widely used and consist of a visual representation in the form of a 1D image composed of parallel bars. The binary coding is straightforward: a black bar represents a 1, and a white bar represents a 0. The sequence of consecutive 0s or 1s determines the thickness of the bars.

978-3-8416-7393-0

Figure 2.20. *A barcode*

The barcode follows the EAN (European Article Number) standard, ensuring compatibility within the European Union. It is divided into two distinct parts by the code 01010 and is framed by the sequence 101 at both the beginning and at the end (see Figure 2.20).

In the EAN standard, there are two types of barcodes: one with 8 digits and one with 13 digits. The structure of these barcodes is defined in Figure 2.21. EAN barcodes provide unique identification of products and items, facilitating their traceability and management in retail systems. Widely used in the retail, logistics and distribution industries, these barcodes offer an efficient method for quickly scanning and reading product information.

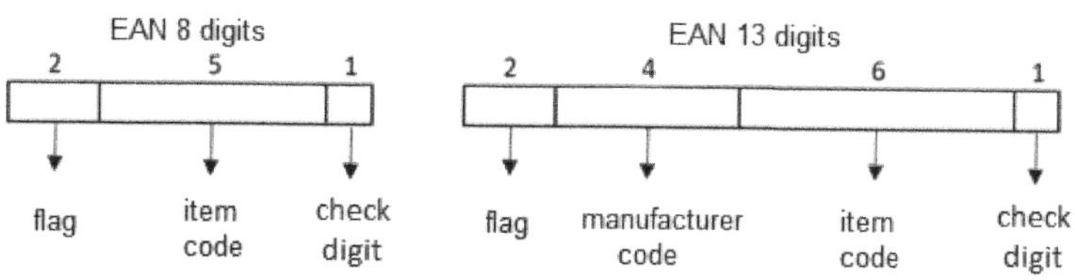

Figure 2.21. *EAN barcode*

Let us examine the method for calculating the check digit in a 13-digit barcode, as illustrated previously. The process is as follows:

1) Each digit of the barcode is multiplied by a multiplicative factor that alternates between 1 and 3, resulting in a series of products corresponding to each digit.

2) Calculate the sum S of these products.

3) Perform a congruence calculation modulo 10 using the equation $Y \equiv S\,[10]$. This involves dividing S by 10 and finding the remainder.

4) If the remainder is 0, the check digit is 0.

5) Otherwise, the check digit is calculated as 10 minus the remainder obtained in the previous step (i.e. $10 - Y$).

This method of calculating the check digit helps verify the integrity of the barcode and detect potential entry or reading errors. It also enhances reliability during the automatic reading of barcodes by scanners and readers (for more details on this process, see Figure 2.22).

digits	9	7	8	3	8	4	1	6	7	3	9	3
multiplicative factor	1	3	1	3	1	3	1	3	1	3	1	3
result	9	21	8	9	8	12	1	18	7	9	9	9
sum	120											

Figure 2.22. *Integrity check*

In Figure 2.22, $Y \equiv 120$ [10] is shown, indicating that Y is congruent to 120 modulo 10. This congruence results in a remainder of 0 [10], making the check digit in this example 0.

2.5.4. *QR codes*

The QR code , which stands for Quick Response, is a two-dimensional barcode developed in Japan in 1994 by Denso, a Toyota subcontractor. It consists of a square image with black and white pixels, each representing a bit of information, known as modules. Designed to be decoded by a specific reader, the QR code is commonly used to provide quick access to a website, usually containing a URL. The QR code has been standardized under the ISO-18004, ensuring compatibility across various readers and systems.

The ability of QR codes to store a large amount of information in a compact format has made them highly popular in various fields, including advertising, marketing, product tracking, event tickets and mobile payments. By scanning the code with a smartphone or compatible reader, users can quickly access the embedded information. This makes QR codes a practical tool for the rapid exchange of information in today's connected world.

Figure 2.23. *Wikipedia QR code (see:* https://en.wikipedia.org)

A QR code is a square matrix composed of modules, which are small squares that can be either black or white. These modules represent binary information, with a white module corresponding to 0 and a black module corresponding to 1.

Versions of a QR code

QR code comes in several versions, each designed to encode different amounts of information, which determines the matrix size. For instance, version 1 features a 21×21 matrix, while version 40 has a 177×177 matrix. With each version increment, the matrix size increases by 4, so the matrix for version 2 is 25×25.

Type of information in a QR code

A QR code includes both encoded information and redundant data. This redundancy enables the recovery of the information intact, even though part of the QR code is damaged, through an error correction process.

This process is divided into four levels:

– Level L: allows information to be read even though 7% of it is illegible.

– Level M: tolerates up to 15% of illegible information.

– Level Q: tolerates up to 25% of illegible information.

– Level H: the highest level of error correction, enabling information to be read even if 30% of it is illegible.

In this case, the higher the error correction level, the greater the amount of data required to be stored in the QR code. This allows for flexibility in balancing storage capacity with the level of robustness against read errors.

Information included in a QR code

The amount of information a QR code can store is influenced by factors such as the code version, error correction level and encoding mode. There are four encoding modes:

1) Numeric mode: represents numbers from 0 to 9. This mode encodes numeric character efficiently to maximize the space available in the QR code.

2) Alphanumeric mode: represents alphanumeric characters using uppercase letters, numbers from 0 to 9, and special symbols such as $, %, *, +, -, '.', /, : , and the space character (sp). It provides greater flexibility for encoding varied information.

3) Byte mode: used by default, this mode allows character encoding according to the ISO-8859-1 standard and is compatible with UTF-8 encoding, allowing the representation of a wide range of characters.

4) Kanji mode: specifically designed for Japanese characters, this mode ensures they are encoded and displayed correctly.

Each encoding mode provides specific advantages in terms of efficiency and compatibility with various types of information. The combination of the QR code's version, error correction level and encoding mode determines its storage capacity and its resistance to reading errors.

Encoding a QR code

The encoding of a QR code involves converting each character of information into a sequence of bits according to one of the four encoding modes. Once encoded, the textual information is represented as a series of bytes. The sequence of bytes is then processed to generate redundancy using the so-called Reed–Solomon error correction process, based on the chosen error correction level.

An optical reader scans both the useful and redundant information. If errors are detected in the useful information, the reader uses the redundant data to perform corrections.

The amount of redundant information is greater at higher error correction levels, such as level H, compared to lower levels such as L. As a result, for a given QR code version, the length of useful information decreases with higher error correction levels. For instance, in binary mode, version 40 can encode up to 2,953 characters (8 bits) at level L, while version 1 can encode only 7 characters at level H. Therefore, the choice of encoding mode, error correction level and QR code version determines both the amount of useful information that can be stored and the code's error correction capacity.

The sequence of bytes obtained by encoding and error correction is arranged in the QR-code matrix according to a specific pattern. For larger QR codes, this sequence of bytes is divided into blocks, which can be interleaved to optimize the distribution of data. In addition, a masking system is used to improve readability and reduce the effects of patterns that could interfere with scanning. The masking process adds complexity to the placement of data with the QR code matrix. The QR code reader must be capable of identifying the mask applied during the encoding process to correctly interpret the information.

Composition of a QR code

A QR code (see Figure 2.24) includes several positioning elements to facilitate reading and orientation. These elements include the following:

– Three square positioning blocks: located at the top left, top right and bottom left of the QR code. These blocks help determine the code's orientation during scanning. They are separated from the rest of the matrix by a white strip called a separator.

– Two dotted lines, known as timing lines: these lines connect the three positioning blocks and assist in synchronizing the reading process.

– An additional marker pattern: for versions greater than 1, there is a smaller alignment block located at the bottom right of the QR code. This (marker) pattern helps in detecting the QR code version.

– A dark module: positioned next to the bottom left positioning block it serves as a reference point to aid in reading the QR code.

These positioning and reference elements are important for ensuring accurate and precise reading of the QR code, as well as for assisting optical readers in recognizing and interpreting the information contained within. In addition, areas reserved for the format information are presented in the QR code, highlighted in blue in Figure 2.24.

The encoding mode used in the QR code is recorded using 4 bits, which enables easy identification. The values corresponding to each mode are as follows:

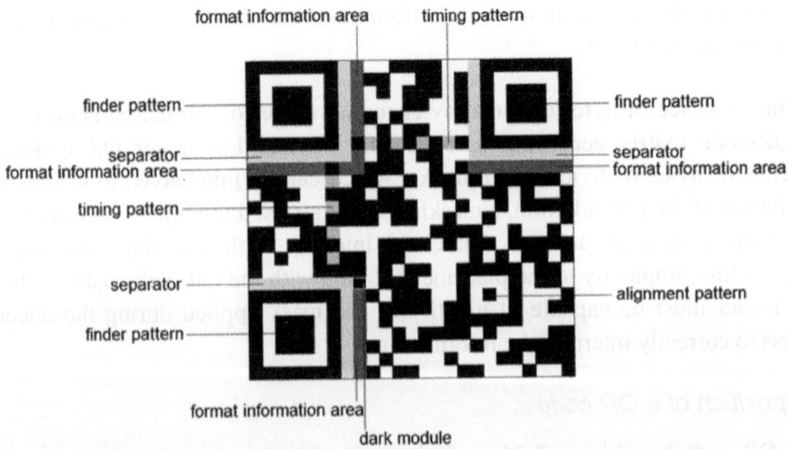

Figure 2.24. *Structure of a QR code. For a color version of this figure, see www.iste.co.uk/cochard/mathematics2.zip*

– Numeric mode: 0001.

– Alphanumeric mode: 0010.

– Byte mode: 0100.

– Kanji mode: 1000.

These values enable QR code readers to quickly identify the encoding mode used and accurately interpret the data contained in the code.

Decoding a QR code

To make understanding QR codes easier, an example will be provided, as their specifications can be quite complex. For a thorough explanation, readers are encouraged to refer to the ISO-18004 standard.

EXAMPLE 2.1.– Consider the example of the QR code shown in Figure 2.25, which is a 25 × 25 matrix corresponding to version 2. To gain further information, the format data, shown in dark blue (1) and light blue (0) in Figure 2.26, will be examined. This format data consists of 15 bits divided into two distinct parts.

Figure 2.25. *QR code to discover*

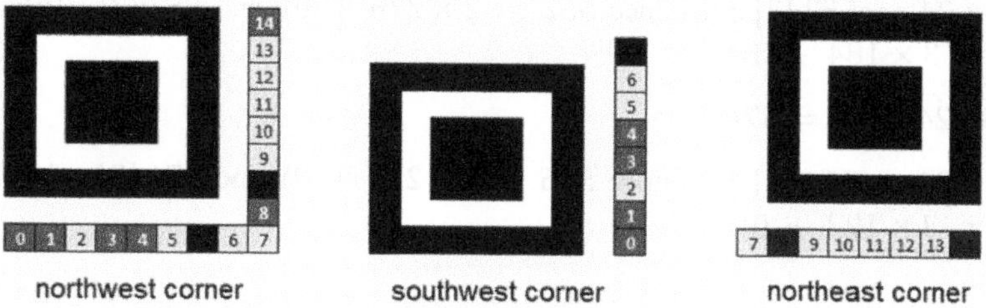

Figure 2.26. *Format data. For a color version of this figure, see www.iste.co.uk/cochard/mathematics2.zip*

The format information can be determined using the table provided in Figure 2.27.

0	1	2	3	4	5	6	7	8	9	10	11	12	13	14
1	1	0	1	1	0	0	0	1	0	0	0	0	0	1

Figure 2.27. *Format information*

According to the standard specifications, the format information codeword corresponds to an error correction level L and a mask 6. In the context of QR codes, a mask is a rule that determines which modules should be inverted (black to white or white to black), while the remaining modules stay unchanged.

Mask 6 is defined by the following rule, based on the coordinates of the modules. In version 2 of the QR code, the module at the top left has coordinates (0, 0), and the module at the bottom right has coordinates (24, 24). For mask rule 6, the rule for a module of coordinates (x, y) is applied as follows:

Calculate $m \equiv xy\,[2]$, calculate $n \equiv xy\,[3]$ and finally determine $z \equiv m + n\,[2]$.

If $z = 0$, the module is inverted; otherwise, it remains unchanged. In practical terms, if the sum $m + n$ is a multiple of 2, the module is inverted according to mask 6. If not, the module remains in its initial state. The mask rule involves division modulo k. In practical terms, if $a \equiv b\ [k]$, a is simply the remainder of the integer division of b by k. Therefore:

– For $23 \times 23 = 529$:

$23 \times 23 = 529\ [2] \equiv 1$ (since $529 = 2 \times 264 + 1$), and $529\ [3] \equiv 1$ (since $529 = 3 \times 176 + 1$)

– For $23 \times 24 = 552$:

$23 \times 24 = 552\ [2] \equiv 0$ (since $552 = 2 \times 276 + 0$), and $552\ [3] \equiv 0$ (since $552 = 3 \times 184 + 0$)

– For $24 \times 24 = 576$:

$24 \times 24 = 576\ [2] \equiv 0$ (since $576 = 2 \times 288 + 0$), and $576\ [3] \equiv 0$ (since $576 = 3 \times 192 + 0$)

This is how mask 6 determines which modules must be modified to ensure a balanced distribution of black and white modules in the QR code, enhancing readability and error detection.

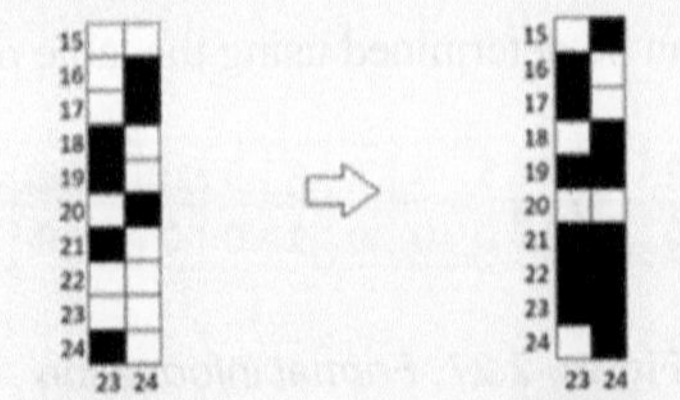

Figure 2.28. *Application of mask 6*

To apply mask 6 (as shown in Figure 2.28):

1) Apply the mask to the modulus (23, 23):

– Calculate $m \equiv 23 \times 23\ [2] = 1$

– Calculate $n \equiv 23 \times 23\ [3] = 1$

– Sum $m + n = 1 + 1 = 2$

– Since $2\ [2] = 0$, the module is inverted.

2) Apply the mask to the modulus (23, 24) and (24, 23):

For both cases:

– Calculate $m \equiv 23 \times 24\ [2] = 0$

– Calculate $n \equiv 23 \times 24\ [3] = 0$

– Sum $m + n = 0 + 0 = 0$

– Since $0\ [2] = 0$, the modules are inverted.

3) Apply the mask to the modulus (24, 24):

– Calculate $m \equiv 24 \times 24\ [2] = 0$

– Calculate $n \equiv 24 \times 24\ [3] = 0$

– Sum $m + n = 0 + 0 = 0$

– Since $0\ [2] = 0$, the module is inverted.

Applying mask 6 twice in a row restores the initial configuration, as the mask is its own inverse.

The QR code matrix contains both "useful" information and redundant data for error correction. For now, let us focus only on the "useful" information, which is recorded starting from the southeast corner (bottom right) of the QR code matrix. The reading pattern proceeds from bottom to top, and then from top to bottom, on columns that are two modules wide (see Figure 2.29). The circulation within a column of width two modules follows the diagram shown in Figure 2.30.

Figure 2.29. *Direction of travel by columns of two modules. For a color version of this figure, see www.iste.co.uk/cochard/mathematics2.zip*

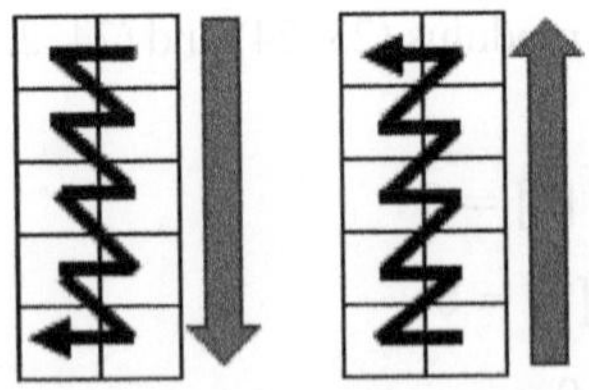

Figure 2.30. *Direction of travel on a column of two modules. For a color version of this figure, see www.iste.co.uk/cochard/mathematics2.zip*

The first block of 4 bits, located in the bottom right of the QR code matrix, represents the encoding mode (see Figure 2.31).

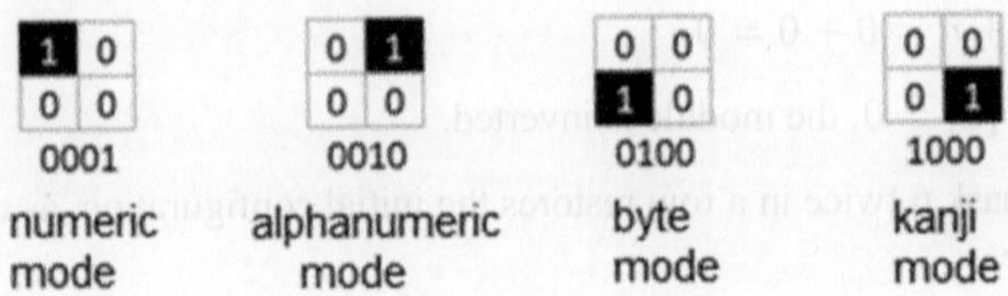

Figure 2.31. *Encoding modes*

EXAMPLE 2.1.– (continued). The representation of the corresponding 4-bit block for the QR-code in question can be observed in Figure 2.32.

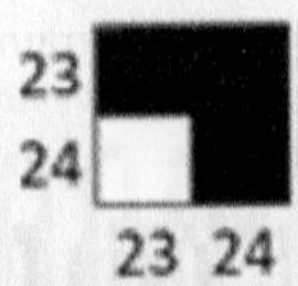

Figure 2.32. *The pattern on which mask 6 is applied*

For the coordinate modules (x, y) involved:

$$23 \times 23 = 529, \ 23 \times 24 = 552, \ 24 \times 23 = 552 \ \text{and} \ 24 \times 24 = 576$$

After applying mask 6 on the QR code, the resulting modification of the modules is shown in Figure 2.33. Mask 6 is a predefined rule that allows certain modules to be altered while leaving others unchanged. In this example, all the modules that have been modified are shown.

Given that the encoding mode used is binary byte mode, the following blocks of information will consist of 8 bits each.

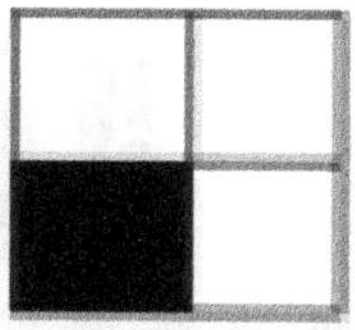

Figure 2.33. *Result of mask application 6*

The next 8-bit block specifies the length of the useful information. After applying mask 6, the result of this block can be seen in Figure 2.34.

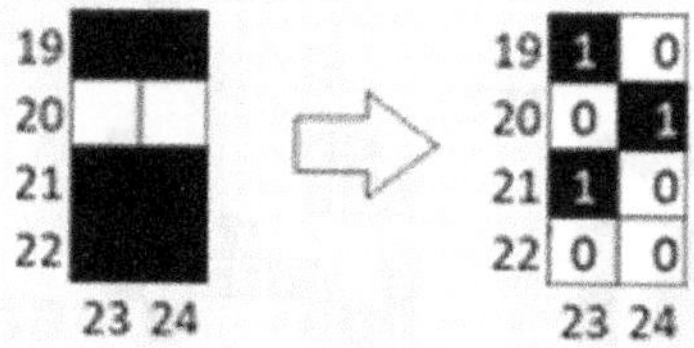

Figure 2.34. *Application of mask 6 to the "length" block*

The 8-bit block, represented by the binary value 00011001, is equivalent to 25 in decimal. This indicates that the length of the useful information is 25 characters.

The next byte, which follows this block, represents the first character of the encoded date. This byte, with the binary value 01101000, corresponds to the character "h" in extended ASCII (see Figure 2.35).

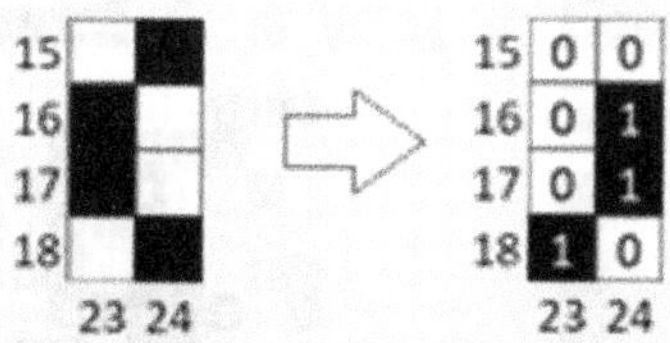

Figure 2.35. *Applying mask 6 to the first character*

This process continues, following the specified navigation direction, until reaching the format information block. The details of this process are shown in Figures 2.36a and 2.36b.

Figure 2.36a. *Before applying mask 6. For a color version of this figure, see www.iste.co.uk/cochard/mathematics2.zip*

Figure 2.36b. *After applying mask 6. For a color version of this figure, see www.iste.co.uk/cochard/mathematics2.zip*

By referring to the extended ASCII code table, the 25 characters can be associated as shown in Figure 2.37.

h	t	t	p	s
01101000	01110100	01110100	01110000	1110011

:	/	/	w	w
01111010	00101111	00101111	01110111	01110111

w	.	u	-	p
01110111	00101101	01110101	00101101	01110000

i	c	a	r	d
01101001	01100011	01100001	01110010	01100100

i	e	.	f	r
01101001	01100101	00101101	01100110	01110010

Figure 2.37. *Transcription of ASCII code*

The information decoded from the QR code is: https://www.u-picardie.fr (the website of the Université de Picardie Jules Verne).

The remaining modules of the QR code are dedicated to error checking and redundancy, using polynomial codes as described in Chapter 10. Manually decoding these modules is highly complex and tedious, so this aspect will not be explored further here. For additional details, refer to the thorough Thonky tutorial available at: https://www.thonky.com/qr-code-tutorial/.

3

Signals and Systems

CONCEPTS COVERED IN THIS CHAPTER.–

In this chapter, the discussion begins with the definition of a signal, along with its characteristics and varieties.

The structure of a signal is then examined, with a particular focus on Fourier analysis for periodic signals.

This chapter concludes with a description of two mathematical tools fundamental to signal analysis and processing: the Dirac distribution and the convolution operation.

References: [PAP 77, MUL 12a, HAY 99, FON 83, COT 97, CHO 17, BRO 82, BER 14, BEL 98].

3.1. Characteristics and varieties

3.1.1. *Introduction*

A signal is a physical medium that carries elements of information. There are various types of signals. For example, sound signals transmit audible information, where the associated physical quantity is the fluctuating air pressure acting on the ear. Similarly, light signals transmit visible information through light waves (electromagnetic waves) captured by the eye. While sound and light signals are common examples, many other types of signals are used across different fields to transmit specific information.

Mathematics for Digital Science 2,
by Gérard-Michel COCHARD and Mhand HIFI.

A signal can be modeled by a simple mathematical representation: it is represented by a function of one or more variables. This mathematical function describes the physical quantity that supports the signal.

EXAMPLE 3.1.– Here are some examples of functions modeling various physical phenomena:

– The function $p(h)$ represents pressure variations p as a function of altitude h, modeling changes in atmospheric pressure at different altitudes.

– The function $T(x, y)$ describes temperature variations T as a function of the coordinates of a point (x, y) on a given flat surface. This can be used to map thermal variations over a specific region.

– The function $f(t)$ describes sound amplitude f as a function of time t, providing a representation of the evolution of sound intensity over time. This function can be used to study and analyze the acoustic properties of a sound signal, such as its envelope, frequency and energy.

Some signals are *deterministic* or *certain*, meaning they are assumed to be known in advance and follow a predictable pattern or function. This is particularly true for signals used in telecommunications, such as digital communication signals or compressed audio and video signals. These deterministic signals can be fully characterized by their waveform and parameters, allowing for precise transmission and reconstruction of information.

On the contrary, some signals are *random*, also known as *stochastic* signals. They occur unpredictably and cannot be predicted with certainty. A common example of a random signal is "noise" in telecommunications.

Noise consists of random variations in amplitude or frequency, resulting from unwanted signals or disturbances, which can degrade communication quality and introduce uncertainty into measurements or data transmissions.

Signals can also be classified into two categories based on their evolution over time:

– *Analog signals:* these are represented by a physical quantity $s(t)$ that varies continuously as a function of time (see Figure 3.1).

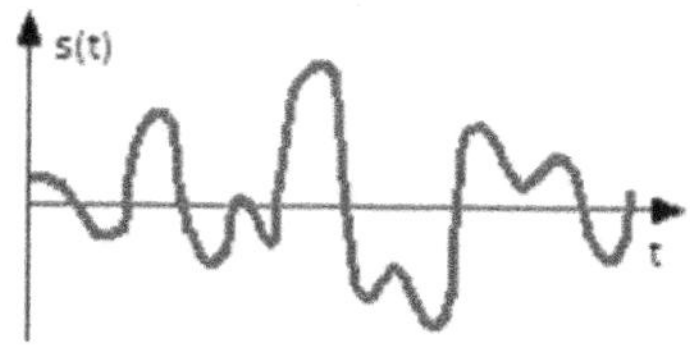

Figure 3.1. *Analog signal. For a color version of this figure, see www.iste.co.uk/cochard/mathematics2.zip*

– *Discrete signals:* these only take on certain specific values at certain times. These signals are also referred to as *digital* (see Figure 3.2).

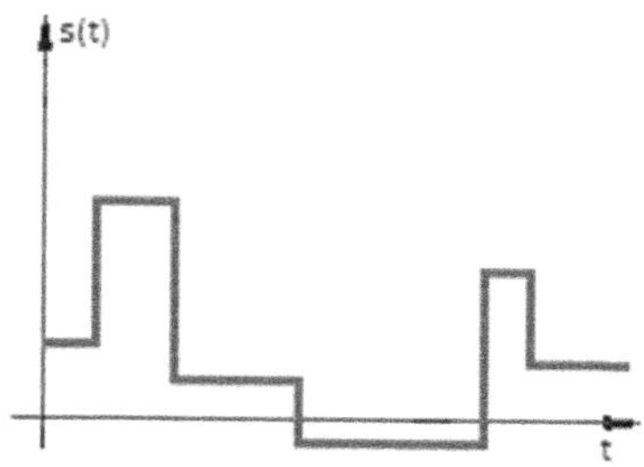

Figure 3.2. *Discrete signal. For a color version of this figure, see www.iste.co.uk/cochard/mathematics2.zip*

3.1.2. *Periodicity*

Some signals have a remarkable property: they repeat regularly over time, as shown in Figure 3.3. These signals are called *periodic signals*. The periodicity of a signal is characterized by a specific value called *period,* denoted by T.

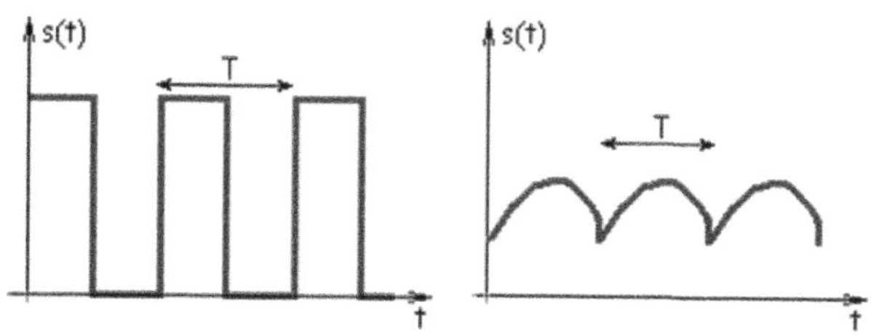

Figure 3.3. *Periodic signals. For a color version of this figure, see www.iste.co.uk/cochard/mathematics2.zip*

The period T of a periodic signal corresponds to the smallest time interval for which the signal remains unchanged. Shifting the signal by an amount equal to its period T results in an exact repetition of the signal pattern. This regular repetition allows for the analysis, synthesis and prediction of signal behavior over an indefinite period of time.

A particularly important signal in the signal theory is the *sinusoidal signal*, represented by the function $s(t) = A\,sin(\omega t)$ function (see Figure 3.4).

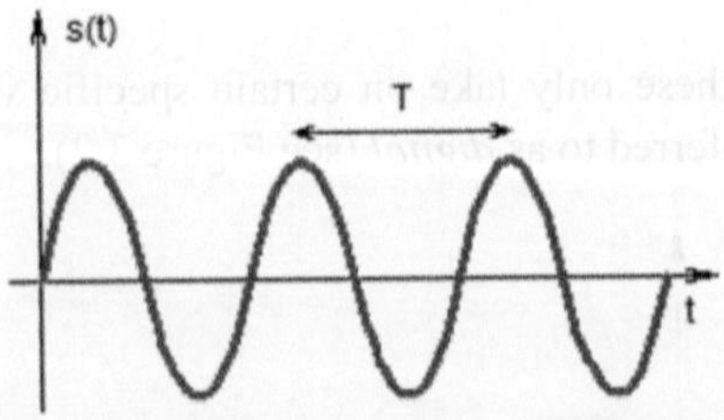

Figure 3.4. *Sinusoidal signal. For a color version of this figure, see www.iste.co.uk/cochard/mathematics2.zip*

"A" is the maximum amplitude, and "ω" is the pulsation (angular frequency), which is related to the period T by the relation $T = \frac{2\pi}{\omega} = \frac{1}{f}$, where f is the frequency.

3.1.3. *Noise*

A signal can be affected by a "parasitic" (an unwanted) signal known as noise, which is superimposed on the nominal signal (see Figure 3.5). This presence of noise in the signal can degrade the quality of the information transmitted by the signal.

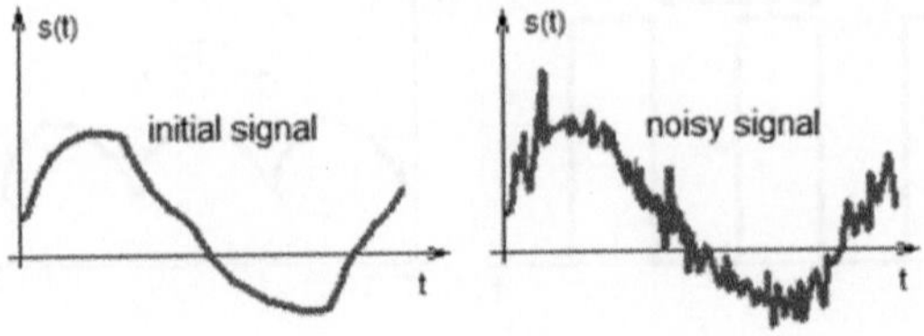

Figure 3.5. *Noise on a nominal signal. For a color version of this figure, see www.iste.co.uk/cochard/mathematics2.zip*

Noise, as shown in Figure 3.5, is a random disturbance that mixes with the signal of interest. It can originate from various sources, including electromagnetic interference, thermal variations, atmospheric disturbances or imperfections in signal capture and transmission devices.

The amount of noise in a signal is generally characterized by the ratio $\frac{signal}{noise} = \frac{P_S}{P_b}$, where P_s and P_b denote the powers (in watts) of the signal and noise, respectively. The signal-to-noise ratio (SNR) is used to assess signal quality by comparing the signal power to the noise power. A higher ratio indicates a stronger signal relative to the noise, reflecting better signal quality. In practice, the signal-to-noise ratio is often expressed in decibels (dB), a logarithmic unit of measurement. The conversion formula is:

$$D = 10 \log_{10}\left(\frac{X_1}{X_2}\right)$$

where D is expressed in decibels (dB), $\log_{10}$ denotes the logarithm to base 10 (decimal logarithm), and $\frac{X_1}{X_2}$ is a ratio of two powers.

EXAMPLE 3.2.– If the signal-to-noise ratio is 30 dB, this means that $\frac{Ps}{P_b} = 1\,000$ because $log_{10}(1000) = 3log_{10}(10) = 3$; the signal power P_s is 1,000 times greater than the noise P_b. In other words, the signal is 1,000 times stronger than the noise.

3.2. Fourier analysis

Throughout this section, only periodic functions are considered. The case of non-periodic functions will be examined later.

3.2.1. *Fourier series expansion*

Consider any periodic function with period T and fundamental frequency f_0. Fourier demonstrated, under certain regularity conditions, that this function can be decomposed into a *superposition of elementary sinusoidal functions with frequencies that are multiples of the base frequency* f_0. The function $s(t)$ can be expressed as:

$$s(t) = \frac{a_0}{2} + \sum_{n=1}^{\infty}[a_n \cos(n\omega_0 t) + b_n \sin(n\omega_0 t)], \text{ where } \omega_0 = \frac{2\pi}{T} = 2\pi f_0,$$

and the coefficients a_n and b_n are given by:

$$a_n = \frac{2}{T}\int_0^T s(t)\cos(n\omega_0 t)\,dt \quad and \quad b_n = \frac{2}{T}\int_0^T s(t)\sin(n\omega_0 t)\,dt$$

The function $s(t)$ can also be expressed in the equivalent form

$$s(t) = \sum_{n=-\infty}^{+\infty} s_n e^{in\omega_0 t} \quad \text{where} \quad s_n = \frac{1}{T}\int_0^T s(t)e^{-in\omega_0 t}\,dt.$$

For $n > 0$, the coefficients can be related to the Fourier series coefficients as follows:

$$s_0 = \frac{a_0}{2} \qquad s_n = \frac{a_n - ib_n}{2} \qquad s_{-n} = \frac{a_n + ib_n}{2} \qquad n > 0$$

The magnitude of the Fourier coefficients is:

$$|s_n|^2 = \frac{a_n^2 + b_n^2}{2}$$

Note that the coefficients here are complex, and that $s_{-n} = s_n^*$, where s_n^* denotes the complex conjugate of s_n.

In conclusion, any periodic signal can be considered as an *infinite superposition of sinusoidal signals*.

3.2.2. Examples

1) Square-wave signal (see Figure 3.6): is a type of periodic waveform that alternates between two levels, typically high and low, with a fixed amplitude and a consistent period.

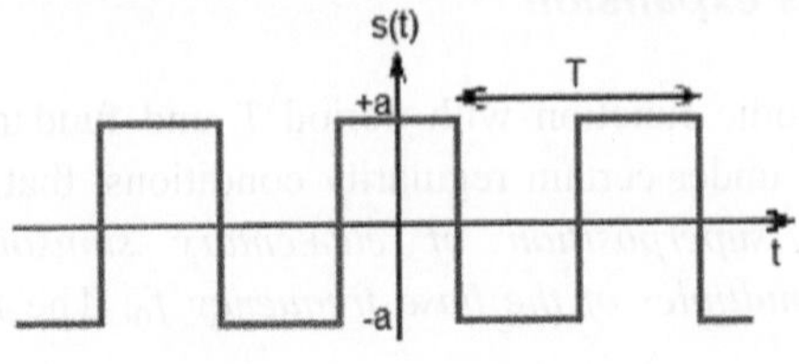

Figure 3.6. *Square-wave signal. For a color version of this figure, see www.iste.co.uk/cochard/mathematics2.zip*

$$s(t) = \frac{4a}{\pi}[cos\omega_0 t - \frac{1}{3}cos3\omega_0 t + \frac{1}{5}cos5\omega_0 t - \frac{1}{7}cos7\omega_0 t + \cdots]$$

2) Triangular signal (see Figure 3.7): is a type of periodic waveform that alternates linearly between a minimum and maximum value, creating a shape that resembles a series of triangles.

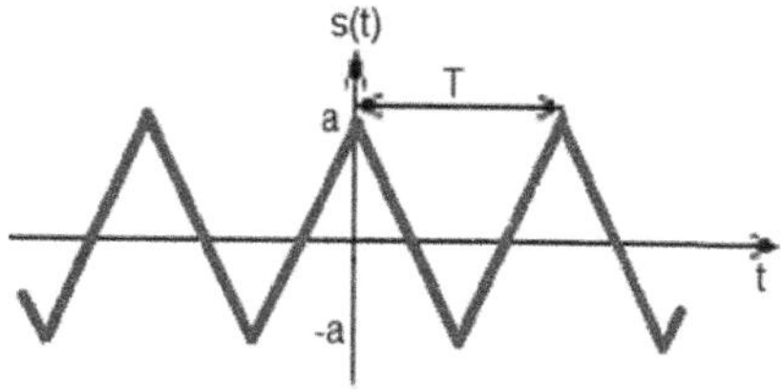

Figure 3.7. *Triangular signal. For a color version of this figure, see www.iste.co.uk/cochard/mathematics2.zip*

$$s(t) = \frac{8a}{\pi^2}\left[cos\omega_0 t + \frac{1}{9}cos3\omega_0 t + \frac{1}{25}cos5\omega_0 t + \cdots\right]$$

3) Sawtooth signal (see Figure 3.8): is a type of periodic waveform that linearly increases or decreases over time, creating a shape resembling the teeth of a saw.

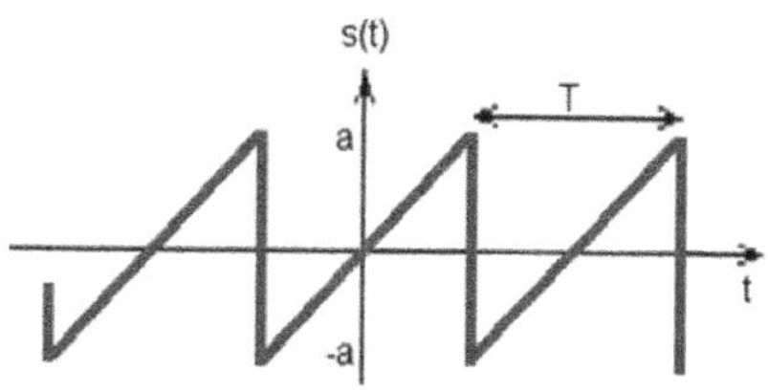

Figure 3.8. *Sawtooth signal. For a color version of this figure, see www.iste.co.uk/cochard/mathematics2.zip*

$$s(t) = \frac{2a}{\pi}\left[sin\omega_0 t - \frac{1}{2}sin2\omega_0 t + \frac{1}{3}sin3\omega_0 t - \frac{1}{4}sin4\omega_0 t + \cdots\right]$$

4) Signal with "spaced bumps": a bump is represented by a cosine function $a\cos(\omega_0 t)$ (see Figure (3.9). The signal consists of these bumps spaced at regular intervals.

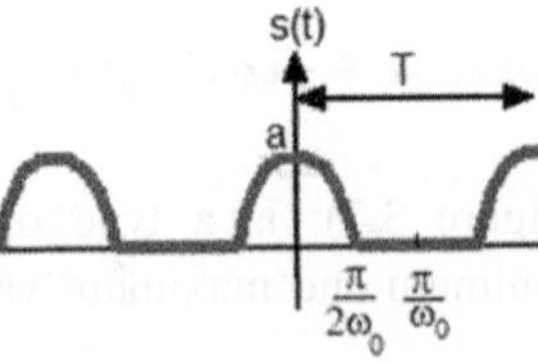

Figure 3.9. *Signal with "spaced bumps". For a color version of this figure, see www.iste.co.uk/cochard/mathematics2.zip*

$$s(t) = \frac{a}{\pi}\left[1 + \frac{\pi}{2}cos\omega_0 t + \frac{2}{3}cos2\omega_0 t - \frac{2}{15}cos4\omega_0 t + \frac{2}{35}cos6\omega_0 t + \cdots\right]$$

5) Signal with "contiguous bumps" (see Figure 3.10): in this case, a bump is represented by a cosine function $a\cos\left(\frac{\omega_0 t}{2}\right)$. The bumps in this signal are contiguous, meaning they touch each other without gaps.

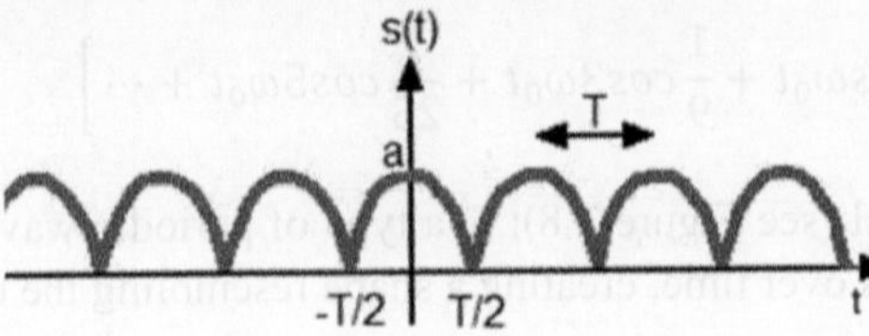

Figure 3.10. *Signal with "contiguous bumps". For a color version of this figure, see www.iste.co.uk/cochard/mathematics2.zip*

$$s(t) = \frac{2a}{\pi}\left[1 + 2\left(\frac{1}{3}cos\omega_0 t - \frac{1}{15}cos2\omega_0 t + \frac{1}{35}cos3\omega_0 t + \cdots\right)\right]$$

6) "Slot" signal (see Figure 3.11): it is characterized by a series of rectangular pulses or slots spaced at regular intervals ($\tau < T$).

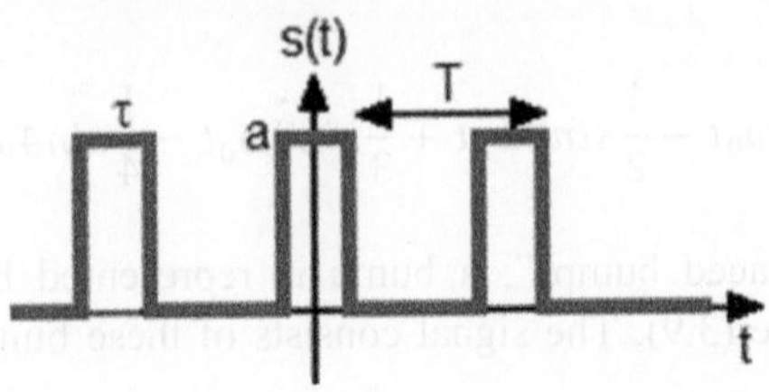

Figure 3.11. *"Slot" signal. For a color version of this figure, see www.iste.co.uk/cochard/mathematics2.zip*

$$s(t) = \frac{a\tau}{T} + \frac{2a}{\pi}[sin\omega_0\frac{\tau}{2}cos\omega_0 t + \frac{1}{2}sin\omega_0\tau cos2\omega_0 t$$
$$+\frac{1}{3}sin3\omega_0\frac{\tau}{2}cos3\omega_0 t + \cdots]$$

3.2.3. *Special cases*

For odd or even functions, the Fourier expansion can be simplified as follows:

– Odd functions: for a function $s(t)$ that is odd, $s(-t) = -s(t)$. The Fourier series expansion of an odd function contains only sine terms:

$$s(t) = \sum_{n=1}^{\infty} b_n \sin(n\omega_0 t),$$

where b_n are the Fourier coefficients for the sine terms.

– Even functions: for a function $s(t)$ that is even, $s(-t) = s(t)$. The Fourier series expansion of an odd function contains only sine terms:

$$s(t) = \frac{a_0}{2} + \sum_{n=1}^{\infty} a_n cos(n\omega_0 t),$$

where b_n are the Fourier coefficients for the sine terms.

3.2.4. *Other development writing*

Another way to express the Fourier series is to include a phase φ_n in the cosine terms of each harmonic. The signal $s(t)$ can be written as:

$$s(t) = A_0 + \sum_{n=1}^{\infty} A_n \cos(n\omega_0 t + \varphi_n)$$

Using the trigonometric identity:

$$\cos(n\omega_0 t + \varphi_n) = \cos(n\omega_0 t)\cos(\varphi_n) - \sin(n\omega_0 t)\sin(\varphi_n)$$

the coefficients a_n and b_n are related to A_n and φ_n as follows:

$$a_0 = 2A_0 \qquad a_n = A_n \cos(\varphi_n) \qquad b_n = -A_n \sin(\varphi_n)$$

Here, A_n represents the amplitudes of the n-th *harmonic*, with A_0 being the amplitude of the fundamental frequency. Specifically, A_1 is the amplitude of the first harmonic, A_2 is the amplitude of the second harmonic, etc.

The unilateral spectrum of the signal is shown in Figure 3.12.

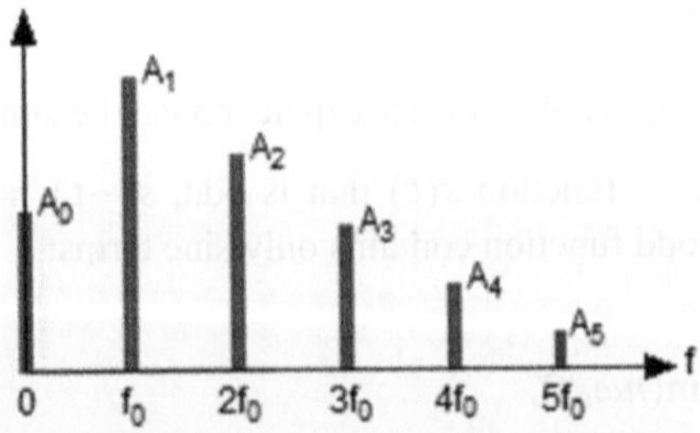

Figure 3.12. *Unilateral spectrum. For a color version of this figure, see www.iste.co.uk/cochard/mathematics2.zip*

The exponential representation of the signal, involving the imaginary argument, leads to the consideration of the bilateral spectrum of the signal (see Figure 3.13).

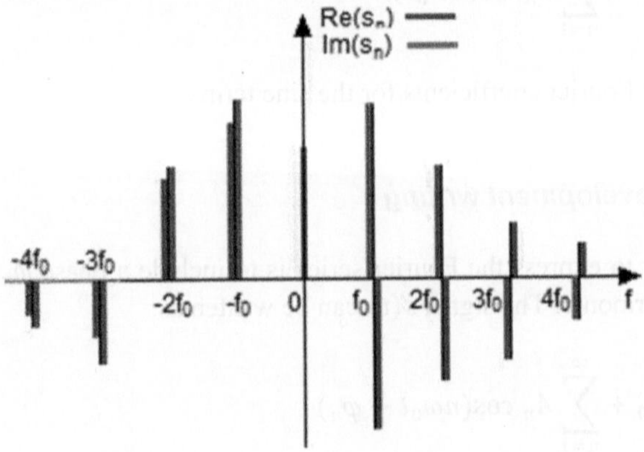

Figure 3.13. *Bilateral spectrum. For a color version of this figure, see www.iste.co.uk/cochard/mathematics2.zip*

3.2.5. *Power*

The quantity $|s(t)|^2$ represents the energy density of the signal. The power transported by the signal over one period T is given by:

$$P = \frac{1}{T}\int_{0}^{T} |s(t)|^2 dt$$

Using the Fourier series expansion:

$$s(t) = \left(\sum_{n=-\infty}^{+\infty} s_n e^{in\omega_0 t}\right)$$

the expression for $|s(t)|^2$ becomes:

$$|s(t)|^2 = \left(\sum_{n=-\infty}^{+\infty} s_n e^{in\omega_0 t}\right)\left(\sum_{m=-\infty}^{+\infty} s_m^* e^{-im\omega_0 t}\right)$$

Expanding this equation:

$$|s(t)|^2 = \sum_{n,m=-\infty}^{+\infty} s_n s_m^* e^{i(n-m)\omega_0 t}$$

Integrating over one period T:

$$\int_{0}^{T} e^{i(n-m)\omega_0 t} dt$$

Since $\omega_0 = \frac{2\pi}{T}$, it follows that $\omega_0 T = 2\pi$. The integral of the complex exponential term over one period is:

$$\int_{0}^{T} e^{i(n-m)\omega_0 t} dt = \int_{0}^{T} e^{i(n-m)\frac{2\pi}{T}t} dt$$

The integral evaluates to T when $n = m$ due to the orthogonality of the complex exponentials, and to 0 otherwise:

$$\int_{0}^{T} e^{i(n-m)\omega_0 t} dt = T\delta_{n,m}$$

where $\delta_{n,m}$ is the Kronecker delta, which equals 1 when $n = m$ and 0 otherwise.

Therefore, the average power P is

$$P = \sum_{n=-\infty}^{+\infty} |s_n|^2$$

3.3. *Dirac distribution*

The Dirac delta distribution $\delta(x)$ is not a function in the traditional sense but a type of generalized function or "pseudo-function"[1]. It is defined by its action on other functions through the integral:

$$f(a) = \int_{-\infty}^{+\infty} f(x)\delta(x-a)dx$$

where $\delta(x-a) = 0$ for $x \neq a$ and acts to extract the value of the function $f(x)$ at a point $x = a$, i.e. $f(a)$.

For the specific case where $a = 0$, this simplifies to

$$f(0) = \int_{-\infty}^{+\infty} f(x)\delta(x)dx$$

In this case, $\delta(x) = 0$ for $x \neq 0$ and effectively picks out the value of $f(x)$ at $x = 0$.

Hence, a fundamental property of the Dirac delta function is:

$$\int_{-\infty}^{+\infty} \delta(x)dx = 1$$

where $\forall x,\ f(x) = 1$. This illustrates that despite $\delta(x)$ being zero at all points except $x = 0$, its integral over the entire real line is equal to 1.

1. The right word is distribution.

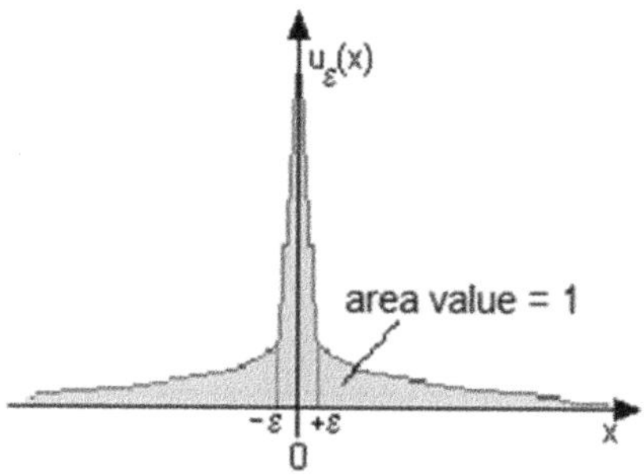

Figure 3.14. *Approximation of the Dirac distribution. For a color version of this figure, see www.iste.co.uk/cochard/mathematics2.zip*

$$\delta(x) = \lim_{\varepsilon \to 0} u_\varepsilon(x)$$

with

$$\int_{-\infty}^{+\infty} u_\varepsilon(x)dx = 1$$

The Dirac delta distribution thus appears as a "function" that would be an infinitely brief impulse at $x = 0$. This shows that $\delta(x)$ can be represented as a distribution that is zero everywhere except at the origin, where it is infinitely large. Conceptually, it can be thought of as the limit of a sequence of functions $u_\varepsilon(x)$ as $\varepsilon \to 0$ (see Figure 3.14). Various functions $u_\varepsilon(x)$ can be used to approximate the Dirac delta distribution. Two common examples are provided below.

EXAMPLE 3.3.– Consider the function $u_\varepsilon(x)$ defined as

$$u_\varepsilon(x) = \begin{cases} \frac{1}{2\varepsilon} & if \ |x| < \varepsilon \\ 0 & otherwise \end{cases}$$

Figure 3.15 shows the representation of the function $u_\varepsilon(x)$. Such a function is a rectangular pulse centered at $x = 0$ with height $\frac{1}{2\varepsilon}$ and width 2ε. As ε approaches 0, $u_\varepsilon(x)$ becomes increasingly narrow and its height grows infinitely, approaching the Dirac delta distribution $\delta(x)$.

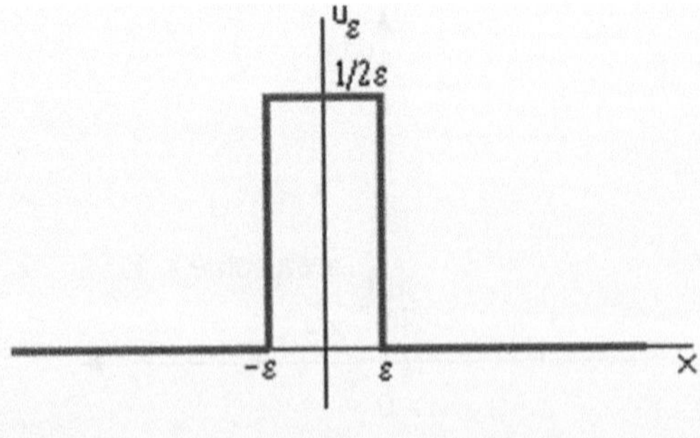

Figure 3.15. *An example of an approximation to the Dirac distribution. For a color version of this figure, see www.iste.co.uk/cochard/mathematics2.zip*

EXAMPLE 3.4.– Consider the function $u_\varepsilon(x)$ defined by

$$u_\varepsilon(x) = \frac{1}{\pi}\frac{\sin\left(\frac{x}{\varepsilon}\right)}{x} = \frac{1}{2\pi}\int_{-\frac{1}{\varepsilon}}^{+\frac{1}{\varepsilon}} e^{ikx}\, dk$$

The graphical representation of $u_\varepsilon(x)$ is shown in Figure 3.16. This function is known as the sinc function in its normalized form. As ε approaches 0, $u_\varepsilon(x)$ becomes a more concentrated impulse centered at $x = 0$, approximating the Dirac delta distribution $\delta(x)$.

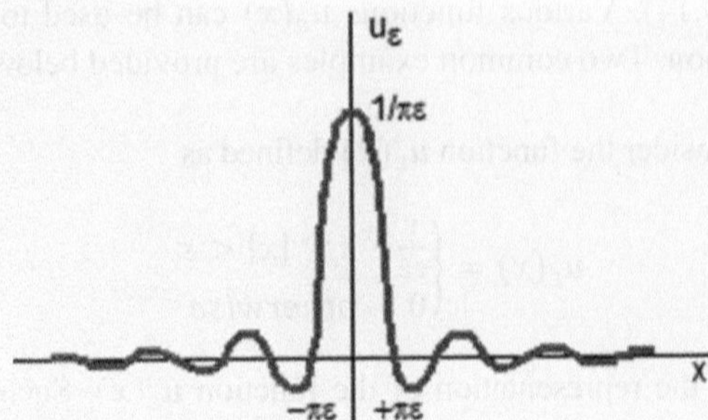

Figure 3.16. *Another example of an approximation to the Dirac distribution. For a color version of this figure, see www.iste.co.uk/cochard/mathematics2.zip*

The previous example shows that an integral representation of the Dirac delta function $\delta(x)$ can be obtained as:

$$\delta(x) = \frac{1}{2\pi} \int_{-\infty}^{+\infty} e^{ikx} dk$$

This relationship is quite useful in various applications.

Below are some of the main properties of the Dirac delta distribution:

1) *Symmetry*: $\delta(x) = \delta(-x)$.

   The Dirac delta function is an even function.

2) *Multiplication by* x: $x\delta(x) = 0$.

   Multiplying the Dirac delta function by x yields zero.

3) *Derivative properties*: $\delta'(-x) = -\delta'(x)$.

   The derivative of the Dirac delta function is an odd function.

4) *Multiplication by x and its derivative*: $x\delta'(x) = -\delta(x)$.

   This relationship involves the derivative of the Dirac delta function.

5) *Scaling property*: $\delta(cx) = \frac{1}{|c|}\delta(x)$ for $c > 0$.

   Scaling the argument of the Dirac delta function scales its magnitude inversely.

The conventional representation of the Dirac delta distribution is shown in Figure 3.17.

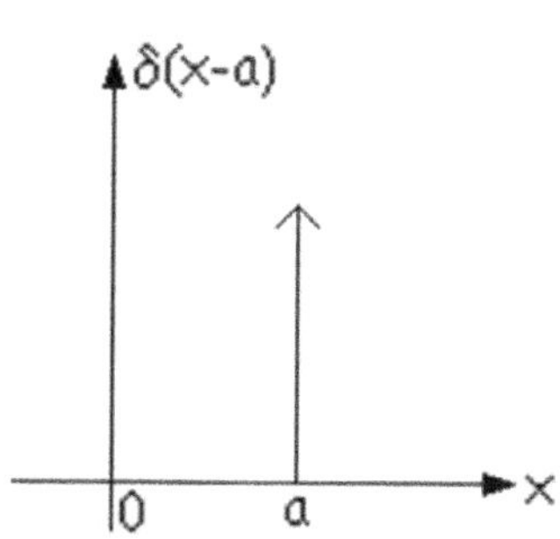

Figure 3.17. *Conventional representation of the Dirac distribution*

3.4. Convolution

3.4.1. *Definition*

Convolution, also known as the *convolution product*, is a mathematical operation used to combine two functions f and g representing signals. It is defined by the following integral:

$$(f * g)(x) = \int_{-\infty}^{+\infty} f(x-t)g(t)dt$$

The convolution product is defined if this integral exists.

The convolution product of two functions f and g can also be expressed as:

$$(f * g)(x) = \int_{-\infty}^{+\infty} f(t)g(x-t)dt$$

To show that both expressions are equivalent, perform a change of variable $u = x - t$. Then,

$$(f * g)(x) = \int_{-\infty}^{+\infty} f(x-t)g(t)dt = \int_{-\infty}^{+\infty} f(u)g(x-u)du$$

Thus, the convolution operation can be represented in either form depending on convenience.

EXAMPLE 3.5.– Perform the convolution of the two functions $f(x)$ and $g(x)$ as defined in Figure 3.18.

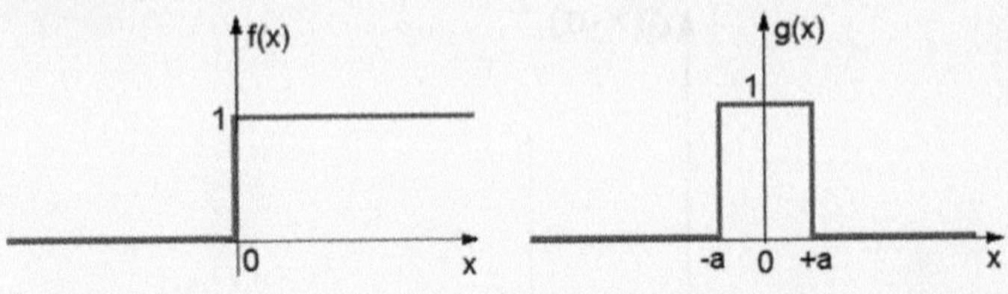

Figure 3.18. *Convolution of the two functions. For a color version of this figure, see www.iste.co.uk/cochard/mathematics2.zip*

$$(f * g)(x) = \int_{-\infty}^{+\infty} f(x-t)g(t)dt = \int_{-\infty}^{+\infty} g(t)dt = \begin{cases} 0 & if\ x < -a \\ x+a & if\ |x| \le a \\ 2a & if\ x > a \end{cases}$$

The result of the convolution is shown in Figure 3.19.

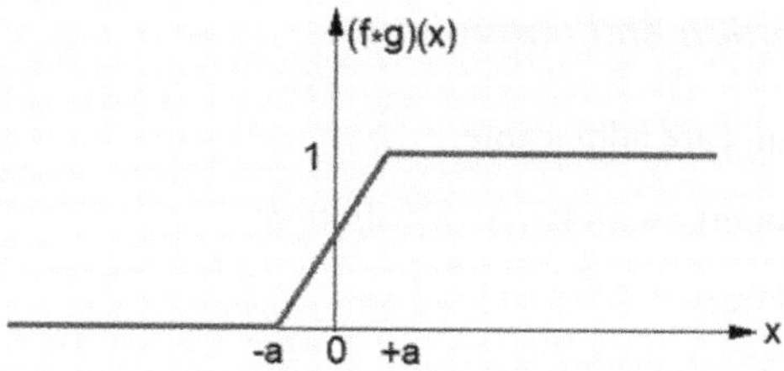

Figure 3.19. *The convolution result. For a color version of this figure, see www.iste.co.uk/cochard/mathematics2.zip*

To provide a graphical interpretation of the convolution $(f * g)$, start by graphing the function $f(-t)$, which is obtained by reflecting the function $f(t)$ across the y-axis.

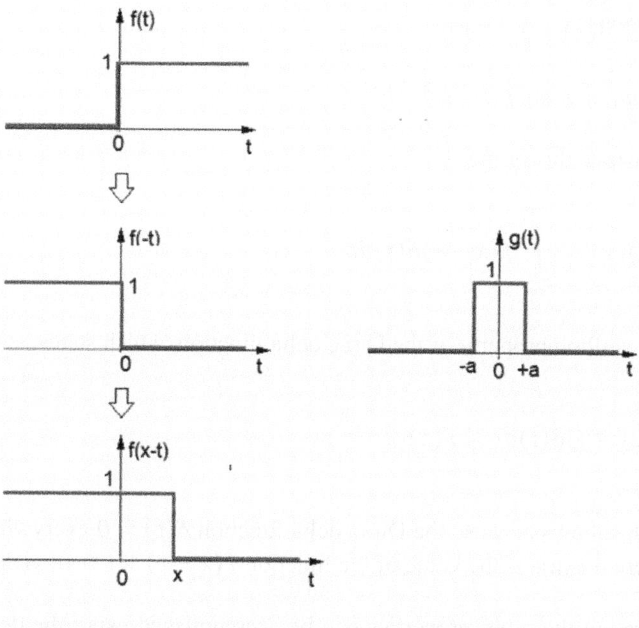

Figure 3.20. *Graphical interpretation of convolution. For a color version of this figure, see www.iste.co.uk/cochard/mathematics2.zip*

Next, graph $f(x-t)$ by shifting the function $f(t)$ horizontally to the right of x units (see Figure 3.20). Then, compute the product $f(x-t) \cdot g(t)$ piece by piece. The convolution $(f * g)$ is obtained by integrating this product over t, which represents the area bounded by the product and the x-axis.

3.4.2. *Dirac distribution and convolution*

The following results are admissible:

1) Convolution property with Dirac distribution:

$$(f * \delta)(x) = f(x).$$

2) Convolution property with shifted Dirac distribution:

$$(f * \delta_a)(x) = f(x-a) \text{ where } \delta_a(x) = \delta(x-a).$$

3) Convolution property with the derivative of Dirac distribution:

$$(f * \delta')(x) = f'(x).$$

4) Convolution property with the n-th order derivative of Dirac distribution:

$$\left(f * \delta^{(n)}\right)(x) = f^{(n)}(x).$$

Demonstration of the first property:

To demonstrate this property,

$$(f * \delta)(x) = \int_{-\infty}^{+\infty} f(x-t)\delta(t)dt$$

Using the shifting property of the Dirac delta function, which states:

$$\int_{-\infty}^{+\infty} f(x-t)\delta(t)dt = f(x)$$

This result follows because the Dirac delta function $\delta(t) = 0$ everywhere except at $t = 0$, where it extracts the value of the function $f(x-t)$ at $t = 0$, yielding $f(x)$.

Demonstrating the other properties can be accomplished using the definition of convolution and the properties of the Dirac distribution.

These results illustrate how convolution with the Dirac distribution or its derivatives can be used to perform useful operations on a function $f(x)$, such as shifting, differentiating or obtaining higher-order derivatives. These properties are widely used in mathematical analysis and physics.

4

z-transforms, Fourier Transforms and Laplace Transforms

CONCEPTS COVERED IN THIS CHAPTER.–

This chapter defines three fundamental mathematical tools for signal analysis.

It begins with the z-transform, covering its properties and application to discrete signals.

Next, it introduces the Fourier transform, which extends the Fourier series from the previous chapter to non-periodic functions. The process of calculating a Fourier transform on a computer requires discretization, which is explained in detail using the fast Fourier transform (FFT) algorithm.

Finally, the chapter presents the Laplace transform and explains its main properties.

Some proofs of theorems are considered relatively complex and are omitted, while more accessible proofs are provided for the reader.

References: [BEL 98, BER 14, BRO 82, COT 97, HAY 99, PAP 77].

Mathematics for Digital Science 2,
by Gérard-Michel COCHARD and Mhand HIFI.

4.1. z-transform

4.1.1. *Definitions and main results*

Consider a discrete signal represented by a sequence $f(n)$. The *z-transform* is defined by the following relation:

$$F(z) = \sum_{n=-\infty}^{+\infty} f(n)z^{-n}$$

where z is a complex variable. This relationship establishes a correspondence between $f(n)$ and $F(z)$. It is important to note that $F(z)$ exists only if the sum converges. According to the theory of complex-variable functions, this sum converges only if z lies within a specific region of the complex plane: $R_1 < |z| < R_2$. The radii R_1 and R_2 depend on the behavior of $f(n)$ as n tends to $-\infty$ or $+\infty$. Consequently, the singularities of $F(z)$ lie outside this annular region.

Specifically, if $f(n) = 0$ for $n < 0$, the z-transform simplifies to:

$$F(z) = \sum_{n=0}^{+\infty} f(n)z^{-n}$$

which implies that the sum contains only negative power terms of z. In this case, the radius of convergence R_1 satisfies $|z| > R_1$.

If $f(n) = 0$ for $n > 0$, then

$$F(z) = \sum_{n=-\infty}^{-1} f(n)z^{-n}$$

where the function contains only terms with positive powers of z. Therefore, the radius of convergence is R_2 such that $|z| < R_2$ (see Figure 4.1).

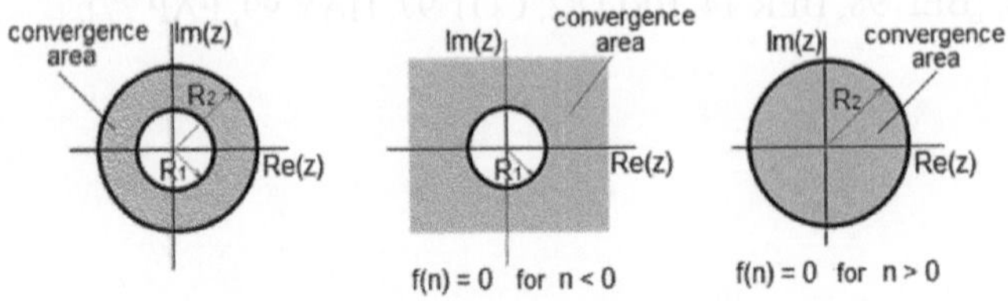

Figure 4.1. *Radii of convergence*

The inverse transform $f(n)$ of $F(z)$ may exist, but its existence and uniqueness depend on the region of convergence.

EXAMPLE 4.1.– Let $f(n) = u(n)$, where $u(n)$ is the unit step function defined as $u(n) = 1$ for $n \geq 0$ and $u(n) = 0$ otherwise. The z-transform of $u(z)$ is given by:

$$U(z) = \sum_{n=-\infty}^{+\infty} u(n) z^{-n} = \sum_{n=0}^{+\infty} \frac{1}{z^n} = 1 + \frac{1}{z} + \frac{1}{z^2} + \frac{1}{z^3} + \cdots$$

This series sums to

$$U(z) = \frac{1}{1-\frac{1}{z}} = \frac{z}{z-1}$$

This expression is valid for $|z^{-1}| < 1$ or $|z| > 1$, which defines the region of convergence.

Interestingly, the function $g(n) = -u(-n-1)$ gives the same transform:

$$U'(z) = -\sum_{n=-\infty}^{+\infty} u(-n-1) z^{-n} = -\sum_{n'=+\infty}^{-\infty} u(n') z^{n'+1} = -\sum_{n'=+\infty}^{0} z^{n'+1}$$

or

$$U'(z) = -[1 + z + z^2 + z^3 + \cdots] + 1 = -\frac{1}{1-z} + 1 = \frac{z}{z-1}$$

provided that $|z| < 1$.

It might seem that $U(z)$ and $U'(z)$ are inverse transforms of $\left(\frac{z}{z-1}\right)$ and that there is no uniqueness of the transform in z. However, this is not the case, as the radii of convergence are different: $|z| > 1$ for $U(z)$ and $|z| < 1$ for $U'(z)$.

The z-transform simplifies the analysis of discrete-time signals and systems by converting them into algebraic functions of a complex z.

When dealing with discrete-time systems, it is often useful to establish relationships between the current value of a sequence $f(n)$ and its previous values, such as $f(n-1)$. The z-transform facilitates this by transforming the difference equations into algebraic equations.

For example, to establish a relationship between $f(n)$ and $f(n-1)$, define $\omega(n) = f(n-1)$. The z-transform of $\omega(n)$, denoted as $\Omega(z)$, is:

$$\Omega(z) = \sum_{k=-\infty}^{+\infty} z^{-k}\omega(k)$$

Substituting $\omega(k) = f(k-1)$, it follows:

$$\Omega(z) = \sum_{k=-\infty}^{+\infty} z^{-k} f(k-1)$$

To simplify, let $k' = k - 1$ (so $k = k' + 1$):

$$\Omega(z) = \sum_{k'=-\infty}^{+\infty} z^{-(k'+1)} f(k') = z^{-1} \sum_{k'=-\infty}^{+\infty} z^{-k'} f(k') = z^{-1}F(z)$$

Here, $F(z)$ is the z-transform of $f(n)$.

Role of z^{-1}: the factor z^{-1} represents a delay of one time step in the discrete-time sequence. Multiplying the z-transform $F(z)$ by z^{-1} corresponds to shifting the sequence $f(n)$ by one unit to the right, resulting in $f(n-1)$.

In block diagrams of digital systems, z^{-1} is used to denote this "delay" operation. Figure 1.2 shows the transition process from signal $f(n)$ to signal $f(n-1)$.

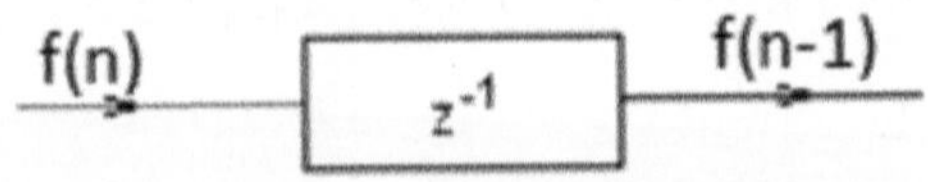

Figure 4.2. *Delay function*

4.1.2. *Application to discrete systems*

Impulse property $\delta(n)$

Let $\Delta(z)$ denote the *z-transform* of $\delta(n)$, where $\delta(n)$ is the unit impulse function. By definition:

$$\Delta(z) = \sum_{n=-\infty}^{+\infty} \delta(n) z^{-n}$$

In this expression, all terms are zero except for $n = 0$ ($\delta(n) = 1$), which gives $\Delta(z) = 1$. Thus, when $z = 1$, it represents a time shift of one unit, graphically symbolized by z^{-1} (see Figure 4.2). This symbol indicates the transition from a signal $f(n)$ to $f(n-1)$.

Similarly, the z-transform $\mathrm{ZT}[\delta(n-1)]$ of $\delta(n-1)$ is defined as:

$$\mathrm{ZT}[\delta(n-1)] = \sum_{n=-\infty}^{+\infty} \delta(n-1)\, z^{-n}$$

All terms of the sum are zero except for $n = 1$ ($\delta(n-1) = 1$ and $z^{-1} = \frac{1}{z}$). Therefore,

$$\mathrm{ZT}[\delta(n-1) = \frac{1}{z} = z^{-1}$$

Similarly,

$$\mathrm{ZT}[\delta(n-2)] = z^{-2} \quad \mathrm{ZT}[\delta(n-3)] = z^{-3} \quad \ldots. \quad \mathrm{ZT}[\delta(n-k)] = z^{-k}$$

Now, consider the z-transform of $x(k)\delta(n-k)$. By definition,

$$ZT[x(k)\delta(n-k)] = \sum_{n=-\infty}^{+\infty} x(k)\delta(n-k)z^{-n} = \sum_{n=-\infty}^{+\infty} x(k)z^{-n}\delta(n-k) = x(k)z^{-k}$$

The z-transform of $x(k)\delta(n-k)$ shows that each signal sample $x(n)$ is multiplied by a decreasing power of z in its $z -$ transform.

h*(*z*) *transfer function

Consider a system L with input signal $e(n) = z^n$. The output signal $s(n)$ can be expressed as a function of the impulse response $h(n)$ using the following equation:

$$s(n) = h(n) * e(n) = \sum_{k=-\infty}^{+\infty} e(k)h(n-k) = \sum_{k=-\infty}^{+\infty} e(n-k)h(k)$$

In this particular case:

$$s(n) = \sum_{k=-\infty}^{+\infty} e(n-k)h(k) = z^n \sum_{k=-\infty}^{+\infty} h(k)z^{-k} = z^n H(z)$$

Here, $H(z)$ represents the z-transform of the impulse response $h(n)$. $H(z)$ is known as the *transfer function* and is crucial for analyzing linear systems. The transfer function directly links the impulse response $h(n)$ to the system's overall behavior.

Let us explore a few examples to determine the transfer function $H(z)$ from $h(n)$.

EXAMPLE 4.2.– Consider the signal $s(n) = \frac{1}{2}[e(n) + e(n-1)]$.

By applying pulses $\delta(n)$ with different values of n, the system's response $h(n)$ can be observed. For this example, the response $h(n)$ is given by:

$$h(n) = \frac{1}{2}[\delta(n) + \delta(n-1)]$$

To determine $h(n)$, evaluate different values of n:

For < 0: $\delta(n) = \delta(n-1) = 0$, so $h(n) = 0$.
For $n = 0$: $\delta(n) = 1$ and $\delta(n-1) = 0$, thus, $h(0) = \frac{1}{2}$.
For $n = 1$: $\delta(n) = 0$ and $\delta(n-1) = 1$, so $h(1) = \frac{1}{2}$.
For > 1: $\delta(n) = \delta(n-1) = 0$, therefore $h(n) = 0$.

Hence, the system response $h(n)$ is non-zero only at $n = 0$ and $n = 1$, where it equals $\frac{1}{2}$.

The z-transform of $h(n)$ is

$$H(z) = \sum_{n=-\infty}^{+\infty} h(n)z^{-n} = \frac{z^0}{2} + \frac{z^{-1}}{2} = \frac{1}{2}(1 + z^{-1})$$

EXAMPLE 4.3.– Let $s(n) = \frac{1}{2}[e(n) + s(n-1)]$

By applying pulses $\delta(n)$ at the input for different values of n, the output $h(n)$ can be observed. The results are as follows:

For $n < 0$: $s(n-1) = 0$ and $\delta(n) = 0$, so $h(n) = 0$.
For $n = 0$: $\delta(n) = 1$ and $s(n-1) = 0$, thus $h(0) = \frac{1}{2}$.

For $n = 1$: $\delta(n) = 0$ and $s(n-1) = h(0) = \frac{1}{2}$, so $h(1) = \frac{1}{4}$.
For $n = 2$: $\delta(n) = 0$ and $s(n-1) = h(1) = \frac{1}{4}$, thus $h(2) = \frac{1}{8}$.
For $n > 0$: $h(n) = \frac{1}{2^{n+1}}$.

The z-transform of $h(n)$ is calculated as follows:

$$H(z) = \sum_{n=-\infty}^{+\infty} h(n)z^{-n} = \frac{z^0}{2} + \frac{z^{-1}}{4} + \frac{z^{-2}}{8} + \cdots = \frac{1}{2}[\frac{z^0}{1} + \frac{z^{-1}}{2} + \frac{z^{-2}}{4} + \cdots]$$

The series in square brackets is a geometric series with the first term of 1 and a (reason) of $\frac{z^{-1}}{2}$. Therefore:

$$H(z) = \frac{1}{2} \cdot \frac{1}{1-\frac{z^{-1}}{2}} = \frac{1}{2-z^{-1}}$$

provided that $|z| > \frac{1}{2}$.

4.2. Fourier transform

4.2.1. *Periodic signals*

The Fourier analysis of periodic signals was described in Chapter 3. The key results are as follows: under certain regularity conditions, a signal $s(t)$ can be decomposed into a sum of elementary sinusoidal functions, where the frequencies are multiples of the base frequency f_0.

The signal $s(t)$ can be expressed as:

$$s(t) = \frac{a_0}{2} + \sum_{n=1}^{\infty} [a_n cos(n\omega_0 t) + b_n sin(n\omega_0 t)]$$

where $\omega_0 = \frac{2\pi}{T} = 2\pi f_0$, and the coefficients are given by:

$$a_n = \frac{2}{T}\int_0^T s(t)\, cos(n\omega_0 t)\, dt \quad \text{and} \quad b_n = \frac{2}{T}\int_0^T s(t)\, sin(n\omega_0 t)\, dt$$

Alternatively, the periodic signal can be represented as:

$$s(t) = \sum_{n=-\infty}^{+\infty} s_n e^{in\omega_0 t} \quad \text{with} \quad s_n = \frac{1}{T}\int_0^T s(t)e^{-in\omega_0 t}dt$$

where

$$s_0 = \frac{a_0}{2} \qquad s_n = \frac{a_n - ib_n}{2} \qquad s_{-n} = \frac{a_n + ib_n}{2} \qquad |s_n|^2 = \frac{a_n^2 + b_n^2}{4} \quad for\ n > 0$$

For a periodic signal, the Fourier series decomposition provides a precise representation of the signal's frequency content.

4.2.2. Non-periodic signals

Let us now consider the transition from periodic to non-periodic signals. Examining the spectrum as a function of frequency, take the example of the periodic function $s(t)$ shown in Figure 4.3. This illustrates how various frequency components contribute to the shape of the signal. The spectrum shows the amplitude of each harmonic as a function of its frequency, revealing how each harmonic component contributes to the overall structure of the signal. Higher-frequency harmonics introduce faster variations to the signal, while the fundamental harmonic (lowest frequency) defines the signal's basic shape.

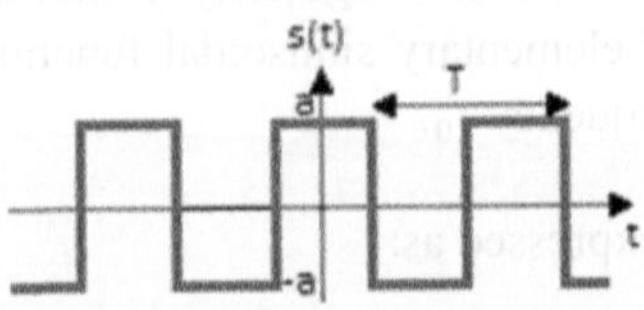

Figure 4.3. *Example of a periodic function. For a color version of this figure, see www.iste.co.uk/cochard/mathematics2.zip*

This spectral analysis allows us to understand how different frequencies combine to create the temporal shape of the periodic signal. It provides an interesting insight into the frequency composition of the signal and highlights the dominant components that define its character. For example, the Fourier series expansion of $s(t)$ is represented by:

$$s(t) = \frac{4a}{\pi}[cos\omega_0 t - \frac{1}{3}cos3\omega_0 t + \frac{1}{5}cos5\omega_0 t - \frac{1}{7}cos7\omega_0 t + \cdots]$$

The frequency spectrum of a periodic signal consists of "lines" at frequencies that are integer multiples of the fundamental frequency f_0. These lines appear at frequencies f_0, $3f_0$, $5f_0$, etc., where f_0 is defined as the reciprocal of the period $T (f_0 = \frac{1}{T})$. Examining the spectrum shows how these lines are distributed relative to the signal period. Varying the period T and plotting the spectrum $|s_n|$ of the corresponding signal clarifies this relationship. Figures 4.4–4.7 show these spectra for different values of T.

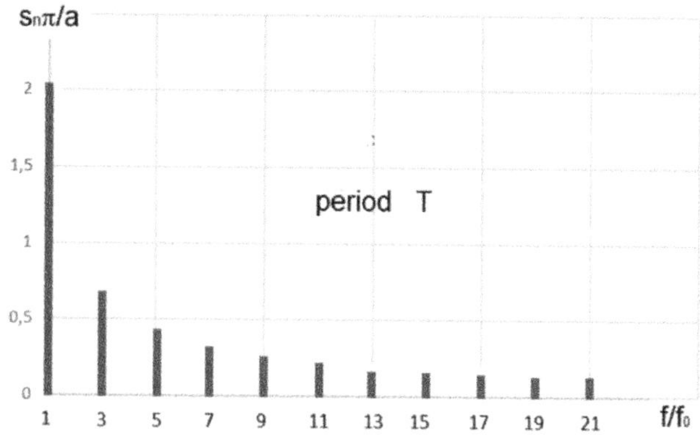

Figure 4.4. *Period T. For a color version of this figure, see www.iste.co.uk/cochard/mathematics2.zip*

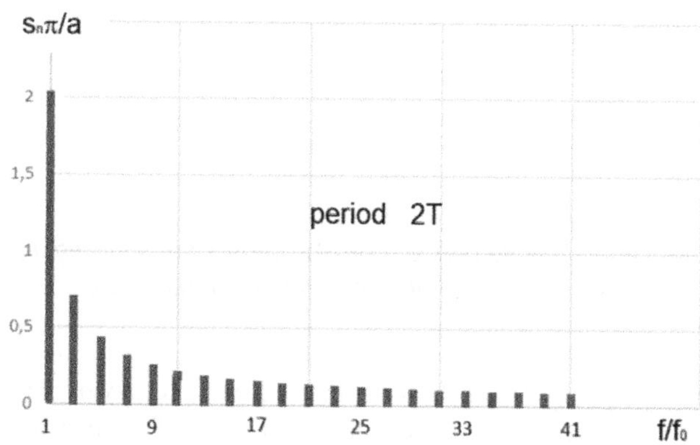

Figure 4.5. *Period 2T. For a color version of this figure, see www.iste.co.uk/cochard/mathematics2.zip*

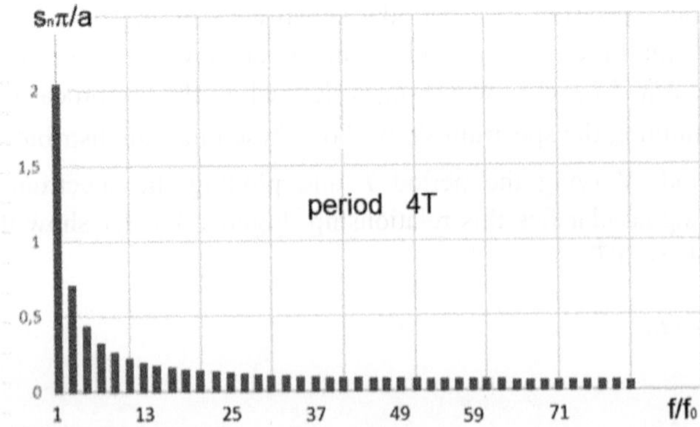

Figure 4.6. *Period 4T. For a color version of this figure, see www.iste.co.uk/cochard/mathematics2.zip*

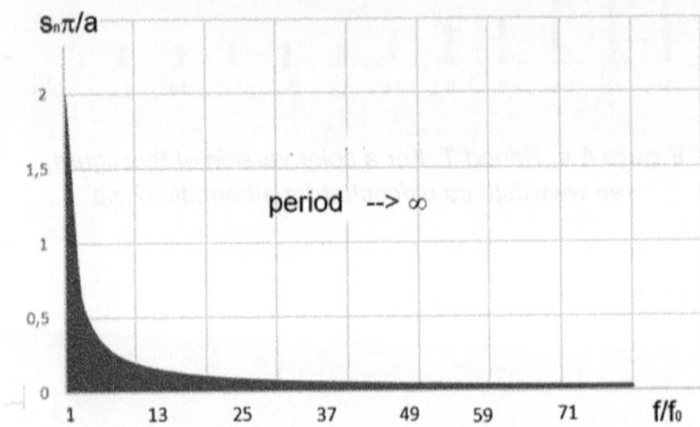

Figure 4.7. *Infinite period. For a color version of this figure, see www.iste.co.uk/cochard/mathematics2.zip*

Examining these spectra reveals how the position and intensity of the harmonic lines change with the signal period. As the period increases, the fundamental frequencies of the harmonic lines decrease, and new lines emerge at lower frequencies. In the extreme case, as the period approaches infinity, the discrete "stick" spectrum of the periodic signal transitions into a continuous spectrum bounded by an envelope curve.

For a non-periodic signal, no specific period can be defined due to the absence of a repetitive pattern. Instead, the period is considered infinite, and the non-periodic signal is represented as a continuous superposition of elementary signals with varying frequencies, contrasting with the discrete representation of periodic signals.

***Fourier transform**:* captures the transition from periodic to non-periodic signals. Consider the Fourier series expansion:

$$s(t) = \sum_{n=-\infty}^{+\infty} s_n e^{in\omega_0 t} \quad \text{where} \quad s_n = \frac{1}{T} \int_{-T/2}^{T/2} s(t) e^{-in\omega_0 t}\, dt$$

As the period T approaches infinity, the signal becomes essentially non-periodic. In this case, the Fourier series expansion shifts to a Fourier transform representation:

$$s(t) = \int_{-\infty}^{+\infty} S(f) e^{2i\pi f t} df \quad \text{where} \quad S(f) = \int_{-\infty}^{+\infty} s(t) e^{-2i\pi f t} dt$$

Here, $S(f)$ represents the Fourier transform of the signal $s(t)$, and the two expressions are inverse of each other. To show this, consider

$$g(u) = \int_{-\infty}^{+\infty} S(f) e^{2i\pi f u} df$$

Substitute the Fourier transform $S(f)$ $(g(u) \; = \; s(u))$:

$$g(u) = \int_{-\infty}^{+\infty} (\int_{-\infty}^{+\infty} s(t) e^{-2i\pi f t} dt) e^{2i\pi f u} df$$

Rearrange the order of integration:

$$g(u) = \int_{-\infty}^{+\infty} s(t) \left(\int_{-\infty}^{+\infty} e^{2i\pi f (u-t)} df \right) dt$$

Change variables by setting $\tau = u - t$:

$$g(u) = \int_{-\infty}^{+\infty} s(u-\tau)\left(\int_{-\infty}^{+\infty} e^{2i\pi f\tau} df\right) d\tau$$

The integral inside the brackets in the Fournier transform of a Dirac delta function:

$$\delta(\tau) = \int_{-\infty}^{+\infty} e^{2i\pi f\tau} df = \frac{1}{2\pi}\int_{-\infty}^{+\infty} e^{iF\tau} dF$$

where $F = 2\pi f$.

Thus,

$$g(u) = \int_{-\infty}^{+\infty} s(u-\tau)\delta(\tau) d\tau = s(u)$$

EXAMPLE 4.4.– The "square" or "square wave" signal, as shown in Figure 4.8, is a fundamental waveform commonly used in computing and digital electronics.

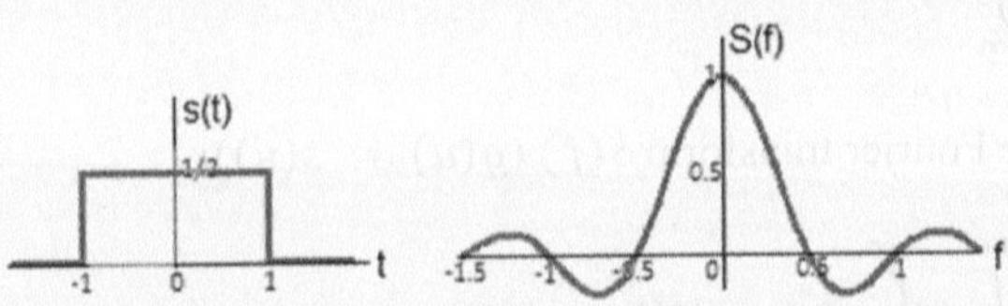

Figure 4.8. *Time and frequency representations of the square-wave signal. For a color version of this figure, see www.iste.co.uk/cochard/mathematics2.zip*

The Fourier transform $S(f)$ of a square wave signal $s(t)$ is calculated as follows:

$$S(f) = \int_{-\infty}^{+\infty} s(t)e^{-2i\pi ft}dt = \frac{1}{2}\int_{-1}^{+1} e^{-2i\pi ft}dt = \frac{e^{2i\pi f} - e^{-2i\pi f}}{4i\pi f} = \frac{\sin(2\pi f)}{2\pi f}$$

	functions	transforms
	$u(t), v(t), s(t)$	$U(f), V(f), S(f)$
linearity	$au(t) + b(v(t)$	$aU(f) + bV(f)$
dilatation in t	$s(at)$	$\frac{1}{\|a\|}S(\frac{f}{a})$
translation in t	$s(t - a)$	$e^{-2i\pi fa}S(f)$
translation in f	$e^{2i\pi at}s(t)$	$S(f - a)$
derivation in t	$s'(t)$	$2i\pi fS(f)$
	$s^{(n)}(t)$	$(2i\pi f)^n S(f)$
derivation in f	$ts(t)$	$-\frac{1}{2i\pi}S'(f)$
	$t^n s(t)$	$\left(-\frac{1}{2i\pi}\right)^n S^{(n)}(f)$

Figure 4.9. *Fourier transform properties*

4.2.3. *Main properties of Fourier transforms*

Without exploring the derivations in detail, the key properties of Fourier transforms are summarized in the table in Figure 4.9. Figure 4.10 provides illustrations of some commonly used Fourier transforms.

s(t)	S(t)
$\Pi_\tau(t) = u\left(t+\frac{\tau}{2}\right) - u(t - \frac{\tau}{2})$	$\frac{\sin(\pi\tau f)}{\pi f}$
$\frac{\sin(\pi\lambda t)}{\pi t}$	$\Pi_\lambda(f) = u\left(f+\frac{\lambda}{2}\right) - u(f - \frac{\lambda}{2})$
$e^{-\pi t^2}$	$e^{-\pi f^2}$
e^{-at^2} with $a > 0$	$\sqrt{\frac{\pi}{a}}e^{-\frac{\pi^2 f^2}{a}}$
$e^{-a\|t\|}$ with $a > 0$	$\frac{2a}{a^2 + 4\pi^2 f^2}$
$\frac{2a}{a^2 + 4\pi^2 t^2}$	$e^{-a\|f\|}$
$\frac{1}{1 + t^2}$	$\pi e^{-2\pi\|f\|}$
$\cos(at^2)$ with $a > 0$	$\sqrt{\frac{\pi}{a}}[\cos\left(-\frac{\pi^2 f^2}{a}\right) - \frac{\pi}{4}]$
$\sin(at^2)$ with $a > 0$	$\sqrt{\frac{\pi}{a}}[\cos\left(-\frac{\pi^2 f^2}{a}\right) + \frac{\pi}{4}]$
$e^{-at}u(t)$	$\frac{1}{a + 2i\pi f}$

Figure 4.10. *Common Fourier transforms*

Now, let us explore some highly useful theorems.

THEOREM 4.1.– (Parseval's): consider a function $f(t)$ and its Fourier transform $F(f)$. The relationships between these two functions are as follows:

$$\int_{-\infty}^{+\infty} |f(t)|^2 dt = \int_{-\infty}^{+\infty} |F(f)|^2 df$$

Proof of Theorem 4.1. Consider the calculation of the following integral:

$\int_{-\infty}^{+\infty} |f(t)|^2 dt.$

Starting with the identity

$$\int_{-\infty}^{+\infty} |f(t)|^2 dt = \int_{-\infty}^{+\infty} f(t)^* f(t) dt$$

this can be rewritten as:

$$\int_{-\infty}^{+\infty} f(t)^* f(t) dt = \int_{-\infty}^{+\infty} f(t) [\int_{-\infty}^{+\infty} F(f) e^{2i\pi ft} df]^* \, dt$$

By switching the order of integration:

$$= \int_{-\infty}^{+\infty} f(t) \left(\int_{-\infty}^{+\infty} F(f)^* e^{-2i\pi ft} df \right) dt$$

$$= \int_{-\infty}^{+\infty} F(f)^* \left(\int_{-\infty}^{+\infty} f(t) e^{-2i\pi ft} dt \right) df$$

This simplifies to:

$$\int_{-\infty}^{+\infty} F(f)^* F(f) df = \int_{-\infty}^{+\infty} |F(f)|^2 df$$

Thus, Parseval's theorem is obtained.

THEOREM 4.2.– (Fourier transform and convolution). The Fourier transform of the convolution of two functions $f(t)$ and $g(t)$ is equal to the product of their Fourier transforms $F(f)$ and $G(f)$.

Proof of Theorem 4.2. The convolution product $(f * g)(t)$ can be written as:

$$c(t) = (f * g)(t) = \int_{-\infty}^{+\infty} f(t-u)g(u)du \quad \text{by definition.}$$

Let $C(f)$, $F(f)$ and $G(f)$ be the Fourier transforms of $c(t), f(t)$ and $g(t)$, respectively. The Fourier transform of $c(t)$ is:

$$C(f) = \int_{-\infty}^{+\infty} c(t)\, e^{-2i\pi ft}dt = \int_{-\infty}^{+\infty}\left(\int_{-\infty}^{+} f(t-u)g(u)du\right)e^{-2i\pi ft}dt$$

By switching the order of integration:

$$C(f) == \int_{-\infty}^{+\infty} g(u)\left(\int_{-\infty}^{+\infty} f(t-u)e^{-2i\pi ft}dt\,)\right)du$$

Letting $\tau = t - u$ to simplify the expression

$$C(f) = \int_{-\infty}^{+\infty} g(u)\left(\int_{-\infty}^{+} f(\tau)e^{-2i\pi f(u+\tau)}dt\,)\right)du.$$

This becomes:

$$C(f) = \left(\int_{-\infty}^{+\infty} g(u)e^{-2i\pi fu}du\right)\left(\int_{-\infty}^{+\infty} f(\tau)e^{-2i\pi f\tau}d\tau\right) = G(f)F(f)$$

which proves Theorem 4.2.

THEOREM 4.3.– The Fourier transform of the product of two functions $f(t)$ and $g(t)$ is equal to the convolution of their Fourier transforms $F(f)$ and $G(f)$.

Theorem 4.3 is the converse of Theorem 4.1.

Proof of Theorem 4.3. Starting with the hypothesis:

$$f(t)g(t) = \int_{-\infty}^{+\infty} F(f)e^{2i\pi ft}df \int_{-\infty}^{+\infty} G(f)e^{2i\pi f't}df'$$

this can be rewritten as:

$$f(t)g(t) = \int_{-\infty}^{+\infty}\int_{-\infty}^{+\infty} F(f)G(f')\, e^{2i\pi(f+f')t}dfdf'$$

Let $u = f + f'$, hence $f' = u - f$. Substituting this into the expression gives:

$$f(t)g(t) = \int_{-\infty}^{+\infty} e^{2i\pi ut}\left(\int_{-\infty}^{+\infty} F(f)G(u-f)df\right)du$$
$$= \int_{-\infty}^{+\infty} e^{2i\pi ut}(F * G)(u)du$$

which is equivalent to:

$$f(t)g(t) = \mathrm{FT}^{-1}((F * G)(t))$$

proving Theorem 4.3.

Figure 4.11 provides a summary of the two previous theorems.

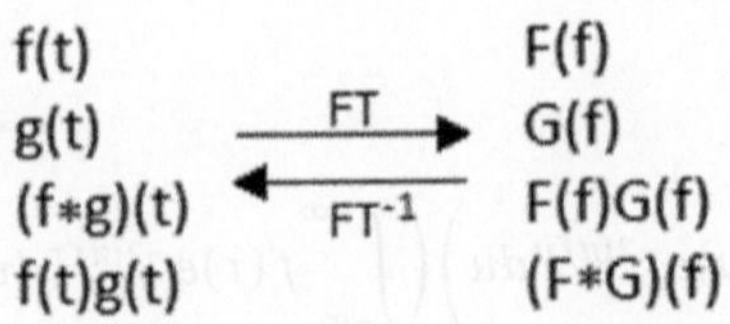

Figure 4.11. *Theorems 4.2 and 4.3*

4.2.4. *Application to analog signals and systems*

The beauty of the Fourier transform lies in its ability to convert a signal from the time domain to the frequency domain. This transformation allows analyzing a signal in both time and frequency domains, providing a deeper understanding of its characteristics.

By applying the Fourier transform, a signal originally defined in the time domain is represented in the frequency domain. This representation highlights the various frequency components that constitute the signal, revealing their amplitudes and frequencies contribute to the overall composition of the signal, offering a clearer picture of its underlying structure.

EXAMPLE 4.5.- To determine the Fourier transform of the sinusoidal signal $f(t) = cos\left(\frac{2\pi\, t}{T}\right)$ with period T (see Figure 4.12), follow these steps.

The Fourier transform $F(f)$ is defined as:

$$F(f) = \int_{-\infty}^{+\infty}\left[f(t) \times e^{(-i2\pi f t)}\right]dt$$

Substitute $f(t) = cos\left(\frac{2\pi t}{T}\right)$ into this definition:

$$F(f) = \int_{-\infty}^{+\infty} \cos(2\pi \frac{t}{T})e^{-2i\pi f t}\, dt$$

Using Euler's formula, $\cos(x) = \frac{e^{ix}+e^{-ix}}{2}$, rewrite $\cos\left(\frac{2\pi\, t}{T}\right)$ as:

$$\cos\left(\frac{2\pi\, t}{T}\right) = \frac{e^{i\frac{2\pi\, t}{T}}+e^{-i\frac{2\pi\, t}{T}}}{2}$$

Thus,

$$F(f) = \int_{-\infty}^{+\infty} \cos\left(2\pi \frac{t}{T}\right) e^{-2i\pi f t}\, dt = \frac{1}{2}\int_{-\infty}^{+\infty} \left[e^{2i\pi\frac{t}{T}} + e^{-2i\pi\frac{t}{T}}\right] e^{-2i\pi f t}\, dt = I_1 + I_2$$

where

$$I_1 = \frac{1}{2}\int_{-\infty}^{+\infty} e^{2i\pi\left(\frac{1}{T}-f\right)t}\, dt \quad and \quad I_2 = \frac{1}{2}\int_{-\infty}^{+\infty} e^{-2i\pi\left(\frac{1}{T}+f\right)t} dt$$

Evaluate each integral using the property of the Dirac delta function:
By setting $k = 2\pi(\frac{1}{T} - f)$, then

$$I_1 = \frac{1}{2}\int_{-\infty}^{+\infty} e^{ikt} dt = \pi\delta\left(2\pi\left(\frac{1}{T} - f\right)\right) = \frac{1}{2}\delta(f - \frac{1}{T})$$

By setting $k' = 2\pi(\frac{1}{T} + f)$, $I_2 = \frac{1}{2}\delta(f + \frac{1}{T})$
This produces a frequency domain spectrum with two distinct lines: $f = \pm \frac{1}{T}$.

$$F(f) = \frac{1}{2}\delta\left(f - \frac{1}{T}\right) + \frac{1}{2}\delta(f + \frac{1}{T})$$

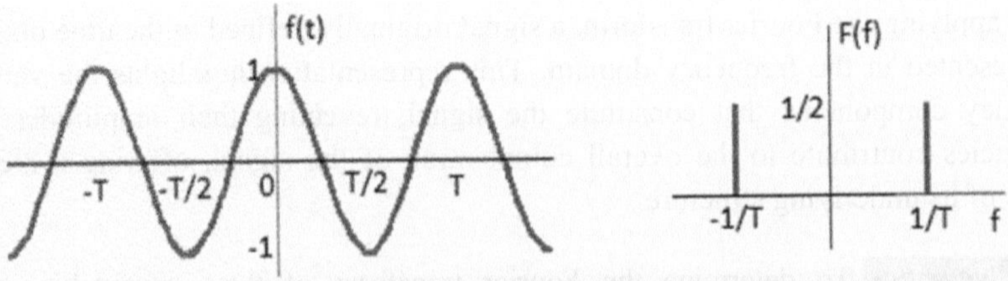

Figure 4.12. *Time and frequency representations of a sinusoidal signal. For a color version of this figure, see www.iste.co.uk/cochard/mathematics2.zip*

EXAMPLE 4.6.– Consider the signal represented by the $f(t)$, which is a "triangle" signal (see Figure 4.13). The Fourier transform $F(f)$ of $f(t)$ can be calculated as follows:

$$F(f) = \int_{-\infty}^{+\infty} f(t)e^{-2i\pi ft}dt = \int_{-a}^{0}(t+a)e^{-2i\pi ft}dt + \int_{0}^{a}(-t+a)e^{-2i\pi ft}dt$$

Since $f(t)$ is piecewise defined over $[-a, a]$, the integral can be split into two parts:

$$F(f) = \int_{-a}^{0}(t+a)e^{-2i\pi ft}dt + \int_{0}^{a}(-t+a)e^{-2i\pi ft}dt$$

This simplifies to:

$$F(f) = \int_{-a}^{0} te^{-2i\pi ft}dt - \int_{0}^{a} te^{-2i\pi ft}dt + a\int_{-a}^{a} e^{-2i\pi ft}dt$$

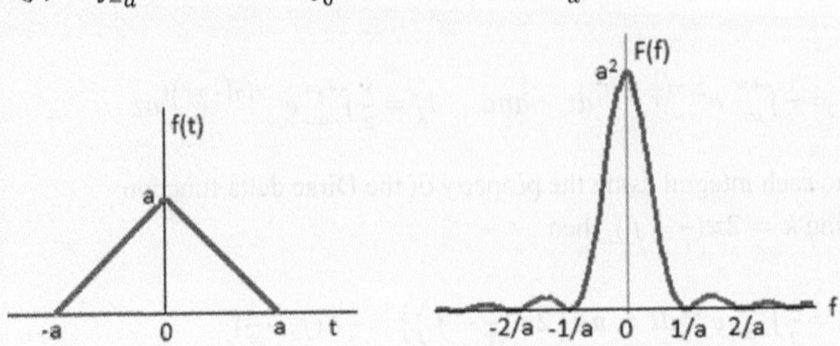

Figure 4.13. *Time and frequency representations of the "triangle" signal. For a color version of this figure, see www.iste.co.uk/cochard/mathematics2.zip*

At this stage, integration by parts can be applied to the first two integrals, while the third is a standard Fourier transform integral for a constant function over the interval $[-a, a]$. In this case,

$$F(f) = -\frac{a}{\pi f}\sin(2\pi fa) + \frac{1}{\pi^2 f^2} sin^2(\pi fa) + \frac{a}{\pi f}\sin(2\pi fa) = a^2(\frac{\sin(\pi fa)}{\pi fa})^2$$

Thus,

$$F(f) = a^2\left(\frac{\sin(\pi fa)}{\pi fa}\right)^2$$

EXAMPLE 4.7.– Consider a signal consisting of an impulse $\delta(t)$, and calculate its corresponding Fourier transform $\Delta(f)$ (see Figure 4.14).

The impulse $\delta(t)$ is defined as:

$$\delta(t) = 0 \text{ for } t \neq 0 \text{ and } \delta(t) = \infty \text{ for } t = 0,$$

with the property that $\int_{-\infty}^{+\infty}\delta(t)\,dt = 1$.

The Fourier transform $\Delta(f)$ of the signal is calculated as follows:

$$\Delta(f) = \int_{-\infty}^{+\infty}\delta(t)e^{-2i\pi ft}dt$$

Since the impulse is concentrated at t=0, this implies

$$\Delta(f) = e^0 = 1$$

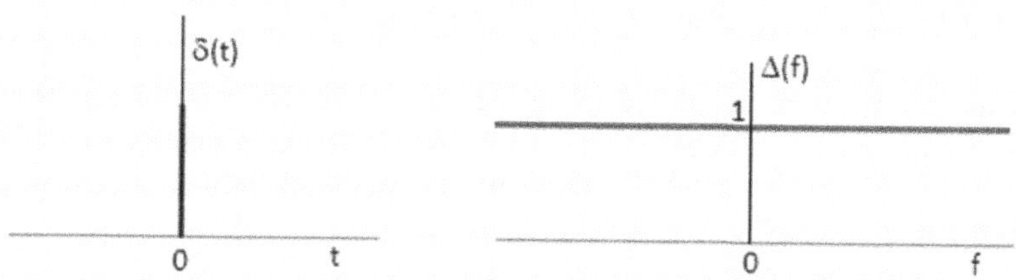

Figure 4.14. *Time and frequency representations of the δ(t) signal. For a color version of this figure, see www.iste.co.uk/cochard/mathematics2.zip*

EXAMPLE 4.8.– Now consider a regular sequence of pulses represented by the signal $f(t)$, as shown in Figure 4.15.

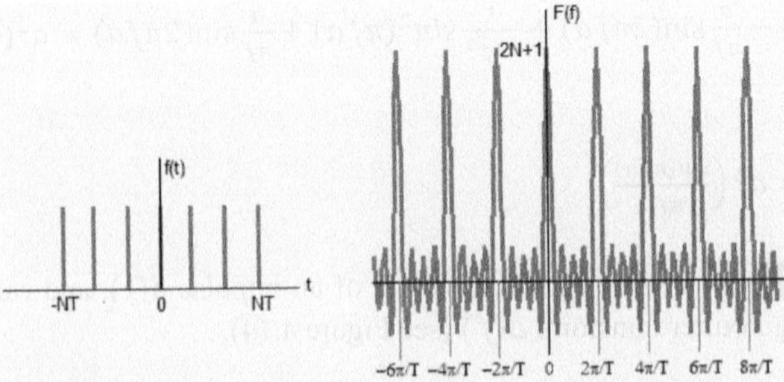

Figure 4.15. *Time and frequency representations of a pulse sequence. For a color version of this figure, see www.iste.co.uk/cochard/mathematics2.zip*

The signal $f(t)$ consists of periodic pulses of width Δ, spaced by a period T. Mathematically, it can be expressed as an infinite sum of shifted pulses $\delta(t)$, where each pulse is shifted by nT for every integer n:

$$f(t) = \sum_{n=-N}^{+N} \delta(t + nT).$$

The Fourier transform of this regular sequence of pulses is:

$$\mathrm{F}(f) = \sum_{n=-N}^{+N} \int_{-\infty}^{+\infty} \delta(t + nT)e^{-2i\pi f t}dt = \sum_{n=-N}^{+N} e^{2i\pi f nT}$$

The expression can be simplified using the formula for the sum of a geometric series. It represents a geometric series with $2N + 1$ terms, a first term $e^{-2i\pi f NT}$, and a common reason of $\mathrm{e}^{2i\pi f T}$. Thus, the Fourier transform is given by (see Figure 4.15):

$$F(f) = e^{-2i\pi f NT}\left(\frac{1-e^{2i\pi f(2N+1)T}}{1-e^{2i\pi f T}}\right) = \frac{\sin[2\pi f\left(N+\frac{1}{2}\right)T]}{\sin(\pi f T)}$$

4.2.5. *Transfer function*

Consider an analog system L to which an input signal $e(t) = e^{(2i\pi f t)}$ is applied. The output signal is given by the convolution of the input signal with the impulse response $h(t)$:

$$s(t) = \int_{-\infty}^{+\infty} e(t-u)h(u)du = \int_{-\infty}^{+\infty} e^{2I\pi f(t-u)}h(u)du$$

This simplifies to:

$$s(t) = e^{2i\pi ft} \int_{-\infty}^{+\infty} h(u)e^{-2i\pi fu}du = H(f)e^{2i\pi ft}$$

where $H(f)$ is the Fourier transform of $h(t)$. Thus, in this particular case, the output signal $s(t)$ can be obtained by multiplying the Fourier transform of the input signal by the transfer function $H(f)$ of the system:

$$s(t) = H(f)e(t)$$

$H(f)$ is commonly referred to as the *transfer function* of the L system.

Since the signal $s(t)$ is the convolution product $(e * h)(t)$, the fundamental property of the Fourier transform can be applied to yield:

$$S(f) = H(f) \times E(f),$$

where $S(f)$ and $E(f)$ represent the Fourier transforms of $s(t)$ and $e(t)$, respectively, and $H(f)$ is the Fourier transform of $h(t)$.

EXAMPLE 4.9.– Consider an analog system for which the impulse response is a "square" signal, as in Example 4.1. In this case, we can express the transfer function as $H(f)$ as follows:

$$H(f) = \frac{\sin(2\pi f)}{2\pi f}$$

EXAMPLE 4.10.– Consider an analog system obeys the following differential equation:

$$s(t) + RC\frac{ds(t)}{dt} = e(t)$$

If $S(f)$ and $E(f)$ are the Fourier transforms of $s(t)$ and $e(t)$, respectively, applying the Fourier transform to the differential equation results in:

$$\begin{array}{ccc} s(t) & \longrightarrow & S(f) \\ s'(t) & \longrightarrow & 2i\pi f S(f) \end{array}$$

Thus, the transformed equation becomes:

$$S(f) + 2i\pi fRCS(f) = E(f)$$

Hence,

$$H(f) = \frac{1}{1+2i\pi fRC}$$

4.2.6. *Autocorrelation and intercorrelation of signals*

Let two signals be represented by the functions $x(t)$ and $y(t)$.

The *autocorrelation function* $R_{xx}(\tau)$ of the signal $x(t)$ is defined by:

$$R_{xx}(\tau) = \int_{-\infty}^{+\infty} x(t)x^*(t-\tau)dt.$$

This function measures the similarity of a signal to itself for different offsets τ. If the signal $x(t)$ is real, then the previous expression simplifies to a convolution product, which is an interesting property. Furthermore, it can be shown that:

$$R_{xx}(\tau) = R_{xx}(-\tau),$$

indicating that the autocorrelation function is an even function. This property shows that the similarity of a signal to itself does not depend on the direction of the time shift.

On the contrary, when the signal is not shifted, i.e. when $\tau = 0$, there is an exact superposition of the signal with itself. In this case, the resemblance is maximal, and it follows that the autocorrelation function $R_{xx}(0)$ is equal to the energy of the signal, which is the integral of the square of the function $x(t)$ over its entire duration:

$$R_{xx}(0) = \int_{-\infty}^{+\infty} |x(t)|^2 dt.$$

Therefore,

$$|R_{xx}(\tau)| \leq |R_{xx}(0)|$$

Similarly, the *intercorrelation function* between signals $x(t)$ and $y(t)$ is defined by:

$$R_{xy}(\tau) = \int_{-\infty}^{+\infty} x(t)y^*(t-\tau)dt.$$

The *power spectral density* of a signal represented by a function $x(t)$ is given by $PDS(f) = |X(t)|^2$, where $X(t)$ is the Fourier transform of $x(t)$. Therefore:

$$|X(f)|^2 = X(f)X^*(f) = \int_{-\infty}^{+\infty} (x * x^*)(t) e^{-2i\pi f t} dt$$

This can be rewritten as:

$$|X(f)|^2 = \int_{-\infty}^{+\infty} (\int_{-\infty}^{+\infty} x(\tau)x^*(\tau - t)\, d\tau) e^{-2i\pi f t} dt = \int_{-\infty}^{+\infty} R_{xx}(t) e^{-2i\pi f t} dt,$$

which shows that the power spectral density (PSD) of $x(t)$ is the Fourier transform of the autocorrelation function $R_{xx}(t)$.

Similarly, the Fourier transform of the intercorrelation function $R_{xy}(\tau)$ between two signals $x(t)$ and $y(t)$ is equal to the product of their conjugate Fourier transforms:

$$FT\{R_{xy}(\tau)\} = X(f)Y^*(f),$$

where $X(f)$ and $Y(f)$ are the Fourier transforms of $x(t)$ and $y(t)$, respectively (see Figure 4.16).

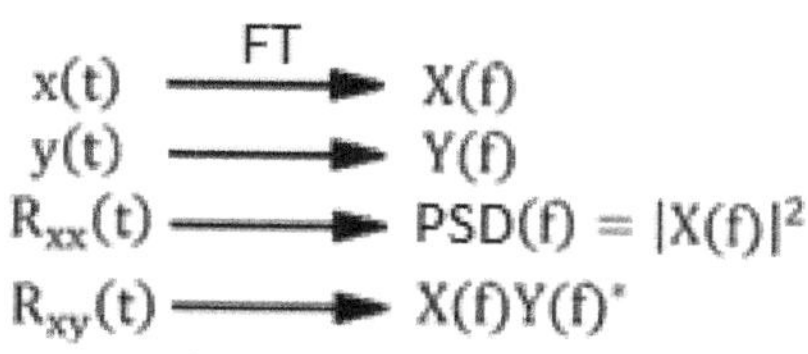

Figure 4.16. *Fourier transform correspondences*

EXAMPLE 4.11.– Consider the square-wave signal represented by the function $x(t)$ in Figure 4.17. Its Fourier transform is given by $X(f)$:

$$X(f) = \int_{-\infty}^{+\infty} x(t) e^{-2i\pi f t} dt.$$

For a square wave of width $2a$ and period T, the Fourier transform can be computed as:

$$X(f) = \int_{-a}^{+a} e^{-2i\pi ft}dt = \frac{1}{-2i\pi f}\left(e^{-2i\pi fa} - e^{2i\pi fa}\right) = \frac{\sin(2\pi fa)}{\pi f}$$

Thus,

$$|X(f)|^2 = \left(\frac{\sin(2\pi fa)}{\pi f}\right)^2$$

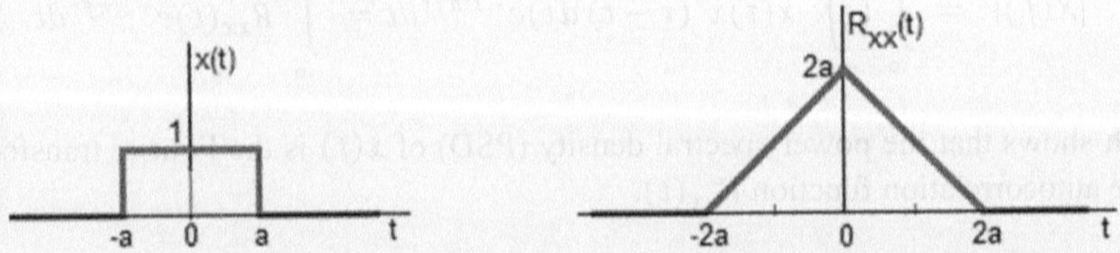

Figure 4.17. *Graphical representations of x(t) and Rxx(t). For a color version of this figure, see www.iste.co.uk/cochard/mathematics2.zip*

In addition, the autocorrelation function $R_{xy}(\tau)$ is given by:

$$R_{xx}(\tau) = \int_{-\infty}^{+\infty} x(t)x(t-\tau)dt = \begin{cases} 0 \ if\ \tau < -2a \\ \tau + 2a \ if\ -2a \leq \tau \leq 0 \\ -\tau + 2a \ if\ 0 \leq \tau \leq 2a \\ 0 \ if\ \tau > 2a \end{cases}$$

The Fourier transform of. $R_{xy}(\tau)$ is

$$\int_{-\infty}^{+\infty} R_{xx}(t)e^{-2i\pi ft}dt = \int_{-2a}^{0}(t+2a)e^{-2i\pi ft}dt + \int_{0}^{2a}(-t + 2a)e^{-2i\pi ft}dt = \frac{\sin^2(2\pi fa)}{\pi^2 f^2}$$

This confirms the expected result.

Note that the total power carried by a signal $x(t)$ can be calculated by integrating the square of the signal over its entire duration. This corresponds to the signal's total energy.

Mathematically, the total power W of the signal $x(t)$ is given by:

$$W = \int_{-\infty}^{+\infty} |x(t)|^2 dt.$$

According to Theorem 4.1 (Parseval's Theorem), this can be expressed in the frequency domain as:

$$W = \int_{-\infty}^{+\infty} |X(f)|^2 df = \int_{-\infty}^{+\infty} PSD(f)df,$$

where $PSD(f)$ denotes the power spectral density of the signal.

Signals can be classified according to their energy content. The total power over a finite interval $[0, \tau]$ is

$$P = \frac{1}{\tau}\int_0^{\tau} |x(t)|^2 dt.$$

When τ tends to infinity ($\tau \rightarrow \infty$):

– If the integral of the power spectral density ($\int_{-\infty}^{+\infty} PSD(f)df$) is finite, this indicates that the signal has finite energy. In this case, the signal power is zero.

– If the integral of the power spectral density is infinite, this indicates that the signal has infinite energy. However, the signal power may still be finite.

4.3. Discrete Fourier transform and fast Fourier transform

4.3.1. *Discrete Fourier transform*

DEFINITION.– Let $f(n)$ be a sequence of N numbers. The *discrete Fourier transform* (DFT) of $f(n)$ is the sequence $F(k)$ defined as:

$$F(k) = \sum_{n=0}^{N-1} f(n)e^{-2i\pi\frac{kn}{N}},$$

where k is an integer ranging from 0 to $N-1$, and $F(k)$ represents the Fourier coefficients corresponding to the frequency component indexed by k.

The inverse discrete Fournier transform is given by:

$$f(n) = \frac{1}{N}\sum_{k=0}^{N-1} F(k)e^{2i\pi\frac{kn}{N}}$$

Demonstration of the inverse transform:

$$A = \frac{1}{N}\sum_{k=0}^{N-1} F(k)e^{2i\pi\frac{kn}{N}} = \frac{1}{N}\sum_{k=0}^{N\_1}\left[\sum_{m=0}^{N-1} f(m)e^{-2i\pi\frac{km}{N}}\right]e^{2i\pi\frac{kn}{N}}$$

according to the definition of the DFT. Reversing the order of summations:

$$A = \frac{1}{N}\sum_{m=0}^{N-1} f(m)\left[\sum_{k=0}^{N-1} e^{2i\pi k\frac{n-m}{N}}\right]$$

Now, evaluate the terms in square brackets. If $n \neq m$, the term $\sum_{k=0}^{N-1} e^{2i\pi k\frac{n-m}{N}}$ forms a geometric series with ratio q, and the sum is:

$$\sum_{k=0}^{N-1} e^{2i\pi k\frac{n-m}{N}} = \frac{1-q^N}{1-q} = \frac{1-e^{2i\pi(n-m)}}{1-q}$$

This simplifies to:

$$\sum_{k=0}^{N-1} e^{2i\pi k\frac{n-m}{N}} = \frac{1-\cos\bigl(2\pi(n-m)\bigr) - i\,\sin\bigl(2\pi(n-m)\bigr)}{1-q} = \frac{1-1}{1-q} = 0$$

For $n = m,$ the sum is:

$$\sum_{k=0}^{N-1} e^{2i\pi k\frac{n-m}{N}} = N$$

Thus, the expression for A reduces to:

$$A = \sum_{m=0}^{N-1} f(n) = f(n)$$

This proves the accuracy of the inverse DFT.

This link between DFT and the continuous Fourier transform lies in the process of discretizing the continuous domain, which is necessary for computation. Continuous variables and infinite sums are replaced by discrete values and finite sums. This involves the following steps:

1) Replace the continuous $s(t)$ by sampled values $s(nT_e)$, where the signal sampled at a frequency $f_e = \frac{1}{T_e}$. This is the discretization process.

2) Limit the number of values to be calculated to a finite number N. The continuous Fourier transform $S(f)$ of the signal $s(t)$:

$$S(f) = \int_{-\infty}^{+\infty} s(t)e^{-2i\pi ft}dt$$

is replaced by the following discrete sum:

$$\bar{S}(f) = \sum_{n=0}^{N-1} s(nT_e)e^{-2i\pi fnT_e}$$

3) The calculation is performed for a finite number of variables of f, which can be chosen as the regular multiples of τ:

$$\bar{S}(k\tau) = \sum_{n=0}^{N-1} s(nT_e)e^{-2i\pi kn\tau T_e}$$

The choice of τ is arbitrary but can be simplified by taking $\tau = \frac{f_e}{N} = \frac{1}{NT_e}$, establishing a link between the time and frequency discretization. This relates both the discretization of t and that of f:

$$\bar{S}(k\tau) = \sum_{n=0}^{N-1} s(n\,T_E)e^{-2i\pi k\frac{n}{N}}$$

Alternatively, by redefining the functions, the expression can be simplified as:

$$\bar{S}(k) = \sum_{n=0}^{N-1} s(n)e^{-2i\pi k\frac{n}{N}}$$

Note that $s(n)$ is limited to N successive values, making it periodic with period N. Consequently, the expression shows that $S(k)$ is also periodic with period N. In the expression above, the DFT formula is recovered.

Let us look at some immediate properties of the DFT:

– Linearity: given two sequences $u(n)$ and $v(n)$ with their discrete Fournier transforms $U(k)$ and $V(k)$, respectively:

$$u(n) \quad \longrightarrow \quad U(k)$$

$$v(n) \quad \longrightarrow \quad V(k)$$

For a linear combination of $u(n)$ and $v(n)$:

$$w(n) = \lambda u(n) + \mu v(n) \quad \longrightarrow \quad W(k) = \lambda U(k) + \mu V(k)$$

λ, μ are constants.

This property shows that the DFT of a linear combination of signals is the same linear combination of their respective DFTs.

– Symmetry: if the sequence f(n) is real, then the DFT satisfies a symmetry property:

$$F(N-k) = F(k)$$

This indicates that the Fourier transform of a real-valued sequence is symmetric in frequency.

– Convolution: for periodic sequences $f(n)$ and $u(n)$ of period N, the discrete convolution $f * u$ is defined as:

$$g(n) = \sum_{k=0}^{N-1} f(k)u(n-k)$$

The resulting sequence $g(n)$ is also of period N, and in the frequency domain, convolution corresponds to multiplication of the discrete Fournier transforms:

$$f(n) \longrightarrow F(k) \quad \text{and} \quad u(n) \quad \longrightarrow \quad U(k)$$

$$f * u \longrightarrow F(k).U(k)$$

This property shows that convolution in the time domain corresponds to point-wise multiplication in the frequency domain.

– Parseval's equality: Parseval's theorem for the DFT relates the total energy in the time domain to that in the frequency domain:

$$\frac{1}{N}\sum_{n=0}^{N-1} |f(n)|^2 = \sum_{k=0}^{N-1} |F(k)|^2$$

This equality demonstrates that the sum of the squared magnitudes of a sequence in the time domain is equal to the sum of the squared magnitudes of its DFT coefficients in the frequency domain.

4.3.2. *Fast Fourier transform (FFT)*

The DFT is defined as follows:

$$F(k) = \sum_{n=0}^{N-1} f(n)e^{-2i\pi\frac{kn}{N}}$$

This expression allows a computer to calculate the value of $F(k)$ for each k, given a fixed number N of points. The direct calculation of the one-dimensional DFT requires a large number of numerical operations. Specifically, for each k, N complex multiplications of $f(n)$ by $e^{-2i\pi k\frac{n}{N}}$ are needed, along with $N-1$ additions of the results. Since there are N values of k, the total number of operations is proportional to N^2.

For example, if $N = 512$, this direct computation would require 262,144 operations. However, using the "FFT", which reduces the computational complexity to $N\log_2(N)$, the number of operations is significantly reduced to just 4,608 operations for $N = 512$.

The Cooley–Tukey algorithm. The Cooley–Tukey algorithm, one of the most well-known FFT algorithms, assumes that $N = 2M = 2^p$ (i.e. consider N as a power of 2). The main idea is to break the DFT summation into two parts by separating even and odd indexed elements of the sequence:

$$\begin{aligned} F(k) &= \sum_{n=0}^{N-1} f(n)e^{-2i\pi\frac{kn}{N}} \\ &= \sum_{m=0}^{M-1} f(2m)e^{-2i\pi\frac{km}{M}} + \sum_{m=0}^{M-1} f(2m+1)e^{-i\pi\frac{k(2m+1)}{M}} \\ &= F_1(k) + e^{-i\pi\frac{k}{M}}F_2(k) \end{aligned}$$

where $M = \frac{N}{2}$ and $\omega_M = e^{-i\pi\frac{1}{M}}$. Defining the two sums as:

$$F_1(k) = \sum_{m=0}^{M-1} f(2m)\omega_M^{2km} \quad and \quad F_2(k) = \sum_{m=0}^{M-1} f(2m+1)\omega_M^{2km}$$

The result can be expressed as:

$$F(k) = F_1(k) + \omega_M^k F_2(k)$$

where $F_1(k)$ corresponds to the even-indexed terms of the sequence $f(n)$, and $F_2(k)$ corresponds to the odd-indexed terms of the sequence $f(n)$.

Splitting the interval: the next step is to split the interval of k into two parts: $[0, M-1]$ and $[M, N-1]$. The DFT for $k+M$ can be related to the DFT for k as follows:

$$F(k+M) = F_1(k+M) + \omega_M^{(k+M)} F_2(k+M)$$

Moreover,

$$\omega_M^{(k+M)} = e^{-i\pi\frac{(k+M)}{M}} = e^{-i\pi} e^{-i\pi\frac{k}{M}} = -e^{-i\pi\frac{k}{M}} = -\omega_M^k$$

$$\omega_M^{2(k+M)m} = e^{-2i\pi\frac{(k+M)m}{M}} = e^{-2i\pi m} e^{-2i\pi\frac{km}{M}} = \omega_M^{2km}$$

Thus,

$$F(k+M) = F_1(k) - \omega_M^k F_2(k)$$

On the interval $[0, M-1]$, the calculations for the points $0, 1, 2, \ldots, (M-1)$ yield $F_1(k)$ and $F_2(k)$, and thus $F(k)$. On the interval $[M, N-1]$, the calculations for the points $M, (M+1), \ldots, (N-1)$, use $F(k+M)$ as defined by the previous expression.

The number of operations required to perform an FFT can be summarized as:

– For multiplications: $\frac{N}{2}\log_2(N)$.

– For additions: $N\log_2(N)$.

EXAMPLE 4.12.– Consider the eight-point discretized signal shown in Figure 4.18.

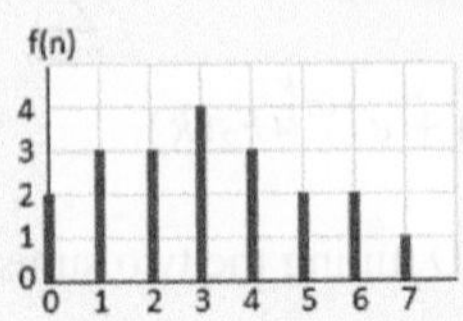

Figure 4.18. *Discretized signal. For a color version of this figure, see www.iste.co.uk/cochard/mathematics2.zip*

Using the FFT, let us compute the values for $F(k)$.
Herein:

$$f(0) = 2,\ f(1) = 3,\ f(2) = 3,\ f(3) = 4,\ f(4) = 3,\ f(5) = 2,$$
$$f(6) = 2,\ f(7) = 1.$$

Splitting the signal into even and odd terms:

1) Even-indexed part $F_1(k)$:

$$F_1(k) = f(0) + f(2)\, e^{-i\pi\frac{k}{2}} + f(4)\, e^{-i\pi k} + f(6)\, e^{-3i\pi\frac{k}{2}}$$

Simplifying:

$$F_1(k) = 2 + 3e^{-i\pi\frac{k}{2}} + 3e^{-i\pi k} + 2e^{-3i\pi\frac{k}{2}} = 2 + 3(-i)^k + 3(-1)^k + 2i^k$$

2) Odd-indexed part $F_2(k)$:

$$F_2(k) = f(1) + f(3)\,(-i)^k + f(5)\,(-1)^k + f(7)\, i^k$$

Simplifying:

$$F_2(k) = 3 + 4(-i)^k + 2(-1)^k + i^k$$

Fourier transform result:

$$\omega_4^k = e^{-i\pi\frac{k}{4}} = (e^{-i\frac{\pi}{4}})^k = (\tfrac{\sqrt{2}}{2}(1-i))^k$$

Calculations for different values of k:

For $k = 0$:
$\omega_4^0 = 1$
$F_1(0) = 2 + 3 + 3 + 2 = 10$
$F_2(0) = 3 + 4 + 2 + 1 = 10$
$F(0) = F_1(0) + F_2(0) = 20$

For $k = 1$:
$\omega_4^1 = \frac{\sqrt{2}}{2}(1-i)$
$F_1(1) = 2 - 3i - 3 + 2i = -1 - i$
$F_2(1) = 3 - 4i - 2 + i = 1 - 3i$
$F(1) = -1 - i + \frac{\sqrt{2}}{2}(1-i)(1-3i) = -1 - \sqrt{2} - i(1 + 2\sqrt{2})$

For $k = 2$:
$\omega_4^2 = -i$
$F_1(2) = 2 - 3 + 3 - 2 = 0$
$F_2(2) = 3 - 4 + 2 - 1 = 0$
$F(2) = 0$

For $k = 3$:
$\omega_4^3 = -\frac{\sqrt{2}}{2}(1 + i)$
$F_1(3) = 2 + 3i - 3 - 2i = -1 + i$
$F_2(3) = 3 + 4i - 2 - i = 1 + 3i$
$F(3) = -1 + i - \frac{\sqrt{2}}{2}(1 + i)(1 + 3i) = -1 + \sqrt{2} + i(1 - 2\sqrt{2})$

For $k = 4$:
$F(4) = F(0 + 4) = F_1(0) - F_2(0) = 10 - 10 = 0$

For $k = 5$:
$F(5) = F(1 + 4) = F_1(1) - \frac{\sqrt{2}}{2}(1 - i)F_2(1) = -1 - i - \frac{\sqrt{2}}{2}(1 - i)(1 - 3i)$
$= -1 + \sqrt{2} - i(1 - 2\sqrt{2})$

For $k = 6$:
$F(6) = F(2 + 4) = F_1(2) + iF_2(2) = 0$

For $k = 7$:
$F(7) = F(3 + 4) = F_1(3) + \frac{\sqrt{2}}{2}(1 + i)F_2(3) = -1 + i + \frac{\sqrt{2}}{2}(1 + i)(1 + 3i)$
$= -1 - \sqrt{2} + i(1 + 2\sqrt{2})$

Efficient calculation of the Fourier transform requires rearranging the sequence $f(n)$ to speed up the algorithm. One effective method is the “reverse bit rule” algorithm. To illustrate this, consider the calculation of the Fourier transform $F(k)$ using eight-point sequence: $\{0, 1, 2, 3, 4, 5, 6, 7\}$. As used in Example 4.12, $M = 4$.

The values $F(k)$ are related to $F_1(k)$ and $F_2(k)$, which correspond to the values $f(2m)$ and $f(2m + 1)$, respectively. For each value of $k, F(k)$ and $F(k + M)$ are derived from $F_1(k)$ and $F_2(k)$, where $k = 0, 1, 2, 3$. This relationship is visualized in the butterfly graph shown in Figure 4.19.

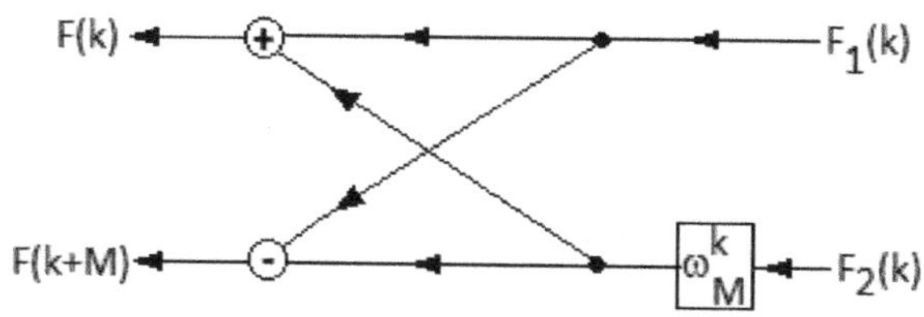

Figure 4.19. *Butterfly graph*

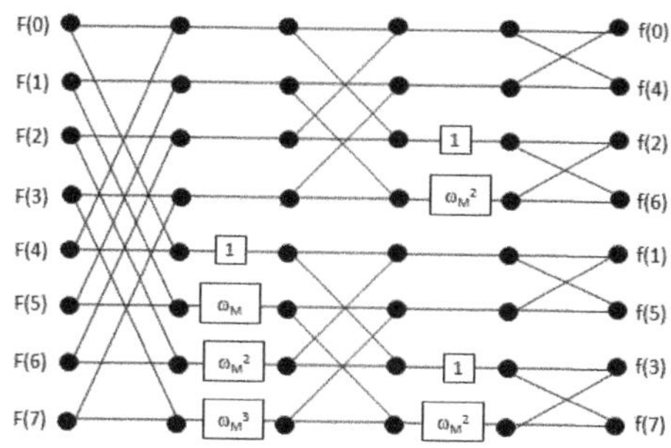

Figure 4.20. *From F(n) to f(n)*

$F_1(k)$ and $F_2(k)$ are linked to the values $f(2m)$ and $f(2m+1)$ through the following matrix expressions:

$$\begin{bmatrix} F_1(0) \\ F_1(1) \\ F_1(2) \\ F_1(3) \end{bmatrix} = \begin{bmatrix} 1 & 1 & 1 & 1 \\ 1 & \omega_M^2 & \omega_M^4 & \omega_M^6 \\ 1 & \omega_M^4 & \omega_M^8 & \omega_M^{12} \\ 1 & \omega_M^6 & \omega_M^{12} & \omega_M^{18} \end{bmatrix} \begin{bmatrix} f(0) \\ f(2) \\ f(4) \\ f(6) \end{bmatrix} \qquad \begin{bmatrix} F_2(0) \\ F_2(1) \\ F_2(2) \\ F_2(3) \end{bmatrix} = \begin{bmatrix} 1 & 1 & 1 & 1 \\ 1 & \omega_M^2 & \omega_M^4 & \omega_M^6 \\ 1 & \omega_M^4 & \omega_M^8 & \omega_M^{12} \\ 1 & \omega_M^6 & \omega_M^{12} & \omega_M^{18} \end{bmatrix} \begin{bmatrix} f(1) \\ f(3) \\ f(5) \\ f(7) \end{bmatrix}$$

Combining the two relationships graphically results in a "butterfly" pattern as shown in Figure 4.20 (intentionally simplified). The structure of the diagram is based on a hierarchical breakdown of the values of $f(n)$.

Initially, the values are divided into two groups based on their indices: $f(2m)$ and $f(2m+1)$. For each of these groups, further divisions are made to form additional groups:

Initial group: $\{f(0), f(1), f(2), f(3), f(4), f(5), f(6), f(7)\}$.

Second-level groups: $\{f(0), f(2), f(4), f(6)\}, \{f(1), f(3), f(5), f(7)\}$.

Third-level groups: $\{f(0), f(4)\}, \{f(2), f(6)\}, \{f(1), f(5\}, \{f(3), f(7)\}$.

n decimal	n binary	reversal	new n decimal
0	000	000	0
1	001	100	4
2	010	010	2
3	011	110	6
4	100	001	1
5	101	101	5
6	110	011	3
7	111	111	7

Figure 4.21. *Renumbering*

In Figure 4.20, obtaining the values $F(k)$ requires organizing the values $f(n)$ according to the inverted bit rule. This rule is visually represented in Figure 4.21, which shows the renumbering of the points $f(n)$ in the sequence $\{0,4,2,6,1,5,3,7\}$. Here is a summary of the FFT algorithm calculation process:

```
Supply of vector f(x) and n
N=2^n //N must be a power of 2
J=1
//renumbering
For I = 1 to N-1
        If I < J
                Swapping f(J) and f(I)
                K=N/2
        EndIf
        While k < J
                J=J-k
                K=k/2
        EndWhile
        J=J+k
EndFor
//calculation
For L = 1 to n
        w=exp(-ip2 )^(L-1)
```

```
        u="i" // i is the imaginary number
        For J=1 to 2^(L-1)
                For I = J to N in 2L steps
                        v=I+2^(L-1)
                        f(v)=f(I)-uf(v)
                        f(I)=f(I)+uf(v)
                EndFor
                U=uw
        EndFor
EndFor
```

There are many other algorithms for calculating the DFT.

4.4. Laplace transform

4.4.1. *Definition*

The *Laplace transform*, like the Fourier transform, is a mathematical tool used for analyzing signals in the frequency domain. However, it offers greater generality than the Fourier transform, as it can be applied to non-periodic signals and dynamic systems.

For a causal signal represented by the function $x(t)$ (i.e. zero for $t < 0$), its Laplace transform $X(p)$ is defined by the following integral:

$$X(p) = \mathrm{LT}\big(x(t)\big) = \int_{0}^{+\infty} x(t)e^{-pt}dt$$

where p is, *in principle*, a complex variable. For practical purpose, p is often considered as a real variable. The definition assumes that the integral converges.

Existence of the Laplace transform: the existence of the Laplace transform depends on the behavior of $x(t)$ at its limits:

– For the limit $t \longrightarrow +\infty$: while $x(t)$ can diverge exponentially, the term $x(t)e^{-pt}$ can remain bounded if $p > p_0$. Here, p_0 is the convergence abscissa, which determines the region in the complex plane where the Laplace transform converges. Specifically, if $x(t)$ grows exponentially, then $e^{-pt}x(t) \longrightarrow e^{-pt}e^{-p_0 t} = e^{-(p-p_0)t}$. This expression remains bounded when $p > p_0$, where p_0 depends on the growth rate of $x(t)$.

– For the limit $t \longrightarrow 0$:, $x(t)$ can possess a singularity at $t = 0$. For instance, if $x(t) = t^n$, the Laplace integral will diverge for $n \leq -1$. The existence of the Laplace transform assumes that the integral converges even in the presence of such singularities.

In the following, it will be assumed that the Laplace integral exists.

Determining the function $x(t)$ from its Laplace transform is not straightforward. However, it is assumed that the inverse Laplace transform exists and yields a unique function $x(t)$.

4.4.2. Properties

Some basic properties of the Laplace transformation are shown in Figures 4.22a and 4.22b.

The additional limits and integrals for the Laplace transform are considered as follows:

– Limits:

$$\lim_{t\to\infty} x(t) = \lim_{p\to\infty} pX(p)$$

Time domain	Laplace domain
ax(t) + by(t)	aX(p)+bY(p)
$x(t-t_0)$	$X(p)e^{-pt_0}$
$X(t)e^{-at}$	X(p-a)
$\frac{dx(t)}{dt}$	pX(p)
$\int_0^t x(u)du$	$\frac{X(p)}{p}$

Figure 4.22a. *Properties of the Laplace transform*

This limit shows that the behavior of $x(t)$ as tt approaches zero can be determined by evaluating $pX(p)$ as p approaches infinity.

– Integral relationship:

$$\int_0^{+\infty} x(t)dt = \int_0^{+\infty} X(p)dp$$

This integral relationship holds provided that the integrals on both sides are convergent.

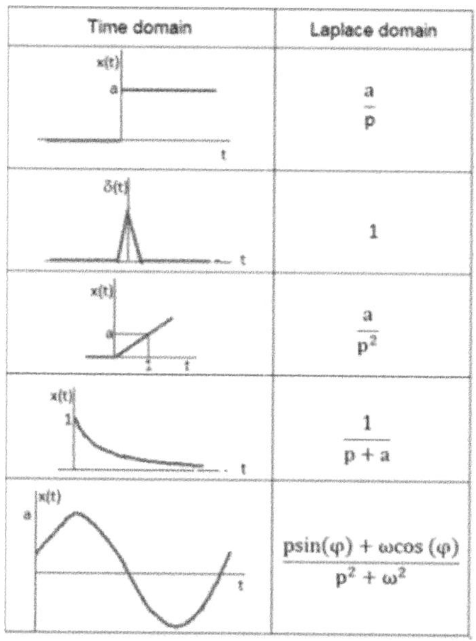

Figure 4.22b. *Laplace transforms of some functions. For a color version of this figure, see www.iste.co.uk/cochard/mathematics2.zip*

4.4.3. *Differential equation*

The Laplace transform facilitates the analysis of filter-type systems by transforming differential equations simpler algebraic equations. This simplifies the process of finding the relationship between the input signal $e(t)$ and the output signal $s(t)$.

Given the Laplace transform $E(p)$ of the input signal $e(t)$ and the Laplace transform $S(p)$ of the output signal $s(t)$, The relationship between these transforms can be expressed through a transfer function $H(s)$. The transfer function is defined as:

$$H(p)E(p) = S(p) \quad \text{or} \quad H(p) = \frac{S(p)}{E(p)}$$

The process of solving the differential equation is simplified by using the Laplace transform. After transforming the differential equation into the Laplace transform domain, the equation becomes algebraic and easier to solve. Once the solution is obtained in the Lagrange domain, the inverse Laplace transform is applied to convert it back to the time domain, providing the solution to the original differential equation.

EXAMPLE 4.13.– Let us consider an RC circuit used as a filter. The differential equation describing this system is:

$$s(t) + RC \cdot \frac{ds(t)}{dt} = e(t)$$

Using the properties of linearity and differentiation, the Laplace transform of $s(t)$ becomes $S(t)$, and the Laplace transform of $\frac{ds(t)}{dt}$ becomes $pS(t)$. This leads to the transformed equation:

$$S(p) + RC \cdot p \cdot S(p) = E(p) \quad \text{hence} \quad H(p) = \frac{1}{1+pRC}$$

Factoring out $S(p)$ on the left-hand side:

$$S(p)\,(1 + RC.p) = E(p) \Rightarrow S(p) = \frac{E(p)}{1+RC.p}$$

The transfer function $H(p)$ relates the output $S(p)$ to the input $E(p)$:

$$H(p) = \frac{S(p)}{E(p)} = \frac{1}{1+RC.p}$$

4.4.4. *Convolution product*

Another interesting property of the Laplace transform is its analogy with the Fourier transform regarding the convolution product. Specifically, the Laplace transform of the convolution product of two functions is equal to the product of their individual Laplace transforms.

THEOREM 4.4.– If $u(t)$ and $v(t)$ are causal functions, and $U(p)$and $V(p)$ are their respective Laplace transforms, then the Laplace transform of the convolution product $(u * v)(t)$ is given by $U(p)V(p)$.

Let $u(t)$ and $v(t)$ be two causal functions, with their Laplace transforms denoted by $U(p)$ and $V(p)$, respectively. Define $G(p)$ as the Laplace transform of the convolution product $(u * v)(t)$. By definition, the convolution product is given by:

$$(u * v)(z) = \int_{-\infty}^{+\infty} u(x - z)v(x)dx$$

$$= \int_{0}^{+\infty} u(x - z)v(x)dx = \int_{0}^{+\infty} u(z - x)v(z)dz$$

since the functions u and v are causal.

To calculate the product of their Laplace transforms $U(p)$ and $V(p)$:

$$U(p) \cdot V(p) = \left(\int_{0}^{+\infty} u(x)e^{-px}dx\right) \cdot \left(\int_{0}^{+\infty} v(y)e^{-py}dy\right)$$

Expanding this product:

$$U(p) \cdot V(p) = \int_{0}^{+\infty}\int_{0}^{+\infty} u(x)v(y)\, e^{-p(x+y)}dx\, dy$$

Changing the order of integration by setting $t = x + y$:

$$U(p)\, V(p) = \int_{0}^{+\infty} v(y)[\int_{0}^{+\infty} u(t - y)e^{-pt}dt]dy = \int_{0}^{+\infty}\left(\int_{0}^{+\infty} v(y)u(t - y)dy\right)e^{-pt}dt$$

In the inner integral, the convolution product of u and v (i.e. $(u * v)(t)$) is obtained:

$$U(p) \cdot V(p) = \int_{0}^{+\infty} (u * v)(t)e^{-pt}dt = \text{LT}[(u * v)]$$

This result shows that the Laplace transform of the convolution product of two causal functions $u(t)$ and $v(t)$ is equal to the product of their Laplace transforms.

For signals, if $e(t)$ is the input signal, $u(t)$ is the output signal, and $h(t)$ is the impulse response of a filter, then taking the Laplace transforms of these signals yields the correspondences shown in Figure 4.23.

e(t) ⟶ E(p)
s(t) ⟶ S(p)
h(t) ⟶ H(p)
s(t) = (h*e)(t) ⟶ S(p)=H(p)E(p)

Figure 4.23. *Laplace transforms*

This correspondence shows that $H(p)$, previously considered as the ratio $\frac{S(p)}{E(p)}$, is indeed the Laplace transform of the impulse response $h(t)$. Taking the inverse Laplace transform provides $h(t)$.

EXAMPLE 4.14.– Returning to the previous example, the transfer function $H(p)$ can be expressed as:

$$H(p) = \frac{\frac{1}{RC}}{p+\frac{1}{RC}}$$

Using the properties of the Laplace transform from the transformation table, the inverse Laplace transform of this function corresponds to:

$$h(t) = \frac{1}{RC} e^{-\frac{t}{RC}}$$

This result represents the impulse response of the system, where RC is the time constant of the circuit.

5

Digitizing an Analog Signal

CONCEPTS COVERED IN THIS CHAPTER.–

Due to considerations related to transport, processing (by computer) and storage, analog signals are converted into digital signals through a process called digitization.

This process consists of several phases, which will be described as follows: sampling, quantization and coding.

References: [BEL 98, BRO 82, COT 97, FON 83, HAY 99, PAP 77].

5.1. Introduction

Digitizing an analog signal generally involves two successive phases, as shown in Figure 5.1.

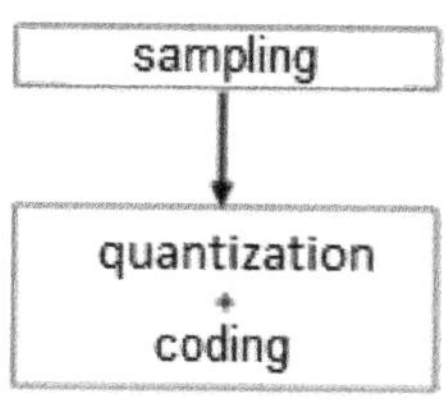

Figure 5.1. *Digitalizing phases*

The first phase is sampling, during which the continuous analog signal is measured at regular time intervals. This process involves taking discrete samples of the signal at specific moments in time. Each sample is represented by a digital value that corresponds to the amplitude of the signal at that instant.

The second phase is quantization, where the sampled values are rounded to discrete levels. These levels are determined by the resolution of the analog-to-digital converter used during digitization process. Higher resolution results in finer quantization levels, allowing for a more accurate representation of the signal.

By combining these two phases, a digital version of the original analog signal is obtained. The digital signal consists of a sequence of discrete values that represent the analog signal on a discrete time and amplitude scale.

Each of these phases will now be examined in detail.

5.2. Sampling

The sampling process involves converting a continuous analog signal $s(t)$ into a discrete signal $s(nT_e)$, where T_e represents the sampling period (and $f_e = \frac{1}{T_e}$ is the sampling frequency). The goal is to periodically measure the signal's amplitude at specific time intervals.

During sampling, the signal is measured at regular intervals of T_e. Each sample corresponds to the signal amplitude at a specific moment in time, resulting in a collection of discrete values that represents the signal at equidistant points.

For the purposes of the analysis, it is assumed that the signal to be sampled has a limited spectrum. This means that the Fourier transform of the signal, denoted as $S(f)$, is zero for frequencies, where $|f| > f_{max}$ (beyond a certain maximum frequency f_{max}). This spectrum limitation is shown in Figure 5.2.

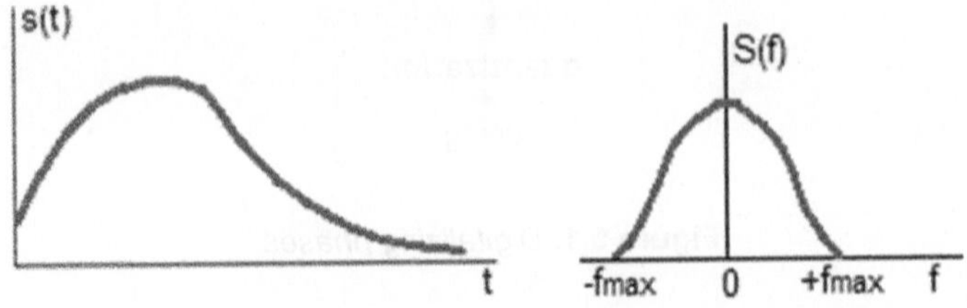

Figure 5.2. *Limited spectrum signal. For a color version of this figure, see www.iste.co.uk/cochard/mathematics2.zip*

Mathematically, sampling involves multiplying the signal by a train of periodic pulses, commonly referred to as a *Dirac comb* $P_{T_e}(t)$. This process is often illustrated graphically, as shown in Figure 5.3.

$$s_e(t) = s(t)P_{T_e}(t) = \sum_{n=-\infty}^{+\infty} s(t)\delta(t - nT_e)$$

Between two measurements, the digitized signal remains constant, resulting in a piecewise constant or "blocky" representation. The shape of the digitized signal is shown in Figure 5.4 and is referred to as a "sample and hold" signal.

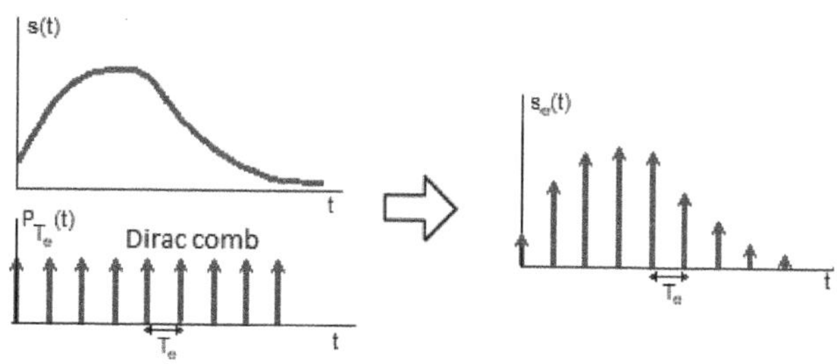

Figure 5.3. *Multiplication by a Dirac comb. For a color version of this figure, see www.iste.co.uk/cochard/mathematics2.zip*

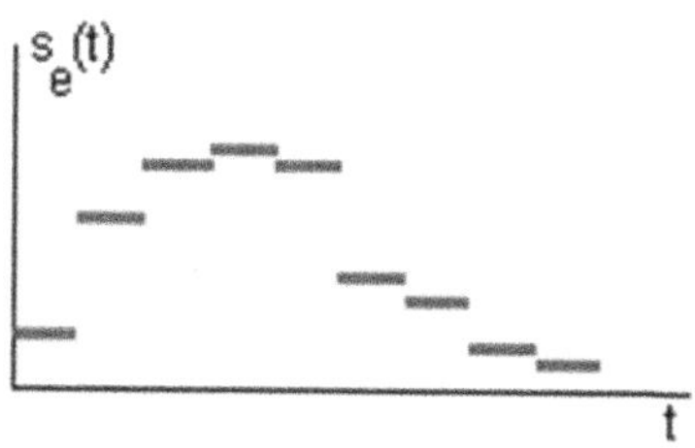

Figure 5.4. *Signal blocking. For a color version of this figure, see www.iste.co.uk/cochard/mathematics2.zip*

The signal is then analyzed in the frequency domain using Fourier transformation. Applying the properties of this transformation yields:

$$s(t) \quad \rightarrow \quad S(f)$$

$$P_{T_e}(t) \quad \rightarrow \quad f_e P_{T_e}(t)$$

$$s(t)P_{T_e}(t) \quad \rightarrow \quad f_e(S * P_{T_e})(f)$$

It is noted that:

$$f_e(S * P_{T_e})(f) = f_e \int_{-\infty}^{+\infty} S(f - \varphi)P_{T_e}(\varphi)d\varphi$$

$$= f_e \sum_{n=-\infty}^{+\infty} \int_{-\infty}^{+\infty} S(f - \varphi)\delta(\varphi - nf_e)d\varphi$$

$$= f_e \sum_{n=-\infty}^{+\infty} S(f - nf_e)$$

Several important aspects of frequency representation should be noted:

NOTE 5.1.– Sampling $s(t)$ results in a periodic frequency spectrum, as shown in Figure 5.5.

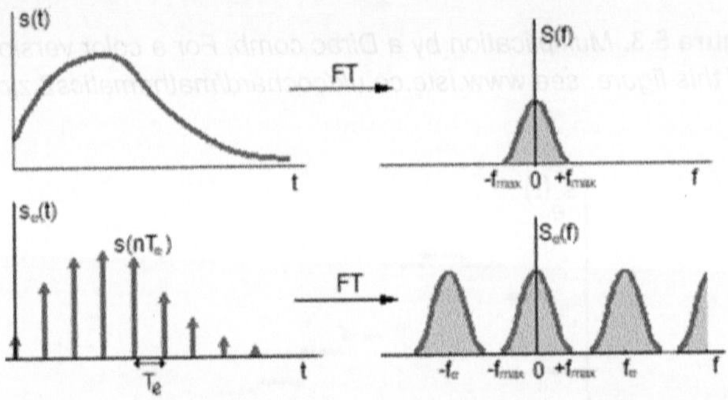

Figure 5.5. *Analog and digital signal spectra. For a color version of this figure, see www.iste.co.uk/cochard/mathematics2.zip*

NOTE 5.2.– The function $S(f)$ is identified by a "gate" function $\Pi_{f_e}(f)$, which acts as a low-pass filter, as shown in Figure 5.6.

$$S(f) = \frac{1}{f_e} S_e(f)\Pi_{f_e}(f)$$

To reconstruct $s(t)$, the inverse Fourier transform (FT) is applied:

$$s(f) = \frac{1}{f_e}\left(S_e * \mathrm{FT}^{-1}(\Pi_{f_e})\right)(f)$$

where

$$\mathrm{FT}^{-1}(\Pi_{f_e})(f) = \frac{\sin(\pi f_e t)}{\pi t}$$

Thus,

$$s(t) = \frac{1}{f_e}\int_{-\infty}^{+\infty} s_e(t-\tau)\frac{\sin(\pi f_e t)}{\pi t}d\tau$$

$$= \frac{1}{f_e}\sum_{n=-\infty}^{+\infty}\int_{-\infty}^{+\infty} s(t-\tau)\delta(t-\tau-nT_e)\frac{\sin(\pi f_e t)}{\pi t}d\tau$$

Therefore,

$$s(t) = \sum_{n=-\infty}^{+\infty} s(nT_e)\frac{\sin(\pi f_e(t-nT_e))}{\pi f_e(t-nT_e)}$$

This shows that $s(t)$ can be reconstructed from the sampled values $s(nT_e)$, where n is an integer representing the sampling instant and T_e is the sampling period.

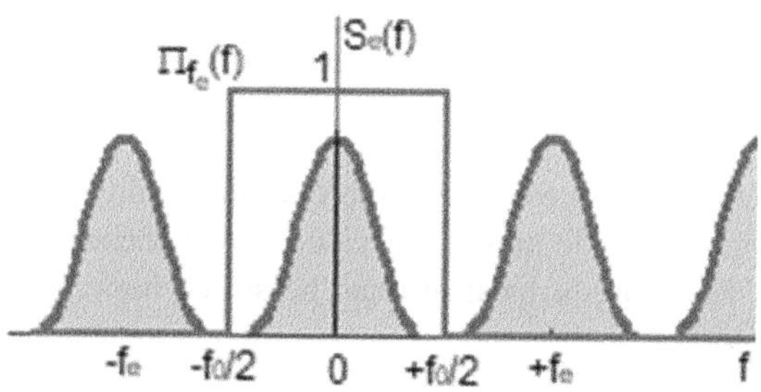

Figure 5.6. *Gate function $\Pi_{f_e}(f)$. For a color version of this figure, see www.iste.co.uk/cochard/mathematics2.zip*

NOTE 5.3.– The previous result holds true if the spectra do not overlap; otherwise, it becomes impossible to "isolate" $S(f)$. When spectra overlap, the situation is as shown in Figure 5.7.

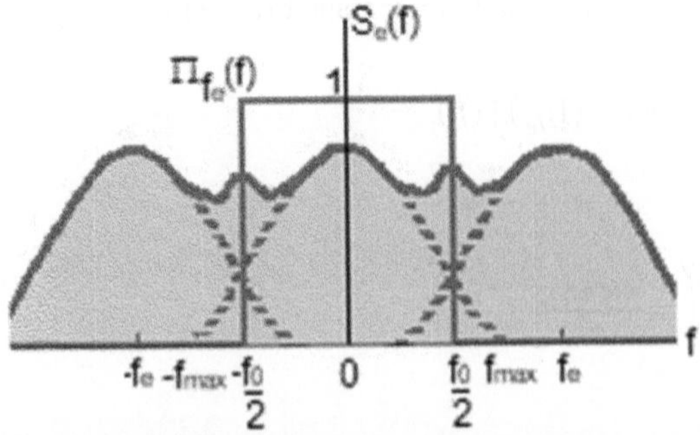

Figure 5.7. *Overlapping spectra. For a color version of this figure, see www.iste.co.uk/cochard/mathematics2.zip*

The validity of the preceding reasoning is ensured by Shannon's theorem, which states that $f_e \geq 2f_{max}$.

The table in Figure 5.8 provides typical values for various sound carriers.

CD audio	44.1 kHz
DVD	48 kHz
Phone	8 kHz
Digital radio	22.5 kHz

Figure 5.8. *Some common values*

Clearly, a higher sampling frequency improves the quality of the digitized signal. However, it also increases the number of values to be encoded, thereby expanding the data volume. Therefore, a balance must be struck between signal quality and data volume.

5.3. Quantization

The values $s(nT_e)$ are mapped to a predefined scale. When this scale is uniform, the graduations are spaced evenly by a step size q, as shown in Figure 5.9.

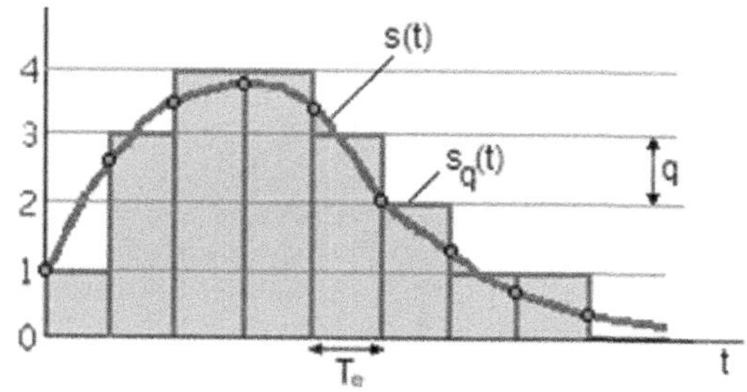

Figure 5.9. *Defining a scale with a q step. For a color version of this figure, see www.iste.co.uk/cochard/mathematics2.zip*

The quantization process involves approximating $s(nT_e)$ to the nearest value m such that:

$$mq - \frac{q}{2} < s(nT_e) < mq + \frac{q}{2}$$

where q is the quantization step size.

The difference $e_q(t) = s(t) - s_q(t)$ represents the quantization error, often referred to as *quantization noise*. To assess the magnitude of this noise, the signal-to-noise ratio is used, defined as P_S/P_B, where P_S is the signal power (in watts) and P_B is the noise power (also in watts).

EXAMPLE 5.1.– Suppose the signal is a sinusoid $y = \frac{5}{4}\sin(t)$. Quantification is performed using the values $0, 0.2, 0.4, 0.6, \ldots, 1.4$ with a quantification step of 0.2, as shown in Figure 5.10.

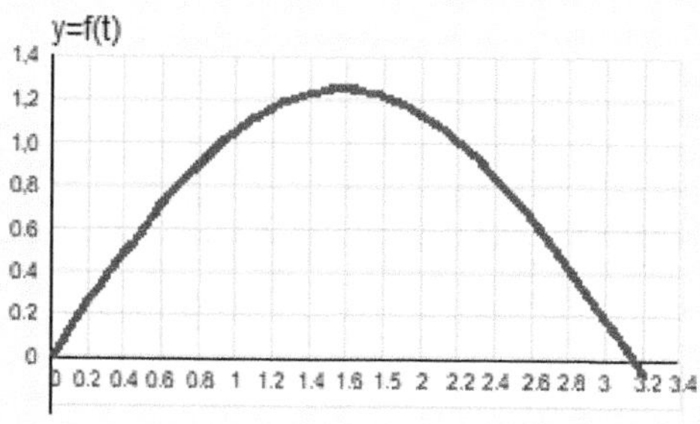

Figure 5.10. *Analog signal. For a color version of this figure, see www.iste.co.uk/cochard/mathematics2.zip*

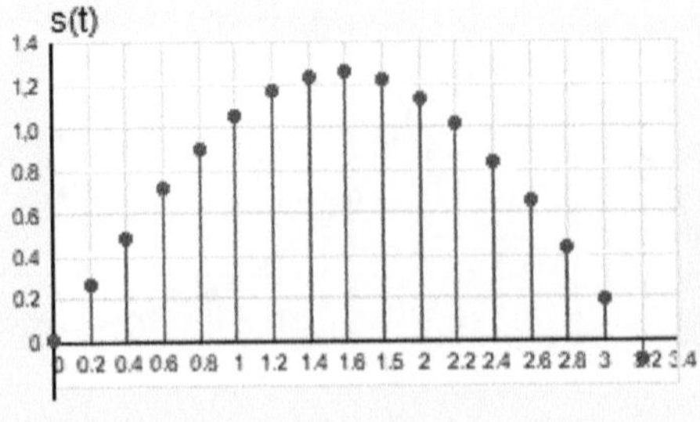

Figure 5.11. *Sampling. For a color version of this figure, see www.iste.co.uk/cochard/mathematics2.zip*

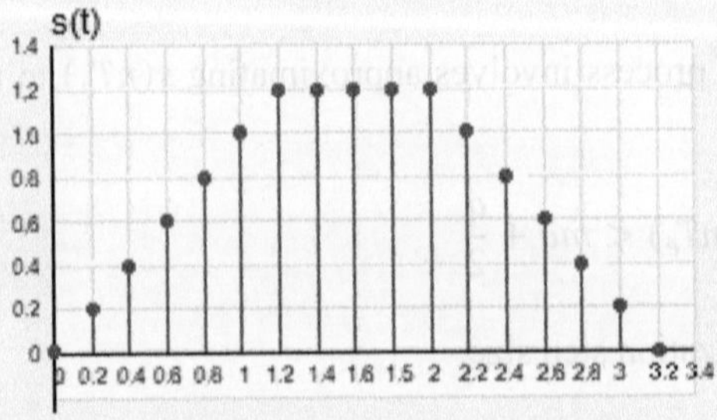

Figure 5.12. *Quantization. For a color version of this figure, see www.iste.co.uk/cochard/mathematics2.zip*

The sampled values $s(t)$ are shown in Figure 5.11. The results of the quantization process are shown in Figure 5.12.

The quantization error, also known as quantization noise, is shown in Figure 5.13.

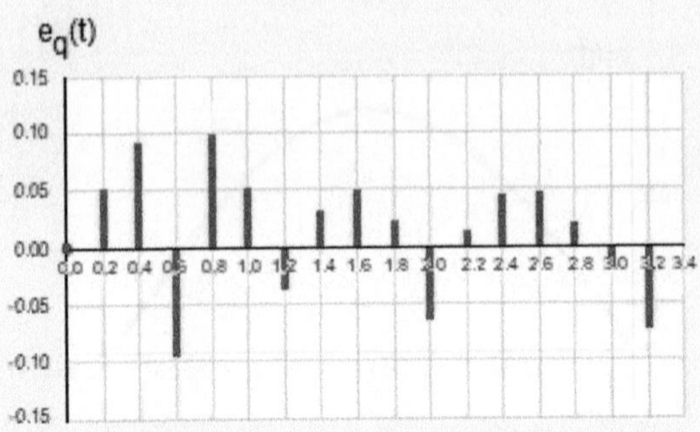

Figure 5.13. *Quantization error. For a color version of this figure, see www.iste.co.uk/cochard/mathematics2.zip*

If the signal ranges between $-V_{max}$ and $+V_{max}$, the signal power P_S can be estimated as:

$$P_S = \frac{V_{max}^2}{2}.$$

When coding with n bits, the quantization divides the signal into $2^n - 1$ intervals of width q. Thus,

$$2V_{max} = (2^n - 1)q \quad \text{and} \quad P_S = \frac{(2^n - 1)^2 q^2}{8} \approx \frac{2^{(2n)} q^2}{8} = 2^{(2n-3)} q^2,$$

assuming n is large.

To express the power of quantization noise, consider the following example. The difference between the sampled value $s(t)$ and the quantized value $s_q(t)$ is within the range $-\frac{q}{2}$ and $+\frac{q}{2}$. Consider the quantization error denoted by $e_q = s(t) - s_q(t)$. This error can be represented as a linear function within the quantization interval, as shown in Figure 5.14.

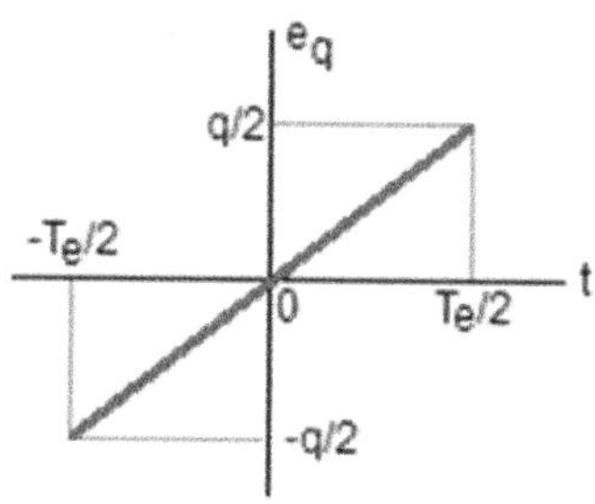

Figure 5.14. *Approximation of the error e_q. For a color version of this figure, see www.iste.co.uk/cochard/mathematics2.zip*

Since $e_q(t) = {}^{qt}/_{T_e}$, the average power of the quantization noise can be calculated as follows:

$$P_B = \frac{1}{T_e} \int_{-T_e/2}^{+T_e/2} e_q^2(t)dt = \frac{q^2}{T_e^2} \int_{-T_e/2}^{+T_e/2} t^2\, dt = \frac{q^2}{3T_e^2}\left(\frac{T_e^2}{8} + \frac{T_e^2}{8}\right) = \frac{q^2}{12}$$

This is an estimate that is usually sufficient. The signal-to-noise ratio

$$P_S/P_B = 3.2^{2n-1}, \text{ in decibels: S/N} = 10\log_{10}\left(\frac{P_S}{P_B}\right) = 6.02n + 1.76$$

5.4. Coding

Coding involves mapping the quantized values $s_q(0), s_q(T_e), s_q(2T_e), s_q(3T_e), \ldots$ to binary combinations.

EXAMPLE 5.2.– Returning to the previous example, the quantization values can be mapped to 3-bit binary numbers as follows:

0.0 → 000 0.2 → 001 0.4 → 010 0.6 → 011 0.8 → 100. 1.0 → 101
1.2 → 110 1.4 → 111

This results in a 3-bit binary encoding of the signal:

000 001 010 100 100 101 110 110 110 110 101 100 010 001 000

The table in Figure 5.15 represents the typical bit depths used for different sound carriers, indicating the precision with which the audio signal is digitized.

CD audio	16 bits
DVD	24 bits
Phone	8 bits
Digital radio	8 bits

Figure 5.15. *Number of bits used*

It is important to recognize that by selecting a smaller step q improves the accuracy of the quantization process, leading to better signal representation. However, this also results in an increased number of values that need to be encoded, which usually means large bit-size words. Consequently, finding an optimal balance between quantization precision and the bit-size of the encoded data is crucial to maintaining efficiency without sacrificing too much quality.

When each quantized value is assigned an independent bit sequence, this method is known as the Pulse Code Modulation (PCM) coding. This was the approach used in the previous example.

Another procedure, known as delta modulation (although this name corresponds to several variants) or differential coding, encodes each value based on its difference from the previous value.

This technique typically reduces the length of the encoded word by taking advantage of the smaller variations between consecutive samples, allowing for more efficient differential quantization.

EXAMPLE 5.3.– Let us revisit to the example of differential coding. The first value is provided "in full": 000 (binary for 0).

The following values are the differences between consecutive values in base 10, as seen in the PCM sequence:

PCM: **0** 1 2 4 4 5 6 6 6 5 4 2 1 0

Differential coding yields the following sequence:

$$\mathbf{0} + 1 + 1 + 2 \quad 0 + 1 + 1 \quad 0 \quad 0 \quad 0 \quad -1 - 1 - 2 - 1 - 1$$

These differences, having an absolute value smaller than 2, require fewer bits to encode. Negative numbers are represented using 2's complement with 3 bits (2 bits for the magnitude and 1 bit for the sign).

The process begins by adding a third bit (0) in front of the positive values. For the negative values, the 0s and 1s are swapped to generate the 1's complement. To obtain the 2's complement, 1 is added to the 1's complement of the negative values:

-2 → 101 → 110 -1 → 110 → 111 0 → 000 → 000 +1 → 001 → 001

+2 → 010 → 010

This results in the sequence:

000 001 001 010 000 001 001 000 000 000 111 111 110 111 111

In this example, there is no reduction in the number of bits per word, as the differences and values are of similar magnitude.

6

Modulation

CONCEPTS COVERED IN THIS CHAPTER.–

Modulation is an effective method for transmitting signals. It involves using a carrier with a simple equation (typically a sinusoid), where the parameters are varied to represent the signal being transmitted.

The three standard types of modulation are:

– amplitude modulation;

– frequency modulation;

– phase modulation.

References: [BEL 98, BER 14, BRO 82, CHO 17, COT 97, FON 83, HAY 99, MUL 12b, MUL 12c, PAP 77].

6.1. Types of modulation

Without modulation, signals are transmitted in their original form, known as baseband transmission, where the signal occupies a frequency band from its minimum to maximum frequency.

Modulation is applied to convert information-carrying signals into forms suited for to the transmission medium, allowing signals to be multiplexed and transmitted efficiently.

Mathematics for Digital Science 2,
by Gérard-Michel COCHARD and Mhand HIFI.

Modulation serves two main objectives:

1) *Adaptation to the transmission medium:* when information is transmitted over a medium, the input signal is modulated using a modulator. This modulation allows the signal to be adapted to the specific characteristics of the medium, such as bandwidth, propagation properties and interference. Modulation optimizes transmission efficiency and ensures better signal quality at reception.

2) *Signal multiplexing:* modulation enables the combination of multiple signals within a single transmission channel, known as multiplexing. Different modulation techniques allow several signals to be transmitted simultaneously over the same medium, optimizing the available bandwidth.

When information is transmitted over a transmission medium, the input signal is modulated using a modulator. At the output, a demodulator recovers (in principle) the original signal. The modulation process is shown in Figure 6.1.

Figure 6.1. *Modulation process*

There are two main types of modulation:

1) *Analog modulation:* in this type of modulation, a "carrier" wave is used, and one of its parameters is adjusted according to variations in the source signal.

2) *Digital modulation:* this type of modulation involves converting the analog signal (if not already digital) into a digital format. This conversion is achieved using techniques such as sampling, quantization and coding (refer to the previous chapter).

Modulation is used to transport a signal and typically uses a *carrier*, which is a simple signal such as a sinusoidal function $c(t)$:

$$c(t) = a \times \cos(\omega_0 t + \varphi)$$

This carrier is characterized by three parameters:

– maximum amplitude a;

– frequency $f_0 = \frac{\omega_0}{2\pi}$;

– phase φ.

Modulation involves varying one or more of these parameters to represent the source signal. In the simplest case, the types of modulation are:

– amplitude modulation: variation of a;

– frequency modulation: variation of f_0;

– phase modulation: variation of φ.

The signal to be modulated is commonly referred to as the *modulating signal.* When combined with the *carrier* signal, or *carrier*, it becomes the *modulated signal.* This distinction reflects the process of altering the original signal to be transmitted over a specific carrier frequency.

6.2. Amplitude modulation

Amplitude modulation was first prominently used in 1921 with an early radio experiment, marking a significant milestone in the history of telecommunications. This modulation method involves varying the amplitude of a carrier signal according to the information to be transmitted.

6.2.1. *Principle*

Given any modulating signal (whether analog or digital) and a sinusoidal carrier $c(t) = a\cos(\omega_0 t + \varphi)$, the modulated signal is expressed as:

$$e(t) = a\big(1 + ks(t)\big)\cos(\omega_0 t + \varphi),$$

where a is the maximum carrier amplitude, k is the modulation coefficient that determines the amplitude of the modulating signal, and ω_0 is the carrier angular frequency.

In this modulation process, only the carrier amplitude is affected by the time-dependent factor $\big(1 + ks(t)\big)$. The coefficient k is a parameter determined by the modulator. The presence of the "1" in the expression signifies that the carrier is preserved. This modulation method is known as double sideband and carrier preserved (DBLPC). In contrast, another amplitude modulation technique is double sideband and carrier suppressed (DBLPS), where the modulation factor is simply $(ks(t))$, resulting in the carrier being suppressed rather than preserved.

In the rest of this study, the assumption will be that the DBLPC technique is used, as shown in Figure 6.2. In this modulated method, the frequency of the modulated signal remains that of the carrier, $f_0 = \frac{\omega_0}{2\pi}$. The shape of the modulated signal, shown in Figure 6.3, closely resembles that of the modulating signal at its extremes, specifically at the envelope level of the signal.

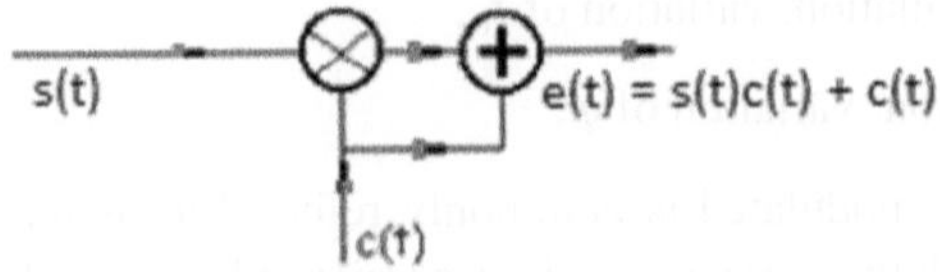

Figure 6.2. *Modulator diagram*

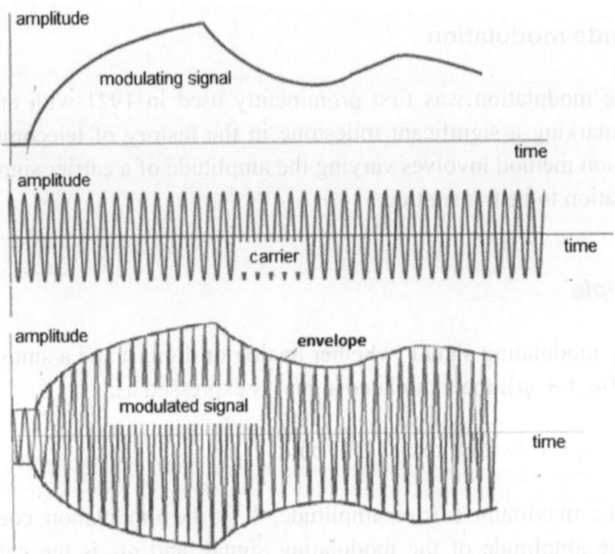

Figure 6.3. *Amplitude modulation. For a color version of this figure, see www.iste.co.uk/cochard/mathematics2.zip*

Consider a special case where the modulating signal is a sinusoidal signal given by $s(t) = A\cos(\Omega_t)$. In this case, the modulated signal can be expressed as follows:

$$e(t) = a[1 + kA\cos(\Omega_t)]\cos(\omega_0 t) = a[1 + m\cos(\Omega_t)]\cos(\omega_0 t)$$

where $m = kA$ is the modulation index.

6.2.2. *Frequency space*

By performing an analysis in the frequency domain using the Fourier transform for non-periodic signals, or using a Fourier series for periodic signals, the frequency components of the modulated signal can be obtained. For a sinusoidal modulating signal, the three-term expansion of the modulated signal, according to the previous expression, is:

$$
\begin{aligned}
e(t) &= a[(1 + m\cos(\Omega_t)]\cos(\omega_0 t) \\
&= a\cos(\omega_0 t) + \frac{am}{2}\cos\big((\omega_0 + \Omega)t\big) \\
&+ \frac{am}{2}\cos\big((\omega_0 - \Omega)t\big)
\end{aligned}
$$

This expression corresponds to a Fourier expansion, revealing three distinct frequencies: $f_0, f_0 + F$, and $f_0 - F$, where $\omega_0 = 2\pi f_0$ and $\Omega = 2\pi F$ (see Figure 6.4).

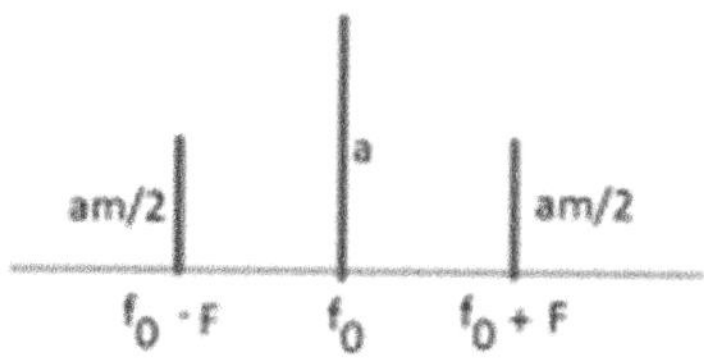

Figure 6.4. *Spectrum of a sinusoidal modulating signal. For a color version of this figure, see www.iste.co.uk/cochard/mathematics2.zip*

The frequency spectrum in this case consists of three discrete components or “lines”. For a periodic modulating signal, the frequency spectrum generally consists of symmetrical lines cantered around the carrier line frequency f_0.

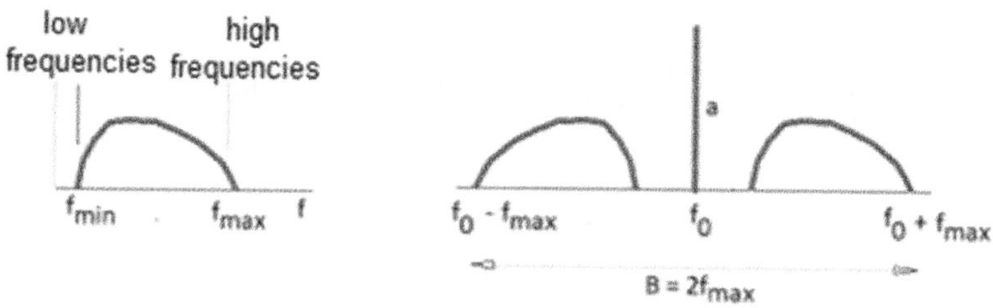

Figure 6.5. *Spectrum for any modulating signal. For a color version of this figure, see www.iste.co.uk/cochard/mathematics2.zip*

For any non-periodic modulating signal, the frequency spectrum can be obtained using the Fourier transform. Generally, this result is shown in Figure 6.5. Carson demonstrated this in 1914.

This diagram extends the result obtained for a sinusoidal modulating signal by describing the spectrum of a modulated signal in more general terms. The spectrum includes a central line corresponding to the carrier frequency and two symmetrical regions that represent the spectrum of the modulating signal. Consequently, the spectrum of a modulated signal features two symmetrical sidebands. The bandwidth occupied by the modulated signal is $B = 2f_{max}$, where f_{max} is the maximum frequency of the modulating signal. In addition, the shape of the sidebands resembles the spectrum of the modulating signal.

6.2.3. Signal strength

Suppose the modulated signal is a sinusoidal radio signal applied to an output antenna, which is considered a simple resistor R. The modulated signal is represented by the voltage across the antenna.

For a sinusoidal signal $x(t) = a\cos(\omega t)$, the root mean square (rms) value X is given by:

$$X^2 = \frac{1}{T}\int_0^T x^2(t)\,dt = \frac{a^2}{T}\int_0^T \cos^2(\omega t)dt$$

$$= \frac{a^2}{2T}\int_0^T [1+\cos(2\omega t)]dt = \frac{a^2}{2} \quad \text{or } X = \frac{a}{\sqrt{2}}$$

Using the trigonometric identity $\cos^2(\omega t) = \frac{1+\cos(2\omega t)}{2}$:

$$X^2 = \frac{a^2}{T}\int_0^T \left[\frac{1+\cos(2\omega t)}{2}\right] dt = \frac{a^2}{2T}\int_0^T [1+\cos(2\omega t)]dt$$

The integral of $\cos(2\omega t)$ over one period T is zero, so:

$$X^2 = \frac{a^2}{2T}\int_0^T 1\,dt = \frac{a^2}{2}$$

Thus, the rms value is:

$$X = \frac{a}{\sqrt{2}}$$

Taking the three-component form of a sinusoidal signal $e(t)$, each signal component contributes to the electrical power. The effective power P_{eff} is given by:

$$P_{eff} = \frac{a^2}{2R} + \frac{a^2m^2}{8R} + \frac{a^2m^2}{8R} = \frac{a^2}{2R}\left(1 + \frac{m^2}{2}\right)$$

Here, m is the modulation index.

6.2.4. *Overmodulation*

It is possible for the upper and lower envelopes of the modulated signal to intersect, creating an intersection zone (see Figure 6.6).

This situation is known as *overmodulation*. It can lead to undesirable problems, particularly in the demodulation circuit, where the response may be impaired. However, this effect can be corrected by using a dynamic control circuit, such as a compressor, to adjust the signal level.

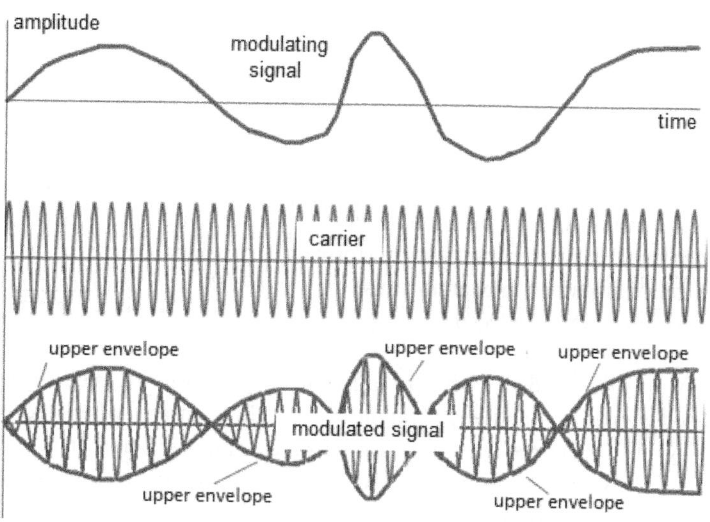

Figure 6.6. *Overmodulation. For a color version of this figure, see www.iste.co.uk/cochard/mathematics2.zip*

6.2.5. *Demodulation*

On reception, recovering the modulating signal can be achieved using different methods. One method is to use a "peak detector", which captures only the upper envelope of the modulated signal, as this envelope mirrors the shape of the modulating signal. It is important to note that corrections are needed if overmodulation occurs.

Another commonly used method involves using a synchronous signal. The modulated signal is expressed as:

$$e(t) = a[1 + ks(t)]\cos(\omega_0 t).$$

In this method, a signal $g(t) = b\cos(\omega_0 t)$ is used, which has the same frequency and phase as the carrier signal (hence the name synchronous signal). By combining the modulated signal by the synchronous signal, the following result is obtained:

$$\begin{aligned} x(t) = e(t)g(t) &= ab[1 + ks(t)]\cos^2(\omega_0 t) \\ &= \frac{ab}{2}[1 + ks(t)][1 + \cos(2\omega_0 t)] \\ &= \frac{ab}{2} + \frac{ab}{2}ks(t) + \frac{ab}{2}\cos(2\omega_0 t) + \frac{abk}{2}s(t)\cos(2\omega_0 t) \end{aligned}$$

The term of interest in the result is the second term, $\frac{ab}{2}ks(t)$. This term represents the modulating signal and can be isolated using a filtering system to recover the modulating signal proportionally.

However, there is an issue due to noise. During transmission, the signal can be affected by various sources of noise, such as electronic agitation, electromagnetic interference and cosmic rays. The performance of signal recovery is assessed using the ratio of carrier power to noise power (P/B) and the modulated signal power-to-noise power ratio (S/B).

As shown in Figure 6.7, when the carrier has minimal noise, recovering a clean demodulated signal is relatively straightforward. In contrast, when the carrier is significantly noisy, the recovery process becomes more difficult. It is shown that in the scenario with minimal noise, the signal-to-noise ratio (S/B) is twice the power-to-noise ratio (P/B) at the demodulator output.

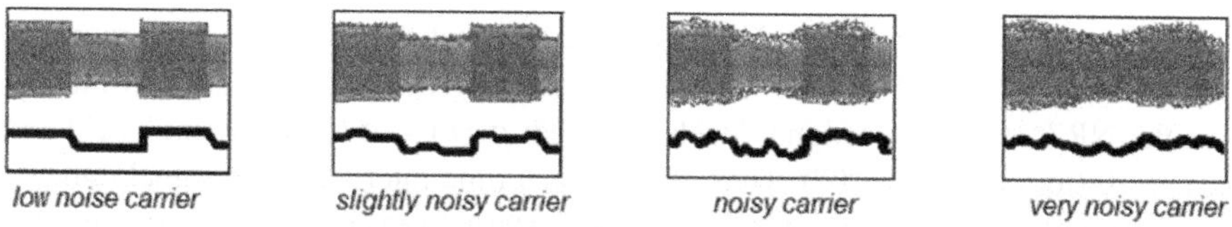

Figure 6.7. *Different types of noise*

6.2.6. *Single sideband*

The single sideband (SSB) technique involves eliminating one of the sidebands, resulting in a reduced total bandwidth. This method is shown in Figure 6.8.

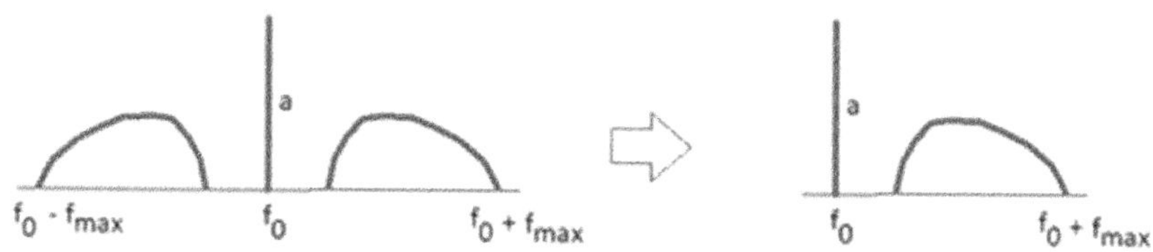

Figure 6.8. *SSB. For a color version of this figure, see www.iste.co.uk/cochard/mathematics2.zip*

6.2.7. *Modulation of a binary signal*

A binary signal can be encoded in various ways. One example is unipolar NRZ ("non-return to zero" unipolar) coding, where a "1" is represented by a voltage a and "0" is represented by zero voltage. This encoding method is shown in Figure 6.9.

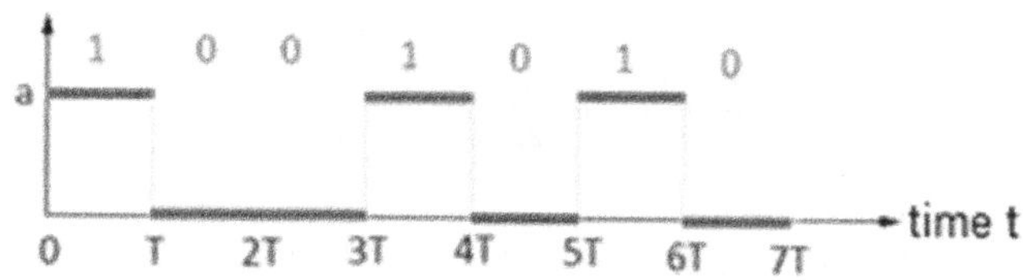

Figure 6.9. *NRZ coding. For a color version of this figure, see www.iste.co.uk/cochard/mathematics2.zip*

The term "non-return to zero" refers to the fact that the signal does not return to zero during the interval $[nT, (n+1)T]$, but instead remains at a constant value. In unipolar NRZ coding, two signal levels are used: $g_1(t) = a$ (to represent a "1") and $g_0(t) = 0$ (to represent a "0"). Thus, during each time interval T, the signal remains at "a" for a "1" and at "0" for a "0". The Fourier transforms of these two signals are:

$$G_0(f) = 0$$

$$G_1(f) = a\int_0^T e^{-2i\pi ft}dt = -\frac{a}{2i\pi f}[e^{-2i\pi fT} - 1]$$
$$= \frac{a}{2i\pi f}e^{-i\pi fT}(e^{+i\pi fT} - e^{-i\pi fT}) = aTe^{-i\pi fT}\frac{\sin(\pi fT)}{\pi fT}$$

The mean power spectral density (MPSD) represents the average power carried by each frequency of the signal. It can be calculated using Bennett's formula (not detailed here):

$$\mathrm{MPSD}(f) = \frac{1}{T}[p_0|G_0(f)|^2 + p_1|G_1(f)|^2] - |p_0G_0(f) + p_1G_1(f)|^2$$
$$+\frac{1}{T^2}[\sum_{n=-\infty}^{n=+\infty} |p_0G_0(f) + p_1G_1(f)|^2\delta(f - \frac{n}{T})]$$

where p_0 and p_1 represent the probabilities of a "0" or a "1", respectively. These are often assumed to be equiprobable: $p_0 = p_1 = \frac{1}{2}$.

In Bennett's formula, two distinct components are recognized: a continuous range of frequencies and a series of discrete lines. When applied to NRZ coding, this formula provides the following expression for the MPSD:

$$\mathrm{MPSD}(f) = \frac{a^2}{4}\frac{\sin^2(\pi fT)}{(\pi fT)^2}[T + \sum_{n=-\infty}^{+\infty} \delta(f - \frac{n}{T})]$$

The second term, involving the Dirac delta distribution function, contributes only at the frequency $f = 0$. This term represents a discrete component that captures the DC (Direct Current) offset of the signal. The overall shape of the MPSD is shown in Figure 6.10.

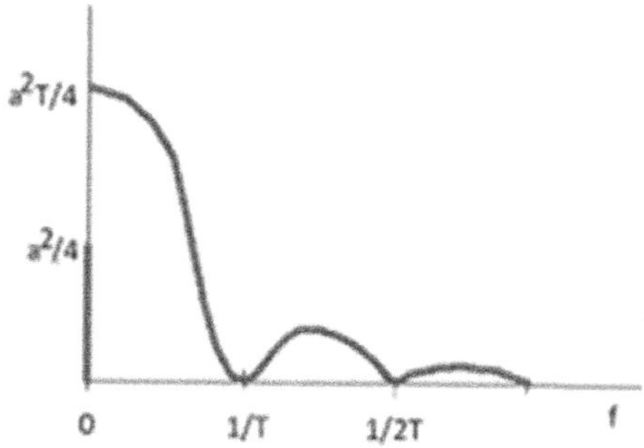

Figure 6.10. *Shape of power spectral density. For a color version of this figure, see www.iste.co.uk/cochard/mathematics2.zip*

The spectrum is not limited (the maximum frequency is theoretically infinite).

When this signal, considered as the modulating signal, is modulated with a sinusoidal carrier, the resulting spectral density distribution exhibits two bands, as shown in Figure 6.11.

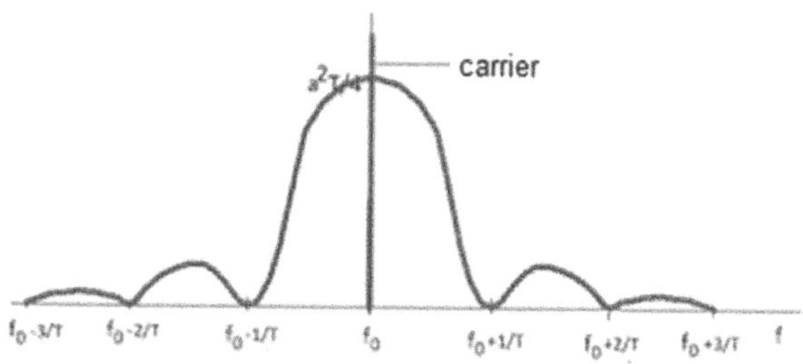

Figure 6.11. *Spectral density distribution. For a color version of this figure, see www.iste.co.uk/cochard/mathematics2.zip*

The extreme parts of the spectrum must be neglected. For instance, frequencies can be considered to range from $f_0 - \frac{4}{T}$ to $f_0 + \frac{4}{T}$, i.e. a width of $B = \frac{8}{T}$. Given a signal bit rate is 4,800 bits/s, the duration of one bit is $T = \frac{1}{4{,}800}$ seconds. The bandwidth B required for the signal is $B = 38.400$ kHz, which is significant.

To tackle the bandwidth limitations of NRZ encoding, alternative encoding methods that use less bandwidth are explored. In addition, filtering the signal can help reduce the frequency range.

Two amplitude modulation techniques for binary signals are as follows:

– ASK (amplitude shift keying): in this technique, both “1” and “0” signals are modulated. The amplitude of the carrier signal is adjusted according to the binary input, allowing for varying signal levels.

– OOK (on–off keying): in this technique, only the “1s” are modulated. The carrier signal is turned on to represent a “1” and turned off to represent a “0”, effectively creating a simpler modulation scheme.

6.3. Frequency modulation

6.3.1. *Principle*

Frequency modulation, developed by Armstrong in 1933, was standardized internationally with the allocation of a dedicated frequency band (88 to 108 MHz) in 1945.

Compared to amplitude modulation, frequency modulation offers several advantages:

– less interference from other transmitters;

– less sensitive to electromagnetic interference and atmospheric radiation;

– avoids the problem of overmodulation.

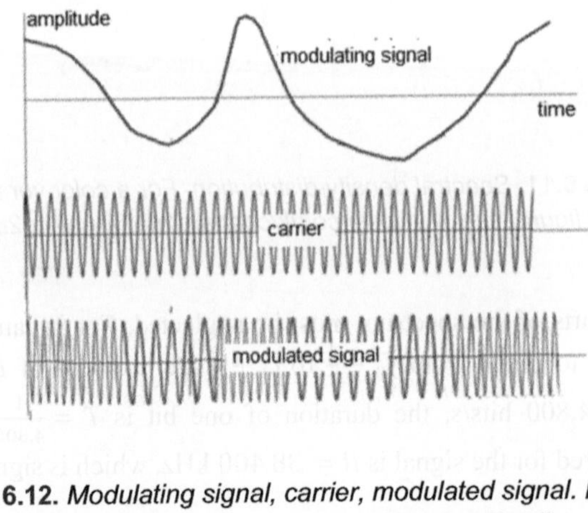

Figure 6.12. *Modulating signal, carrier, modulated signal. For a color version of this figure, see www.iste.co.uk/cochard/mathematics2.zip*

In frequency modulation, a frequency carrier f_0 is used, and the modulating signal $s(t)$ affects the instantaneous frequency of the carrier. The instantaneous frequency of the modulated signal is given by:

$$f(t) = f_0 + ks(t).$$

Here, k is the modulation index, which determines the extent to which the modulating signal influences the carrier frequency. The amplitude of the modulated signal remains constant and corresponds to the carrier amplitude. If $s(t) > 0$, the frequency $f(t)$ exceeds f_0, while if $s(t) < 0$, the frequency $f(t)$ is less than f_0 (see Figure 6.12).

The modulated signal can be expressed as:

$$e(t) = a\cos\big(\varphi(t)\big),$$

where a represents the carrier amplitude. The phase $\varphi(t)$ of the modulated signal is related to the instantaneous frequency $f(t)$ through its time derivative:

$$\omega(t) = 2\pi f(t) = \frac{d\varphi(t)}{dt}$$

Thus, the phase can be written as:

$$\varphi(t) = \omega_0 t + \theta(t)$$

where ω_0 is the angular frequency of the carrier. Consequently, the instantaneous frequency is:

$$\omega(t) = \frac{d\varphi(t)}{dt} = \omega_0 + \frac{d\theta(t)}{dt}$$

and the corresponding frequency is:

$$f(t) = f_0 + \frac{1}{2\pi}\frac{d\theta(t)}{dt}$$

Given that $f(t) = f_0 + ks(t)$, the phase $\varphi(t)$ can be derived as follows:

$$\frac{d\theta(t)}{dt} = 2\pi ks(t)$$

Integrating this equation:

$$\theta(t) = 2\pi k \int_0^t s(u)du$$

Therefore, the expression for the modulated signal is:

$$e(t) = \text{a} \cos\left(\omega_0 t + 2\pi k \int_0^t s(u)du\right)$$

In this expression, $ks(t)$ represents the *frequency* deviation introduced by the modulating signal. If the modulating signal $s(t)$ reaches an absolute maximum value s_{max}, the frequency of the modulated signal can vary between $(f_0 - ks_{max})$ and $(f_0 + ks_{max})$. This variation, $\Delta f = ks_{max}$, is commonly referred to as the *frequency excursion.*

6.3.2. Case of a sinusoidal signal

For a sinusoidal modulating signal, the frequency expression $f(t)$ is given by:

$$f(t) = f_0 + k\, cos(\Omega t) = f_0 + \Delta f\, cos(\Omega t)$$

where $\Delta f = k$ is the frequency deviation.

The phase $\theta(t)$ is defined by:

$$\theta(t) = 2\pi k \int_0^t \cos(\Omega u)\, du = 2\pi k \frac{\sin(\Omega t)}{\Omega}$$

Thus, the expression of the modulated signal is:

$$e(t) = \text{a}\cos\left(\omega_0 t + \Delta f \frac{sin(\Omega t)}{F}\right)$$

where $\Omega = 2\pi F$.

The quantity $(\Delta f / F)$ is known as the *modulation index.*

6.3.3. *Spectrum*

For any modulating signal $s(t)$, calculating the frequency spectrum of the modulated signal can be complex.

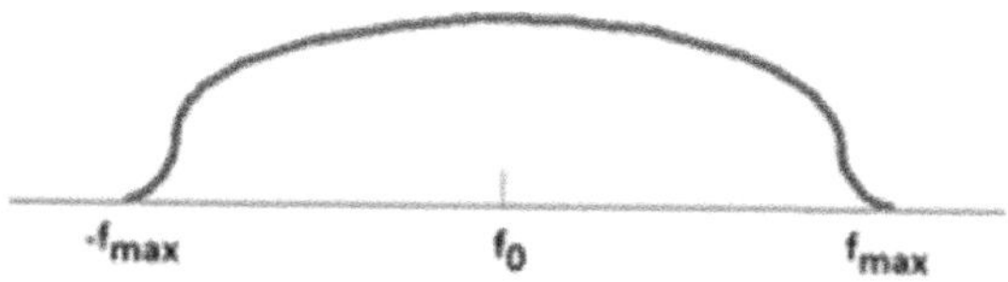

Figure 6.13. *General shape of the modulated signal spectrum. For a color version of this figure, see www.iste.co.uk/cochard/mathematics2.zip*

In general, the spectrum of the modulated signal is symmetrical with respect to the carrier frequency f_0, as shown in Figure 6.13. This symmetry is characterized by the presence of upper and lower sidebands, which are symmetrical around the carrier frequency.

Although the bandwidth of a modulated signal can theoretically be infinite, practical considerations allow for a bounded estimate. When the spectrum of the modulating signal is limited, Carson's formula provides an approximate estimate of the bandwidth. According to Carson's formula:

$$B = 2(\Delta f + f_{max})$$

where B is the bandwidth of the modulated signal, Δf is the frequency deviation (the maximum variation of the modulated signal's frequency from the carrier frequency), and f_{max} is the maximum frequency of the modulating signal.

Expanding the expression of the modulated signal for a sinusoidal modulating signal yields a series of terms involving trigonometric functions:

$$\begin{aligned} e(t) &= \mathrm{a}\cos\big(\omega_0 t + m\sin(\Omega t)\big) \\ &= \mathrm{a}\cos(\omega_0 t)\cos\big(m\sin(\Omega t)\big) \\ &\quad - \mathrm{a}\sin(\omega_0 t)\sin(m\sin(\Omega t)) \end{aligned}$$

The terms $\cos\big(m\sin(\Omega t)\big)$ and $\sin\big(m\sin(\Omega t)\big)$ can be evaluated using Bessel's functions $J_n(m)$. This result, shown in Figure 6.14, is derived from mathematical studies of these functions:

$$\cos(m\sin(\Omega t)) = J_0(m) + 2J_2(m)\cos(2\Omega t) + 2J_4(m)\cos(4\Omega t) + \cdots$$

$$\sin(m\sin(\Omega t)) = 2J_1(m)\sin(\Omega t) + 2J_3(m)\sin(3\Omega t) + 2J_5(m)\sin(5\Omega t) + \cdots$$

Using trigonometric relationships

$$\cos(x)\ \cos(y) = \frac{1}{2}[\cos(x+y) + \cos(x-y)]$$

and

$$\sin(x)\ \sin(y) = \frac{1}{2}[\cos(x-y) - \cos(x+y)]$$

The expansion of the modulated signal $e(t)$ can be derived as:

$$e(t) = a\cos(\omega_0 t)\,[J_0(m) + 2J_2(m)\cos(2\Omega t) + 2J_4(m)\cos(4\Omega t) + \cdots] - a\sin(\omega_0 t)[2J_1(m)\sin(\Omega t) + 2J_3(m)\sin(3\Omega t) + 2J_5(m)\sin(5\Omega t) + \cdots]$$

Therefore,

$$e(t) = aJ_0(m)\cos(\omega_0 t) - 2aJ_1(m)\sin(\omega_0 t)\sin(\Omega t) + 2aJ_2(m)\cos(\omega_0 t)\cos(2\Omega t) - 2aJ_3(m)\sin(\omega_0 t)\sin(3\Omega t) + 2aJ_4(m)\cos(\omega_0 t)\cos(4\Omega t) - 2aJ_5(m)\sin(\omega_0 t)\sin(5\Omega t) + \cdots$$

Rewriting using trigonometric relations:

$$e(t) = aJ_0(m)\cos(\omega_0 t) - aJ_1(m)[\cos(\omega_0 - \Omega)\,t - \cos(\omega_0 + \Omega)\,t] + aJ_2(m)[\cos(\omega_0 + 2\Omega)\,t + \cos(\omega_0 - 2\Omega)\,t] - aJ_3(m)[\cos(\omega_0 - 3\Omega)\,t - \cos(\omega_0 + 3\Omega)\,t] + aJ_4(m)[\cos(\omega_0 + 4\Omega)\,t + \cos(\omega_0 - 4\Omega)\,t\} + \cdots$$

This expansion, similar to a Fourier series, reveals the presence of harmonics ranging from $-\infty$ to $+\infty$

$$\ldots, f_0 - nF, \ldots, f_0 - 2F, f_0 - F, f_0, f_0 + F, f_0 + 2F, \ldots, f_0 + nF, \ldots$$

Thus, the spectrum of the modulated signal contains an infinite number of lines. The amplitude of each line is given by:

$$|aJ_n(m)|$$

where a is the amplitude of the carrier and $J_n(m)$ is a Bessel function of the first kind (see Figure 6.14).

As the number of harmonics (i.e. n) increases, the magnitude of $J_n(m)$ decreases, making higher harmonics less significant. In practical applications, the analysis typically focuses on the main harmonics and neglects higher-order harmonics to simplify the characterization of the spectrum.

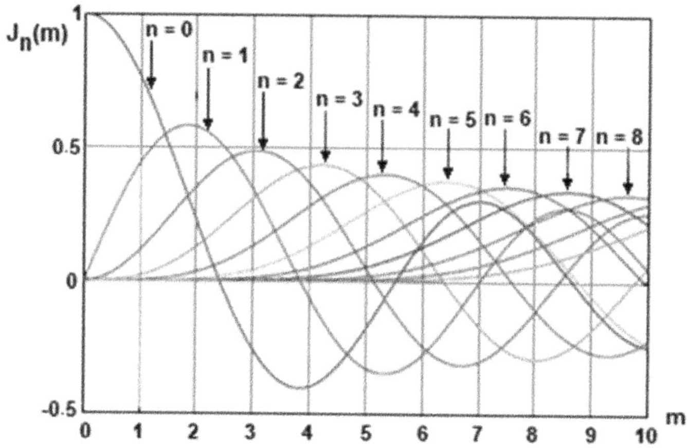

Figure 6.14. *First Bessel functions. For a color version of this figure, see www.iste.co.uk/cochard/mathematics2.zip*

The frequency spectrum of $e(t)$ has a pattern similar to that shown in Figure 6.15. This structure illustrates the harmonic distribution of the modulated signal, with each line representing a distinct frequency component.

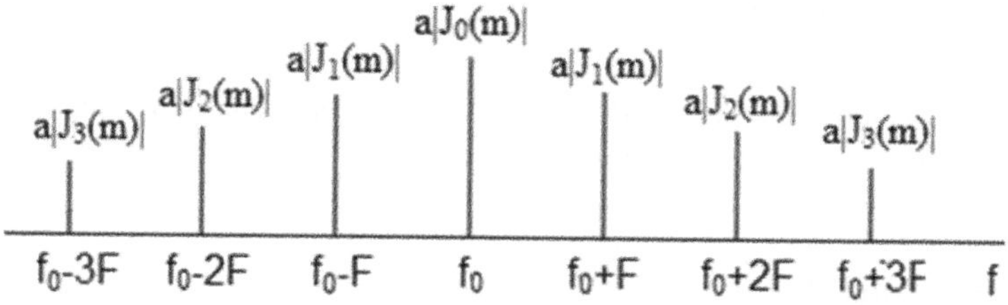

Figure 6.15. *Harmonics of the modulated signal. For a color version of this figure, see www.iste.co.uk/cochard/mathematics2.zip*

The amplitude of each harmonic line in the spectrum depends on both the index n and the modulation index coefficient m. The indicative table in Figure 6.16 shows the relative amplitudes of harmonic lines as functions of these parameters.

In general, as the value of the index n increases, the amplitude of the harmonic lines decreases. This means that higher-order harmonics contribute less to the total power of the modulated signal.

	m			
n	1	2	5	10
0	0.77	0.22	0.18	0.25
1	0.44	0.58	0.33	0.04
2	0.12	0.35	0.46	0.25
3	0.02	0.13	0.36	0.06
4		0.03	0.39	0.22
5			0.26	0.23
6			0.13	0.01
7			0.05	0.22
8			0.02	0.32
9				0.29
10				0.20
11				0.01

Figure 6.16. *Values of* $|J_n(m)|$

6.3.4. *Signal power*

Given the shape of the modulated signal $e(t)$, whose amplitude is a, the power of the signal applied to an output antenna (considered as a simple resistor R) can be calculated as follows. Since frequency modulation affects only the phase of the signal, the amplitude of the signal remains "a", and the power is given by:

$$P_{eff} = \frac{\left(\frac{a}{\sqrt{(2)}}\right)^2}{R} = \frac{a^2}{2R}.$$

This relationship holds in the ideal case where the antenna is perfectly matched and there is no distortion or interference.

6.3.5. *FSK modulation*

Frequency modulation can be applied to an NRZ-coded binary signal, where "1" corresponds to a value U and "0" corresponds to a value $-U$ (as shown in

Figure 6.17). In this case, the instantaneous frequency of the modulated signal is altered based on the binary signal's value:

– When the binary signal is "1", the frequency of the modulated signal is increased by Δf.

– When the binary signal is "0", the frequency of the modulated signal is decreased by Δf.

This method of modulation ensures the frequency of the signal corresponds to the variations in the binary signal, resulting in a frequency shift that represents each bit of the date.

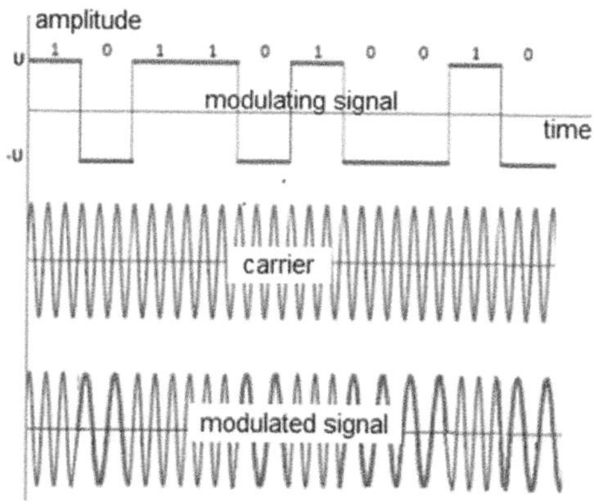

Figure 6.17. *FSK modulation. For a color version of this figure, see www.iste.co.uk/cochard/mathematics2.zip*

The modulation scheme described is frequency-shift keying (FSK) modulation, where two distinct frequencies f_1 and f_2 are used to represent binary symbols. For FSK, the frequency deviation Δf is defined as $\Delta f = kU$, and the modulated signal can be expressed as:

$$e(t) = a\cos[\omega_0 t \pm \Delta f t] = a\cos[(\omega_0 \pm 2\pi\Delta f)t]$$

Here, the frequency f_1 corresponds to the pulsation $(\omega_0 - 2\pi\Delta f)$ or the frequency $(f_0 - \Delta f)$ and the frequency f_2 corresponds to the pulsation $(\omega_0 + 2\pi\Delta f)$ or frequency $(f_0 + \Delta f)$.

To determine the spectral density of the modulated signal, Bennett's formula is used:

$$\mathrm{DSPM}(f) = \frac{1}{T}[p_0|G_0(f)|^2 + p_1|G_1(f)|^2] - |p_0G_0(f) + p_1G_1(f)|^2 + \frac{1}{T^2}[\sum_{n=-\infty}^{+\infty} |p_0G_0(f) + p_1G_1(f)|^2\,\delta\left(f - \frac{n}{T}\right)]$$

In this formula:

– p_1 and p_0 are the probabilities of observing a "1" and a "0" in the binary signal. Assuming equal probabilities, $p_1 = p_0 = \frac{1}{2}$.

– $G_1(f)$ and $G_0(f)$ are the Fourier transforms of the modulating signal for "1" and "0", respectively.

The Fourier transforms are calculated as follows:

$$G_1(f) = U\int_0^T e^{-2i\pi ft}dt = -\frac{U}{2i\pi f}[e^{-2i\pi fT} - 1] = Ue^{-i\pi fT}\frac{\sin(\pi fT)}{\pi f}$$

$$G_0(f) = -Ue^{-i\pi fT}\frac{\sin(\pi fT}{\pi f}$$

Therefore, the power spectral density (DSP) of the FSK-modulated signal is given by:

$$DSP(f) = \frac{1}{T}U^2\frac{\sin^2(\pi fT)}{\pi^2f^2} = TU^2\frac{\sin^2(\pi fT)}{\pi^2f^2T^2}$$

This spectral density shows two symmetrical bands around the carrier frequency f_0. As shown in Figure 6.18, the spectrum consists of two distinct frequency bands corresponding to the modulated frequencies $f_0 - \Delta f$ and $f_0 + \Delta f$, with the central peak and surrounding side frequencies distributed according to the shape of the function, $\left(\frac{\sin^2(\pi fT)}{\pi^2f^2}\right)$.

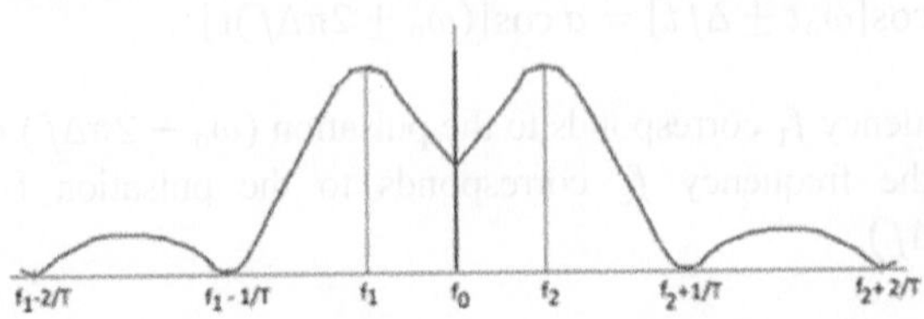

Figure 6.18. *Representation of DSP(f). For a color version of this figure, see www.iste.co.uk/cochard/mathematics2.zip*

The separation between the two frequencies f_1 and f_2 in FSK is $(2\Delta f)$. This separation affects the modulation and bandwidth of the signal in the following ways:

– If Δf is too small, the sidebands corresponding to f_1 and f_2 may overlap significantly. This overlap can cause interference between the two frequencies, making it challenging to distinguish them and potentially degrading the quality of the received signal.

– If Δf is too large, the frequency separation between to f_1 and f_2 increases, resulting in a wider frequency range for the signal. This broader bandwidth can lead to inefficient use of the available spectrum and may cause interference with other signals, impacting spectral efficiency.

In multi-level FSK modulation, such as four-level FSK, multiple frequencies are used to represent different levels of information. For instance, with four distinct frequencies f_1, f_2, f_3, and f_4, different bit combinations ("00", "01", "10", "11") are assigned to these frequencies. This approach allows the transmission of more bits per symbol, enhancing data transmission efficiency and increasing the capacity of the communication system.

6.4. Phase modulation

6.4.1. *Principle*

In phase modulation, the carrier signal $(a\cos(\omega_0 t))$ with frequency $f_0 = \frac{\omega_0}{2\pi}$ is modified by altering its phase in response to the modulating signal $s(t)$. The modulated phase involves modifying the carrier phase by adding a proportional term to the modulating signal. Thus, the modulated phase $\varphi(t)$ is given by:

$$\varphi(t) = \omega_0 t \;\longmapsto\; \omega_0 t + h\, s(t)$$

where h is a constant that controls the extent of the phase deviation introduced by the modulating signal. Consequently, the phase-modulated signal can be expressed as:

$$e(t) = a\cos\big(\omega_0 t + h.s(t)\big) = a\cos\big(\varphi(t)\big).$$

In phase modulation, this phase alteration effectively translates into frequency modulation. Specifically, the instantaneous angular frequency $\omega(t)$ of the modulated signal is:

$$\omega(t) = \frac{d\varphi(t)}{dt} = \omega_0 + h\frac{ds(t)}{dt}$$

which implies that the instantaneous frequency $f(t)$ of the signal is:

$$f(t) = f_0 + \frac{h}{2\pi}\frac{ds(t)}{dt}$$

Thus, phase modulation results in a frequency shift proportional to the derivative of the modulating signal, where:

$$\Delta f(t) = \frac{h}{2\pi}\frac{ds(t)}{dt}$$

represents the frequency deviation induced by the modulation.

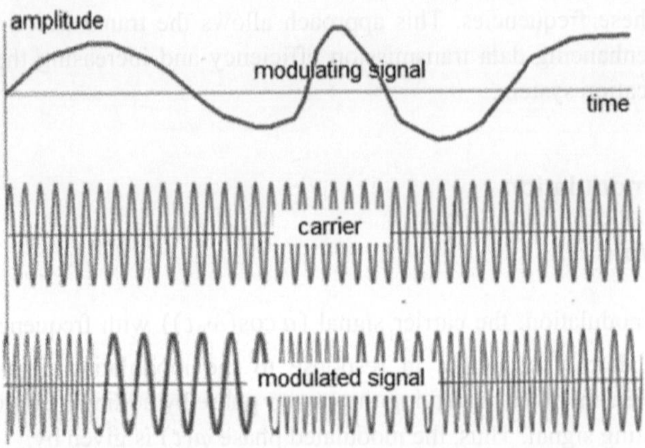

Figure 6.19. *Shape of the modulated signal. For a color version of this figure, see www.iste.co.uk/cochard/mathematics2.zip*

Figure 6.19 shows the shape of the modulated signal resulting from phase modulation. It shows how the frequency of the modulated signal changes in response to variations in the modulating signal. More precisely, the frequency of the modulated signal increases when the modulating signal rising and decreases when the modulating signal is falling. This behavior reflects the relationship where Δf represents the deviation induced by phase modulation. The value of Δf is related to the rate of change (slope) of the modulating signal $s(t)$.

6.4.2. *PSK modulation*

PSK (Phase-Shift Keying) modulation is a popular technique for transmitting digital signals by altering the phase of the carrier signal. In PSK modulation, different digital symbols are represented by distinct phases of the carrier signal. The general expression for a PSK-modulated signal is given by

$$e(t) = a\cos(\omega_0 t + \theta),$$

where "a" is the amplitude of the carrier signal, ω_0 is the angular frequency (pulsation) of the carrier signal, t is the time, and θ is the phase of the modulated signal.

The value θ is determined by the digital symbol being transmitted. For a PSK scheme with m different phases, the phase θ can be expressed as:

$$\theta = a_k \frac{2\pi}{m}$$

where $a_k = 0,1, \dots, (m-1)$. Here, m represents the number of distinct phase states, and a_k is the index corresponding to the phase of the current symbol.

Figure 6.20. *Polar diagram*

In the case of 2-PSK (Binary Phase-Shift Keying, BPSK), only two symbols, "0" and "1", are transmitted. The phase of the carrier signal is shifted by π radians depending on the symbol. The PSK-modulated signal for 2-PSK can be written as:

$$e(t) = a\cos(\omega_0 t + a_k\pi)$$

where $a_0 = 0$ for "1" and $a_1 = 1$ for "0".

Thus, the signal becomes:

$$e(t) = \pm\, a\cos(\omega_0 t)$$

This means that for "0" and "1", the carrier signal is either in phase or out of phase by π, which flips the signal's polarity.

The phase shifts in 2-PSK can be visualized on a simple polar diagram (see Figures 6.20 and 6.21), where the two possible phase states (0 and π) are represented on a unit circle.

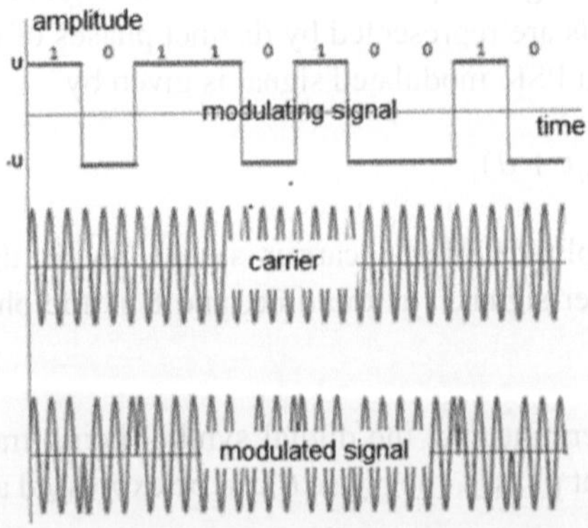

Figure 6.21. *Modulated signal shape. For a color version of this figure, see www.iste.co.uk/cochard/mathematics2.zip*

In the case of 4-PSK modulation (also called Quadrature Phase-Shift Keying, QPSK), the signal can represent four different digital symbols: 00, 01, 10 and 11. The expression of the modulated signal for 4-PSK is given by:

$$e(t) = a\cos\left(\omega_0 t + a_k \frac{\pi}{2}\right),$$

where "a" is the amplitude of the carrier signal, ω_0 is the pulsation of the carrier signal, t is the time, and a_k represents the different digital symbols: $a_0 = 0\ (for\ 00)$, $a_1 = 1\ (for\ 01)$, $a_2 = 2\ (for\ (10)$ and $a_3 = 3\ (for\ 11))$.

This function can be written as follows

$$e(t) = a\cos\left(\omega_0 t + a_k \frac{\pi}{2}\right)$$

$$= \begin{cases} a\cos(\omega_0 t) \text{ for } a_k = 0 \\ a\cos\left(\omega_0 t + \frac{\pi}{2}\right) = a\sin(\omega_0 t) \quad \text{for } a_k = 1 \\ a\cos(\omega_0 t + \pi) = -a\sin(\omega_0 t) \quad \text{for } a_k = 2 \\ a\cos\left(\omega_0 t + \frac{3\pi}{2}\right) = -\sin(\omega_0 t) \quad \text{for } a_k = 3 \end{cases}$$

Given $\theta = a_k \frac{\pi}{2}$, the modulated signal can be expressed as:

$$e(t) = \text{a}\cos(\theta)\cos(\omega_0 t) - \text{a}\sin(\theta)\sin(\omega_0 t)$$

which is of the form

$$e(t) = a\, I \cos(\omega_0 t) - a\, Q \sin(\omega_0 t).$$

where $I = \cos(\theta)$ is the in-phase component and $Q = \sin(\theta)$ is the quadrature component.

The first term, characterized by I, is in phase with the carrier, while the second term, characterized by Q, is in quadrature with the carrier. This is because $\sin(\omega_0 t) = \cos(\omega_0 t - \frac{\pi}{2})$, meaning the quadrature term is shifted by $\frac{\pi}{2}$ relative to the carrier.

Figure 6.23 shows the phase shifts in an (I, Q) diagram.

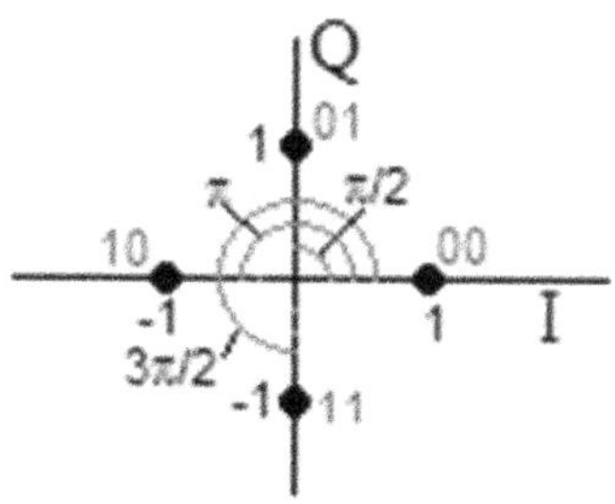

Figure 6.22. *4-PSK modulation. For a color version of this figure, see www.iste.co.uk/cochard/mathematics2.zip*

The expression of $e(t)$, with its in-phase (I) and quadrature (Q) components, allows for definition of the QPSK modulation. By choosing values of I and Q as ± 1, four possible states can be represented, as shown in Figure 6.23 (the constellation diagram):

– State 11: $I = 1,\ Q = 1.$

– State 10: $I = -1,\ Q = 1.$

– State 00: $I = -1,\ Q = -1.$

– State 01: $I = 1,\ Q = -1.$

These four states correspond to phase shifts $\theta = \frac{\pi}{4}, \frac{3\pi}{4}, \frac{7\pi}{4}, \frac{9\pi}{4}$.

For example, for state 10 ($I = -1,\ Q = 1$):

$$s(t) = -a\cos(\omega_0 t) - a\sin(\omega_0 t) = -a\left[\cos(\omega_0 t) - \cos\left(\omega_0 t + \frac{\pi}{2}\right)\right]$$
$$= 2a\sin\left(\omega_0 t + \frac{\pi}{4}\right)\sin\left(-\frac{\pi}{4}\right) = -a\sqrt{2}\sin\left(\omega_0 t + \frac{\pi}{4}\right)$$
$$= a\sqrt{2}\cos(\omega_0 t + \frac{3\pi}{4})$$

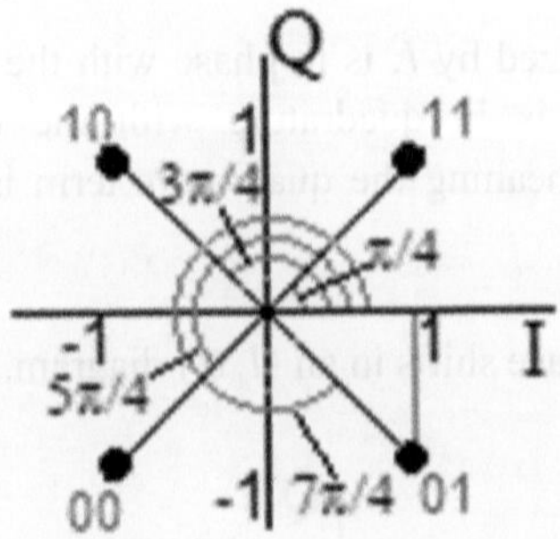

Figure 6.23. *QPSK modulation. For a color version of this figure, see www.iste.co.uk/cochard/mathematics2.zip*

7

Filtering

CONCEPTS COVERED IN THIS CHAPTER

Filtering is a common operation in signal processing used to isolate or eliminate specific frequency components, or simply to "clean up" a signal.

First, analog signal filtering is examined, followed by digital signal filtering.

Throughout this chapter, the z-transform, Fourier transform and Laplace transform, as defined in the previous chapter, are used.

References: [BEL 98, BER 14, BRO 82, CHO 17, COT 97, FON 83, HAY 99, PAP 77].

7.1. Definitions and reminders

A *filter* is a linear system that transforms one input signal $e(t)$ into another output signal $s(t)$ (see Figure 7.1).

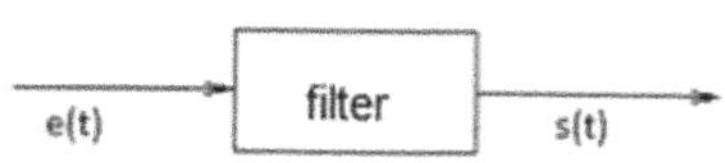

Figure 7.1. *Schematic diagram of a filter*

A filter is a specialized system used to process signals for two main purposes:

1) Cleaning a signal to remove unwanted noise, i.e. eliminating undesired frequencies.

Mathematics for Digital Science 2,
by Gérard-Michel COCHARD and Mhand HIFI.

2) Selecting specific frequency components from a signal.

In both cases, filtering involves frequency selection.

7.1.1. *Discrete signals*

If the signal is discrete, the input signal is denoted by $e(n)$, and the output signal is denoted by $s(n)$, where n represents the sample number.

If the input signal is a pulse $\delta(n)$, such that $\delta(n) = 1$ if $n = 0$ and $\delta(n) = 0$ otherwise, the output signal is the impulse response $h(n)$ (see Figure 7.2). This results in the following relationship:

$$s(n) = h(n) * e(n) = \sum_{k=-\infty}^{+\infty} e(k)h(n-k) = \sum_{k=-\infty}^{+\infty} e(n-k)h(k)$$

Figure 7.2. ***Impulse response***

In discrete signal processing, the transformation into z-domain has been introduced. For a signal $f(n)$, the z-transform is given by:

$$F(z) = \sum_{n=-\infty}^{+\infty} f(n)z^{-n}$$

The z-transform of the impulse response $h(n)$ is the transfer function $H(z)$, expressed as:

$$H(z) = \sum_{n=-\infty}^{+\infty} h(n)z^{-n} = \frac{S(z)}{E(z)}$$

where $E(z)$ and $S(z)$ are the z-transforms of $e(n)$ and $s(n)$, respectively.

7.1.2. *Analog signals*

If the signal is analog, let $e(t)$ and $s(t)$ represent the input and output signals, respectively. Suppose the input signal is a pulse $\delta(t)$ (Dirac distribution); as mentioned earlier, the output signal $h(t)$ is called the impulse response (see Figure 7.3). The relationship is given by:

$$s(t) = (e * h)(t) = \int_{-\infty}^{+\infty} e(\tau)h(t-\tau)d\tau = \int_{-\infty}^{+\infty} e(t-\tau)h(\tau)d\tau$$

Figure 7.3. *Impulse response*

In analog signal processing, it is convenient to use the Fourier transform. If a signal is represented by a function $x(t)$, its Fourier transform is given by:

$$X(f) = \int_{-\infty}^{+\infty} x(t)e^{-2i\pi ft}dt$$

If the input signal $e(t)$ and the output signal $s(t)$ have Fourier transforms $E(f)$ and $S(f)$, respectively, then the transfer function $H(f)$ is defined as

$$H(f) = \frac{S(f)}{E(f)}$$

where $H(f)$ is simply the Fourier transform of the impulse response $h(t)$.

Another useful transformation is the Laplace transform. If $x(t)$ is a causal signal (i.e. $x(t) = 0$ for $t < 0$), its Laplace transform $X(p)$ is defined by:

$$X(p) = \int_{0}^{\infty} x(t)e^{-pt}dt$$

If $H(p), E(p)$ and $S(p)$ are the Laplace transforms of the impulse response $h(t)$, the input signal $e(t)$ and the output signal $s(t)$, respectively, then:

$$H(p) = \frac{S(p)}{E(p)}$$

The same notation is used for both the Fourier and Laplace transforms, but there is no ambiguity because the variable in the Fourier transform f, while in the Laplace transform, it is p.

7.2. Analog filtering

7.2.1. *General information*

Filters can be classified in various ways:

– *Linear* filters:

A linear filter respects a linear combination of input signals.

If $e_1(t)$ and $e_2(t)$ are the input signals with outputs $s_1(t)$ and $s_2(t)$, then for a signal $\lambda e_1(t) + \mu e_2(t)$, the output will be $\lambda s_1(t) + \mu s_2(t)$. Only linear filters will be considered here. It was mentioned earlier that if $h(t)$ is the filter's impulse response, then the output $s(t)$ is expressed by the convolution product $e * h$, such that

$$s(t) = (e * h)(t) = \int_{-\infty}^{+\infty} e(\tau)h(t-\tau)d\tau$$

– *Active* and *passive* filters:

An active filter provides power gain (such as an amplifier or transistor), while a passive filter does not provide any power gain (such as a resistor or inductor).

– *Causal* filter:

A causal filter satisfies the property that $e(t) = 0$ for $t < 0$, which leads to $s(t) = 0$ for $t < 0$.

7.2.2. *Common filters*

Based on the transfer function $H(f)$, which describes the frequency behavior of a filter, four commonly used filter types can be characterized: high-pass, low-pass, band-pass and notch filters.

These filters are designed to meet specific requirements for separating and modifying the frequencies of an electrical signal.

Figure 7.4 shows these filter categories, comparing "ideal" filters on the left with real filters on the right.

It is important to note that, in practice, obtaining filters that operate in a completely ideal manner is difficult.

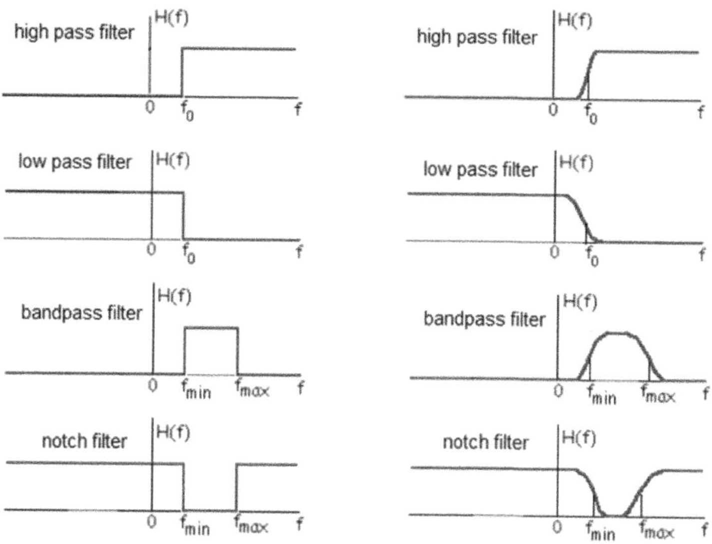

Figure 7.4. *Filter categories. For a color version of this figure, see www.iste.co.uk/cochard/mathematics2.zip*

EXAMPLE 7.1.– Frequency selection:

Consider the signal:

$$e(t) = sin(2\pi f_0 t) + \frac{1}{2} sin(4\pi f_0 t).$$

This signal consists of two sinusoidal components: one with a fundamental frequency f_0 and the other with a double frequency $2f_0$.

The first component has an amplitude of 1, while the second has an amplitude of $\frac{1}{2}$. Figure 7.5 shows this signal graphically.

The calculation of the Fourier transform leads to the following result:

$$E(f) = -i[\delta(f_0 - f) - \delta(f_0 + f) + 0.5\delta(2f_0 - f) - 0.5\delta(2f_0 + f)$$

The spectrum (see Figure 7.6) clearly shows the two frequency components of $e(t)$.

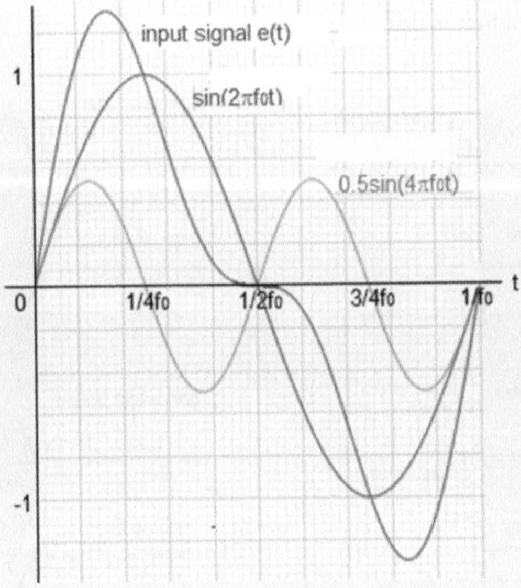

Figure 7.5. *e(t) signal. For a color version of this figure, see www.iste.co.uk/cochard/mathematics2.zip*

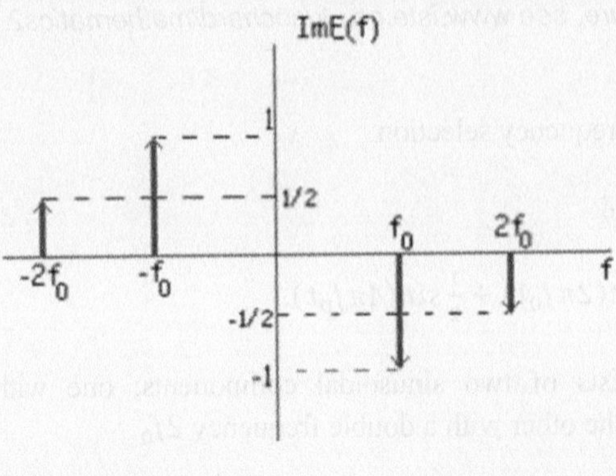

Figure 7.6. *Signal spectrum. For a color version of this figure, see www.iste.co.uk/cochard/mathematics2.zip*

Let us apply a low-pass filter with the transfer function is $H(f)$ (see Figure 7.7)

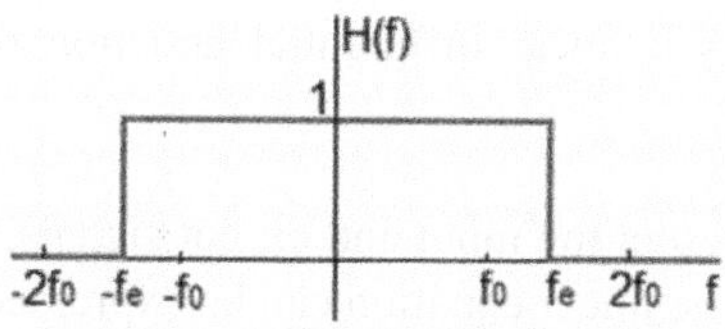

Figure 7.7. *Transfer function. For a color version of this figure, see www.iste.co.uk/cochard/mathematics2.zip*

From the relation $S(f) = E(f) \times H(f)$, the spectrum $S(f)$ can be deduced (see Figure 7.8):

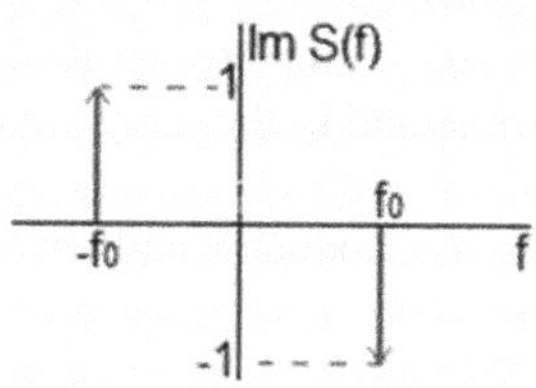

Figure 7.8. *S(f) spectrum. For a color version of this figure, see www.iste.co.uk/cochard/mathematics2.zip*

It can be observed that the filter has eliminated the frequency component $2f_0$ (see Figure 7.9).

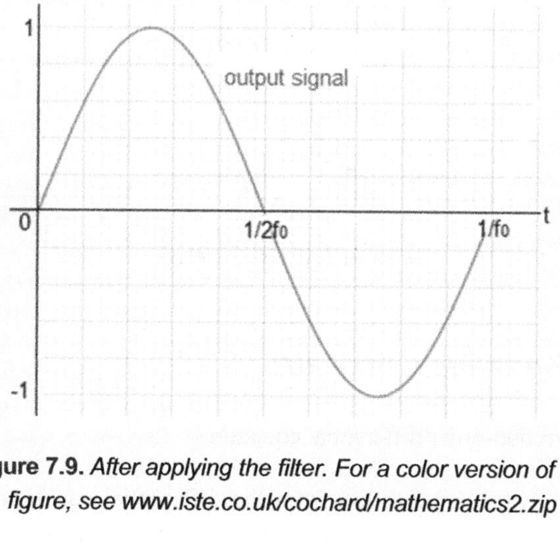

Figure 7.9. *After applying the filter. For a color version of this figure, see www.iste.co.uk/cochard/mathematics2.zip*

7.2.3. *Differential equations and transfer functions*

A linear system obeys a linear differential equation that links the input and output signals.

Let $e(t)$ and $s(t)$ represent the input and output signals, respectively, of a linear system acting as a filter. The linear equation can be expressed in general form as:

$$a_0 s(t) + a_1 \frac{ds(t)}{dt} + a_2 \frac{d^2 s(t)}{dt^2} + \cdots + a_n \frac{d^n s(t)}{dt^n} = b_0 e(t) + b_1 \frac{de(t)}{dt} + b_2 \frac{d^2 e(t)}{dt^2} + \cdots + b_m \frac{d^m e(t)}{dt^m}$$

This equation involves the successive derivatives of $s(t)$ and $e(t)$ with respect to time. The coefficients $a_0, a_1, a_2, \ldots, a_n$ are associated with the terms involving $s(t)$, while $b_0, b_1, b_2, \ldots, b_m$ correspond to the terms involving $e(t)$.

EXAMPLE 7.2.– Consider a resistor, inductor and capacitor (RLC) circuit defined by the diagram in Figure 7.10.

According to the laws of electricity:

$$e(t) = Ri(t) + L\frac{di(t)}{dt} + s(t)$$

$$i(t) = C\frac{ds(t)}{dt}$$

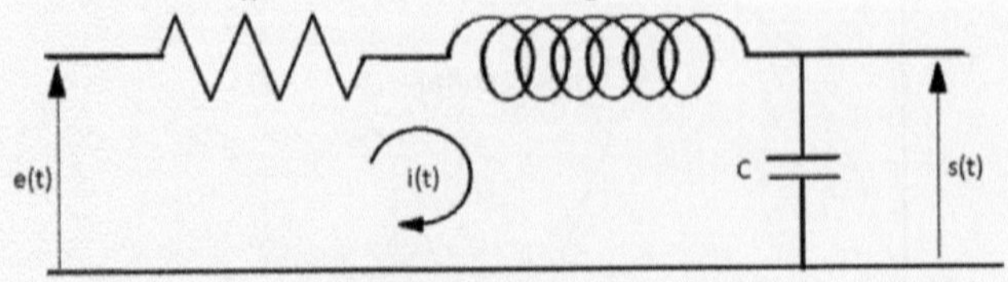

Figure 7.10. *RLC circuit*

or

$$LC\frac{d^2 s(t)}{dt^2} + RC\frac{ds(t)}{dt} + s(t) = e(t)$$

This leads to a second-order differential equation.

It is straightforward to transition from the differential equation to the transfer function $H(p)$ using the Laplace transformation. The transformation converts a differential equation into an algebraic equation in the Laplace domain. Thus, under the Laplace transformation, the preceding differential equation becomes:

$$a_0S(p) + a_1pS(p) + \cdots + a_np^nS(p) \\ = b_0E(p) + b_1pE(p) + \cdots + b_mp^mE(p)$$

In this equation, $S(p)$ and $E(p)$ represent the Laplace transforms of the output signal $s(t)$ and input signal $e(t)$, respectively. The terms p, p^2 ,..., p^m represent the differentiation operations in the Laplace domain, and the coefficients a_0, a_1, a_2, ..., a_n and b_0, b_1, b_2, ..., b_m are associated with these terms. Hence, the transfer function $H(p)$ is given by:

$$H(p) = \frac{S(p)}{E(p)} = \frac{b_mp^m + \cdots + b_1p + b_0}{a_nn + \cdots + a_1p + a_0}$$

The transfer function $H(f)$ in the frequency domain can be obtained by replacing p with $2i\pi f$:

$$H(f) = \frac{S(t)}{E(f)} = \frac{b_m(2i\pi f)^m + \cdots + b_1(2i\pi f) + b_0}{a_n(2i\pi f)^n + \cdots + a_1(2i\pi f) + a_0}$$

The impulse response $h(t)$ can then be calculated using the inverse Fourier transform.

It is often useful to compute the *index response* $h_0(t)$, which represents the output signal when the input signal is the unit step function $u(t)$ defined by:

$$u(t) = 0 \text{ if } t < 0 \text{ and } u(t) = 1 \text{ if } t > 0.$$

Using the convolution relation:

$$h_0(t) = \int_{-\infty}^{+\infty} u(t - \tau)h(\tau)d\tau$$

Since $u(t - \tau) = 0$ when $t - \tau < 0$ (i.e. $\tau > t$) and $u(t - \tau) = 1$ when $\tau < t$, this simplifies to:

$$h_0(t) = \int_{-\infty}^{t} h(\tau)d\tau$$

allowing for the calculation of $h_0(t)$ from $h(t)$.

EXAMPLE 7.3.– Returning to Example 7.1, from the differential equation, the transfer function $H(p)$ is obtained as:

$$H(p) = \frac{1}{LCp^2+RCp+1}$$

The denominator has the discriminant $\Delta = R^2C^2 - 4LC$. Three cases are distinguished:

Case 1: if $\Delta > 0$, there are two real roots p_1 and p_2. Thus, $H(p)$ can be expressed as:

$$H(p) = \frac{1}{LC(p-p_1)(p-p_2)} = \frac{1}{LC}\left[\frac{\alpha}{p-p_1} + \frac{\beta}{p-p_2}\right].$$

The coefficients α and β can be easily obtained, with $\alpha = -\beta = \frac{LC}{\sqrt{\Delta}}$. Hence,

$$H(p) = \frac{1}{\sqrt{\Delta}}\left[\frac{1}{p-p_1} - \frac{1}{p-p_2}\right].$$

The transition from $H(p)$ to $H(f)$ is accomplished by replacing p with $2i\pi f$:

$$H(f) = \frac{1}{\sqrt{\Delta}}\left[\frac{1}{2i\pi f-p_1} - \frac{1}{2i\pi f-p_2}\right]$$

Calculating $|H(f)|$ yields the filter's frequency characteristic. Assuming that $LC = R^2C^2/8$ and considering various values of RC, the resulting curves indicate that the filter behaves like a low-pass filter at certain frequencies (see Figure 7.11).

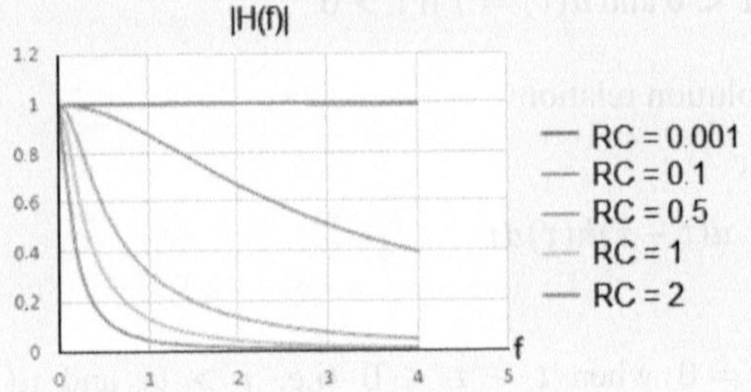

Figure 7.11. *Representation of $|H(f)|$. For a color version of this figure, see www.iste.co.uk/cochard/mathematics2.zip*

The impulse response $h(t)$ is the inverse Fourier transform of $H(f)$. In the Fourier transform table, the correspondence is shown in Figure 7.12, where $u(t)$ is the unit step (or Heaviside) function:

$u(t) = 0$ if $t < 0$ and $u(t) = 1$ if $t > 0$.

Using the inverse Fourier transform, the impulse response can be expressed as:

$$h(t) = \frac{1}{\sqrt{\Delta}}[e^{p_1 t} - e^{p_2 t}]u(t)$$

$$\frac{1}{2i\pi f + a} \xrightarrow{FT^{-1}} e^{-at}u(t)$$

Figure 7.12. *Fourier transform*

It is also interesting to calculate the index response $h_0(t)$ for $t > 0$ (see Figure 7.13).

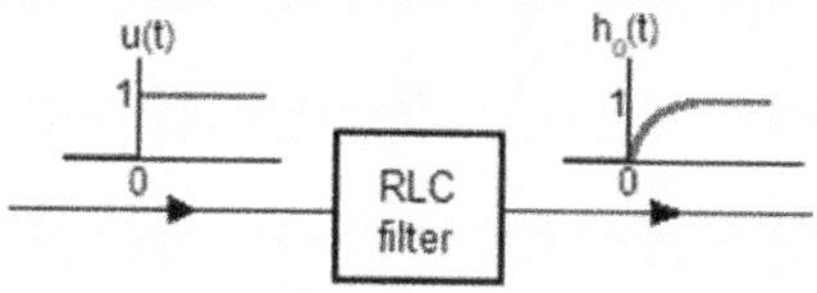

Figure 7.13. *$u(t)$ and $h_0(t)$. For a color version of this figure, see www.iste.co.uk/cochard/mathematics2.zip*

$$h_0(t) = \int_{-\infty}^{t} h(\tau)d\tau = \frac{1}{\sqrt{\Delta}}\int_{-\infty}^{t}[e^{p_1\tau} - e^{p_2\tau}]u(\tau)d\tau = 1 + \frac{e^{p_1 t}}{p_1\sqrt{\Delta}} + \frac{e^{p_2 t}}{p_2\sqrt{\Delta}}$$

Since $h_0(t)$ is zero for $t < 0$, it can therefore be calculated as follows:

$$h_0(t) = \left[1 + \frac{e^{p_1 t}}{p_1\sqrt{\Delta}} + \frac{e^{p_2 t}}{p_2\sqrt{\Delta}}\right]u(t)$$

Assuming $LC = R^2C^2/8$ and considering various values of RC, the resulting curves are shown in Figure 7.14.

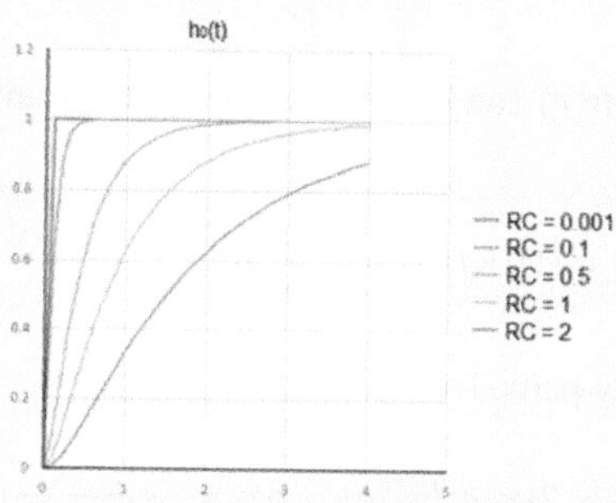

Figure 7.14. *Curves $h_0(t)$. For a color version of this figure, see www.iste.co.uk/cochard/mathematics2.zip*

The "jump" signal deteriorates as RC increases (resulting in the suppression of high frequencies).

Case 2: if $\Delta = 0$, then $R^2C^2 = 4LC$, leading to a double real root $p_0 = -\frac{2}{RC}$. The transfer function in terms of p, $H(p)$, is given by

$$H(p) = \frac{1}{LC(p-p_0)^2} = \frac{1}{R^2C^2(p-p_0)^2}$$

This results in the frequency transfer function $H(f)$:

$$H(f) = \frac{4}{R^2C^2}\frac{1}{(2i\pi f-p_0)^2} = -\frac{1}{2i\pi}\frac{4}{R^2C^2}\frac{d}{df}\left[\frac{1}{2i\pi f-p_0}\right\}$$

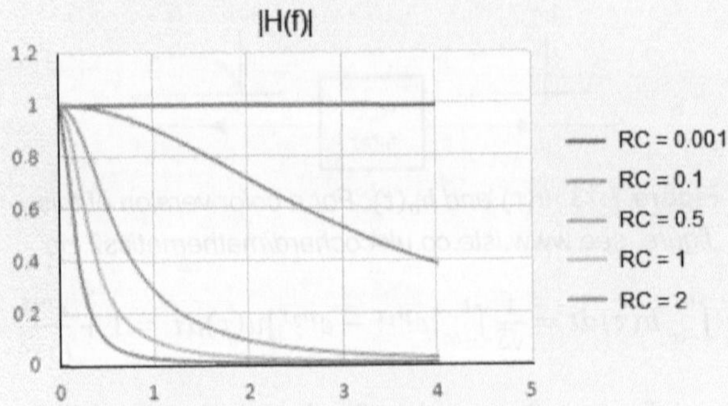

Figure 7.15. $|H(f)|$ *curves. For a color version of this figure, see www.iste.co.uk/cochard/mathematics2.zip*

The magnitude $|H(f)|$ indicates the frequencies selected by the filter (see Figure 7.15). It is evident that for large values of RC, the filter operates as a low-pass filter.

The impulse response $h(t)$ can be obtained through the inverse Fourier transform of $H(f)$:

$$h(t) = \int_{-\infty}^{+\infty} H(f)e^{2i\pi ft}df = -\frac{2}{i\pi R^2C^2}\int_{-\infty}^{+\infty}\frac{d}{df}\left[\frac{1}{2i\pi f-p_0}\right]e^{2i\pi ft}\,df$$

Applying integration by parts, let:

$$u = e^{2i\pi ft} \quad du = 2i\pi te^{2i\pi ft}dt \quad dv = \frac{d}{df}\left[\frac{1}{2i\pi f-p_0}\right]df \quad v = \frac{1}{2i\pi f-p_0}$$

$$h(t) = \frac{2}{i\pi R^2C^2}\left\{\left[\frac{e^{2i\pi ft}}{2i\pi f-p_0}\right]_{-\infty}^{+\infty} - 2i\pi t\int_{-\infty}^{+\infty}\frac{e^{2i\pi ft}}{2i\pi f-p_0}df\right\}$$

$$= -\frac{2}{i\pi R^2C^2}\{\lim_{F\to\infty}\left[\frac{e^{2i\pi ft}}{2i\pi f - p_0}\right]_{-F}^{+F} - 2i\pi t e^{p_0 t}\}$$

$$= -\frac{2}{i\pi R^2C^2}\left\{\lim_{F\to\infty}\frac{2\cos(2i\pi Ft)}{2i\pi F} - 2i\pi t e^{p_0 t}\}\right\} = \frac{4}{R^2C^2} t e^{p_0 t} u(t)$$

This expression can be used to compute the index response:

$$h_0(t) = \int_{-\infty}^{t} h(\tau)d\tau \text{ for } t > 0 \text{ and } h_0(t) = 0 \text{ for } t < 0.$$

For $t > 0$,

$$h_0(t) = \int_{-\infty}^{t} h(\tau)d\tau = \frac{4}{R^2C^2}\int_0^t \tau e^{p_0\tau}d\tau$$

Using the integration by parts again, let:

$$u = \tau \qquad du = d\tau \qquad dv = e^{p_0\tau}d\tau \qquad v = \frac{1}{p_0}e^{p_0\tau}$$

$$h_0(t) = \frac{4}{R^2C^2p_0}\{te^{p_0 t} - \frac{1}{p_0}(e^{p_0 t} - 1)\} = 1 - \left(1 + \frac{2t}{RC}\right)e^{-\frac{2t}{RC}}$$

Thus, (see Figure 7.16)

$$h_0(t) = \left[1 - \left(1 + \frac{2t}{RC}\right)e^{-\frac{2t}{RC}}\right]u(t), for\ all\ t$$

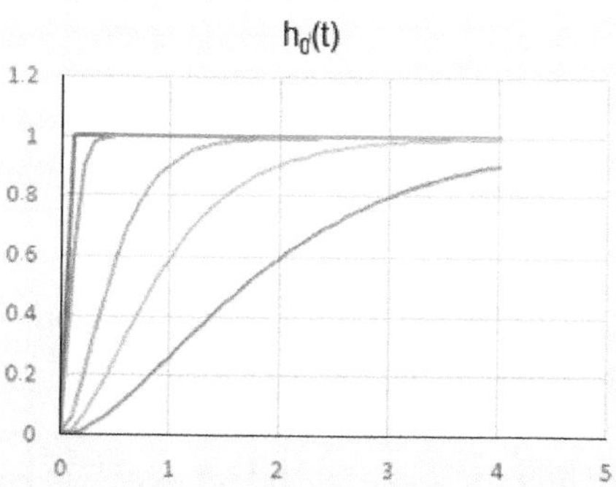

Figure 7.16. *s $h_0(t)$ curves. For a color version of this figure, see www.iste.co.uk/cochard/mathematics2.zip*

Case 3: if $\Delta < 0$ or $LC > R^2C^2/4$, there are two conjugate complex roots given by:

$$q_{1,2} = \frac{-RC \pm i\sqrt{-\Delta}}{2LC}.$$

The transfer function $H(p)$ can be expressed as:

$$H(p) = \frac{1}{LCp^2+RCp+1} = \frac{1}{LC} \cdot \frac{1}{p^2+\frac{R}{L}p+\frac{1}{LC}} = \frac{1}{i\sqrt{-\Delta}}\left[\frac{1}{p-q_1} - \frac{1}{p-q_2}\right]$$

This leads to the frequency transfer function:

$$H(f) = \frac{1}{i\sqrt{-\Delta}}\left[\frac{1}{2i\pi f-q_1} - \frac{1}{2i\pi f-q_2}\right]$$

The following curves illustrate the behavior of $|H(f)|$ for $LC = \frac{R^2C^2}{2}$ and several values of RC.

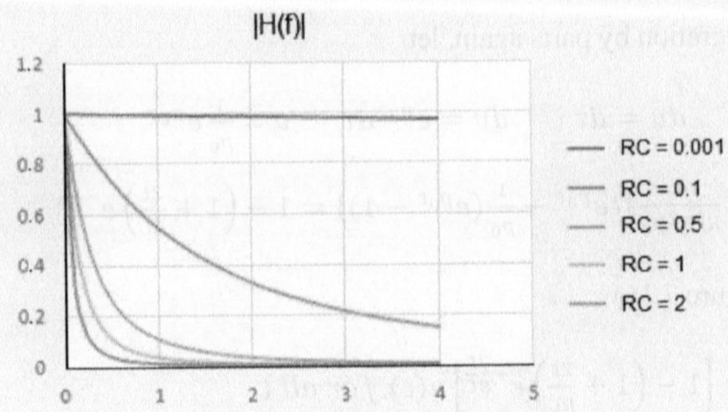

Figure 7.17. *$|H(f)|$ curves. For a color version of this figure, see www.iste.co.uk/cochard/mathematics2.zip*

The RLC system behaves like a low-pass filter, effectively eliminating high frequencies for large values of RC.

The impulse response $h(t)$ is derived as the inverse Fourier transform of $H(f)$. According to the Fourier transform table, the following relationship holds:

$$h(t) = \frac{1}{i\sqrt{-\Delta}}[e^{q_1 t} - e^{q_2 t}]u(t) = \frac{1}{i\sqrt{-\Delta}} e^{-\frac{R}{2L}t}\left[e^{i\frac{\sqrt{-\Delta}t}{2LC}} - e^{-i\frac{\sqrt{-\Delta}t}{2LC}}\right]u(t) = \frac{2}{\sqrt{-\Delta}} e^{-\frac{R}{2L}t} \sin\left(\frac{\sqrt{-\Delta}}{2LC}t\right)u(t)$$

With *h*(*t*), the index response $h_0(t)$ is calculated as follows:

$$h_0(t) = \int_{-\infty}^{t} h(\tau)d\tau = \frac{2}{\sqrt{-\Delta}}\int_0^t e^{-\frac{R}{2L}\tau}\sin\left(\frac{\sqrt{-\Delta}}{2LC}\tau\right)d\tau \text{ for } t > 0$$

and

$$h_0(t) = 0, \text{ for } t < 0$$

For $t > 0$, $h_0(t)$ can be written as:

$$h_0(t) = \frac{2}{\sqrt{-\Delta}} \int_0^t e^{-\frac{R}{2L}\tau} \frac{e^{i\frac{\sqrt{-\Delta}\tau}{2LC}} - e^{-i\frac{\sqrt{-\Delta}\tau}{2LC}}}{2i} d\tau$$

which, after integration, can be rewritten as:

$$h_0(t) = \frac{1}{i\sqrt{-\Delta}} \{ \frac{e^{-\frac{R}{2L}+i\frac{\sqrt{-\Delta}}{2LC}} - 1}{-\frac{R}{2L}+i\frac{\sqrt{-\Delta}}{2LC}} - \frac{e^{-\frac{R}{2L}-i\frac{\sqrt{-\Delta}}{2LC}} - 1}{-\frac{R}{2L}-i\frac{\sqrt{-\Delta}}{2LC}} \}$$

As $\left(-\frac{R}{2L} + i\frac{\sqrt{-\Delta}}{2LC}\right)\left(-\frac{R}{2L} - i\frac{\sqrt{-\Delta}}{2LC}\right) = \frac{1}{LC}$, then:

$$h_0(t) = \frac{LC}{i\sqrt{-\Delta}} \left\{ \frac{R}{2L} e^{-\frac{R}{2L}t} \left[e^{i\frac{\sqrt{-\Delta}t}{2LC}} - e^{-i\frac{\sqrt{-\Delta}t}{2LC}} \right] - i\frac{\sqrt{-\Delta}}{2LC} e^{-\frac{R}{2L}t} \left[e^{-i\frac{\sqrt{-\Delta}t}{2LC}} + e^{i\frac{\sqrt{-\Delta}t}{2LC}} \right] + i\frac{\sqrt{-\Delta}}{LC} \right\} = 1 - e^{-\frac{R}{2L}t} \left[\cos\left(\frac{\sqrt{-\Delta}}{2LC}t\right) + \frac{RC}{\sqrt{-\Delta}} \sin\left(\frac{\sqrt{-\Delta}}{2LC}t\right) \right]$$

and therefore, for all t (see Figure 7.18):

$$h_0(t) = \left\{ 1 - e^{-\frac{R}{2L}t} \left[\cos\left(\frac{\sqrt{-\Delta}}{2LC}t\right) + \frac{RC}{\sqrt{-\Delta}} \sin\left(\frac{\sqrt{-\Delta}}{2LC}t\right) \right] \right\} u(t)$$

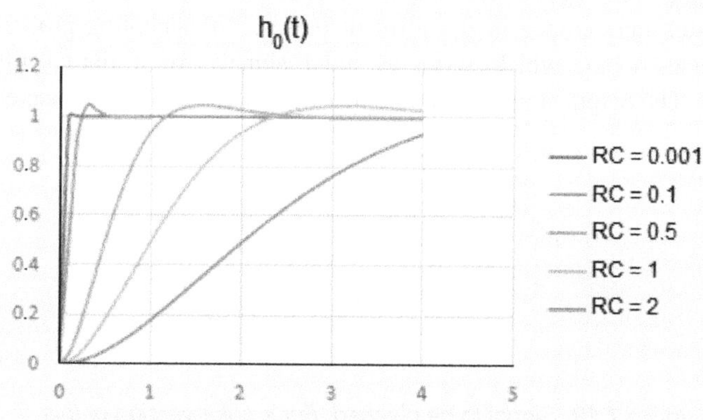

Figure 7.18. *Curves $h_0(t)$. For a color version of this figure, see www.iste.co.uk/cochard/mathematics2.zip*

7.3. Digital filtering

7.3.1. *General information*

Digital filtering, like analog filtering, can possess properties of linearity, causality and time invariance, which are assumed in the following discussion. In addition to these general properties, various specific properties of digital filters include:

– *Stable filter:* a filter is considered stable if it satisfies the following condition: when the input signal amplitude is bounded ($e(t) < B$), the amplitude of the output signal also remains bounded ($s(t) < B'$). In other words, a bounded input signal yields an equally bounded output signal. A stable, causal filter is referred to as *feasible.*

– *Finite impulse response* (FIR) *filter*: an FIR filter is one whose output depends only on the current input and does not rely on previous outputs. This means that calculating the output at a given time does not require storing or recalling past output values.

– *Infinite impulse response* (IIR) *filter*: an IIR filter is one whose output depends on the current input as well as on the previous outputs. This implies that calculating the output at a given moment requires considering past output values.

EXAMPLE 7.4.– Cleaning a discrete signal. Suppose a sinusoidal signal is sampled, which contains interference in the form of pulses. The sampling process yields discrete signals $e(n)$, which serve as input signals for a digital filter (see Figure 7.19). The output signal from this filtering process will be the discrete $s(n)$.

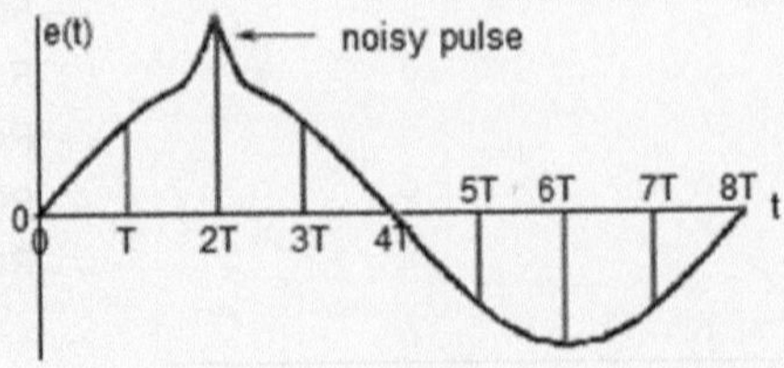

Figure 7.19. *Signal to be cleaned. For a color version of this figure, see www.iste.co.uk/cochard/mathematics2.zip*

The sampled signal is represented as $e(n)$ with $n = 0, 1, 2, \ldots$ (which corresponds to $e(nT)$). To eliminate interference, the moving average method, commonly used in statistics for analyzing seasonal phenomena, can be applied:

$s(n) = \frac{1}{3}[e(n-1) + e(n) + e(n+1)]$

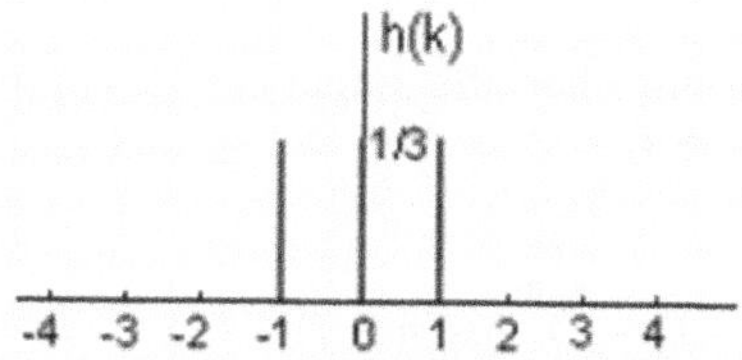

Figure 7.20. *Impulse response. For a color version of this figure, see www.iste.co.uk/cochard/mathematics2.zip*

Three coefficients can be defined as follows:

$$h(-1) = 1/3 \quad h(0) = 1/3 \quad h(1) = 1/3$$

Using these coefficients, the previous expression can be rewritten as:

$$s(n) = h(1)e(n-1) + h(0)e(n) + h(-1)e(n+1)$$

Thus, the output signal can be expressed in terms of a summation:

$$s(n) = \sum_{k=-1}^{+1} h(k)e(n-k)$$

This indicates the convolution product corresponding to the filter effect defined by the impulse response $h(k)$ (see Figure 7.20).

7.3.2. *Difference equation*

Generally speaking, the output of a digital filter *s(n)* depends on several elements:

1) Current input $e(n)$: this is the input signal at the moment n, which is directly processed by the filter.

2) The *N* previous entries $e(n-1), e(n-2), \ldots, e(n-N)$: these values represent the history of *N* previous input samples at time n. Filters that depend solely on the current input and previous samples (e.g. FIR filters) use only these past samples to compute the output.

3) Previous outputs $s(n-1), s(n-2), \ldots, s(n-M)$: these values represent the history of *M* previous output samples at time *n*. Filters that depend on previous

outputs (e.g. IIR filters) incorporate these past samples in addition to the current input to calculate the output.

For a linear time-invariant filter, the output can be expressed with the following linear expression:

$$s(n) = \sum_{i=0}^{p} a_i e(n-i) - \sum_{j=1}^{p} b_j s(n-j)$$

Therefore, the coefficients a_i *and* b_j are constants, and $p = \max(N, M)$. This can also be rearranged to:

$$\sum_{i=0}^{p} a_i e(n-i) = \sum_{j=0}^{p} b_j s(n-j) \quad \text{with} \quad b_0 = 1$$

This equation is referred to as the *difference equation*. If all coefficients $b_j (for\ j \neq 0)$ are zero, the filter is *non-recursive*; otherwise, it is *recursive*.

Solving the difference equation involves finding the expression for $s(n)$. The approach is analogous to solving differential equations with constant coefficients, and the principle is outlined without demonstration here.

The *solution* $s(n)$ can be expressed as:

$$s(n) = s_n(n) + s_p(n)$$

where $s_h(n)$ is the solution of the homogeneous equation (the equation without input terms):

$$s(n) + \sum_{j=1}^{p} b_j s(n-j) = 0$$

and $s_p(n)$ is a particular solution of the complete equation.

EXAMPLE 7.5.– Let $s(n) - \frac{1}{4}s(n-2) = e(n)$ be the difference equation. First, the solution to the homogeneous equation is determined:

$$s_h(n) - \frac{1}{4}s_h(n-2) = 0.$$

The characteristic polynomial is obtained by replacing $s_h(n)$ with z^n:

$$\left(z-\frac{1}{2}\right)\left(z+\frac{1}{2}\right)=0,$$

resulting in two roots: $\frac{1}{2}$ and $-\frac{1}{2}$. This leads to the solution:

$$s_h(n)=B_1\left(\frac{1}{2}\right)^n+B_2\left(-\frac{1}{2}\right)^n,$$

where B_1 and B_2 are constants determined by initial conditions.
Next, a constant is assumed for the particular solution:

$$s_p(n)=C.$$

Substituting this into the difference equation yields:

$$e(n)=\frac{3}{4}C.$$

Thus, for an input function $e(n)=1$, the output function becomes:

$$s_p(n)=\frac{4}{3}.$$

The particular solution is:

$$s_p(n)=\frac{4}{3}.$$

Ultimately, the general solution to the difference equation is:

$$s_n(n)=\frac{4}{3}+B_1\left(\frac{1}{2}\right)^n+B_2\left(-\frac{1}{2}\right)^n$$

Constants B_1 and B_2 can be determined based on the initial conditions of the system.

7.3.3. *Transfer function*

The difference equation leads to the transfer function $H(z)$. Applying the z-transform to both sides of the equation provides:

$$\sum_{n=-\infty}^{+\infty}[\sum_{j=0}^{p}b_js(n-j)]z^{-n}=\sum_{n=-\infty}^{+\infty}\left[\sum_{i=0}^{p}a_ie(n-i)\right]z^{-n}$$

This can also be expressed by inverting the sums:

$$\sum_{j=0}^{p} b_j[\sum_{n=-\infty}^{+\infty} s(n-j)z^{-n}] = \sum_{i=0}^{p} a_i \left[\sum_{n=-\infty}^{+\infty} e(n-i)z^{-n}\right]$$

or in another form:

$$\sum_{j=0}^{p} b_j z^{-j} \left[\sum_{n=-\infty}^{+\infty} s(n-1)z^{-(n-j)}\right] = \sum_{i=0}^{p} a_i z^{-i} [\sum_{n=-\infty}^{+\infty} e(n-1)z^{-(n-i)}]$$

Thus,

$$\sum_{j=0}^{p} b_j z^{-j} \left[\sum_{m=-\infty}^{+\infty} s(m)z^{-m}\right] = \sum_{i=0}^{p} a_i z^{-i} \left[\sum_{m=-\infty}^{+\infty} e(m)z^{-m}\right]$$

Defining $E(z)$ and $S(z)$ as the z-transforms of $e(n)$ and $s(n)$, respectively, the equation can be rewritten as:

$$S(z)\sum_{j=0}^{p} b_j z^{-j} = E(z)\sum_{i=0}^{p} a_i z^{-i}$$

From this, the transfer function $H(z)$ is given by:

$$H(z) = \frac{S(z)}{E(z)} = \frac{\sum_{i=0}^{p} a_i z^{-i}}{\sum_{j=0}^{p} b_j z^{-j}}$$

Note that $H(z)$ is expressed as a rational fraction in z^{-1}. To determine $H(z)$, substitute $e(n) = z^n$ and $s(n) = z^n H(z)$ into the difference equation.

EXAMPLE 7.6.– Consider the difference equation:

$$6s(n) + 5s(n-1) + s(n-2) = e(n).$$

By performing the indicated replacement, the transfer function $H(z)$ is derived as follows:

$$H(z) = \frac{1}{6+5z^{-1}+z^{-2}}$$

EXAMPLE 7.7.– Consider $s(n) = \frac{1}{2}[e(n) + e(n-1)$ that corresponds to an "averaging" filter. The transfer function can be determined as follows. Given the linearity of the z-transform, the expression for $S(z)$ is:

$$S(z) = \frac{1}{2}[E(z) + z^{-1}E(z)]$$

From the previous expression, the transfer function can be deduced:

$$H(z) = \frac{1}{2}[1 + z^{-1}]$$

EXAMPLE 7.8.– Consider $s(n) = \frac{1}{2}[e(n) + s(n-1)]$.
Taking the z-transform of this expression yields:

$$S(z) = \frac{1}{2}[E(z) + z^{-1}S(z)].$$

Rearranging this equation leads to the deduction of the transfer function $H(z)$:

$$H(z) = \frac{S(z)}{E(z)} = \frac{\frac{1}{2}[E(z)+z^{-1}S(z)]}{E(z)}$$

Solving for $H(z)$ gives:

$$H(z) = \frac{1}{2-z^{-1}}$$

EXAMPLE 7.9.– Consider $s(n) = e(n) + 4s(n-1)$.
Taking the z-transform of this expression gives:

$$S(z) = E(z) + 4z^{-1}S(z).$$

Rearranging this equation leads to the deduction of the transfer function:

$$H(z) = \frac{1}{1-4z^{-1}}$$

From the difference equation, the transfer function $H(z)$ can be derived. The converse also holds true.

EXAMPLE 7.10.– Given the transfer function:

$$H(z) = \frac{1+z^{-2}}{1-4z^{-1}+z^{-2}} = \frac{S(z)}{E(z)},$$

the expression can be rearranged to obtain $S(z)$:

$$S(z) = (1 + z^{-2})E(z) + (4z^{-1} - z^{-2})S(z)$$

By isolating $S(z)$:

$$S(z)\left(1 - (4z^{-1} - z^{-2})\right) = (1 + z^{-2})E(z)$$

Taking the inverse z-transform results in:

$$s(n) = e(n) + e(n-2) + 4s(n-1) - s(n-2)$$

For this expression, the coefficients can be identified as follows:

a_i and b_j: $a_0 = 1, a_1 = 0, a_2 = 1, b_1 = -4$ and $b_2 = 1$.

7.3.4. *Filter stability*

An important consequence of $H(z)$ in the z-domain is its ability to be factored. The transfer function can be expressed as:

$$H(z) = C\frac{\prod_{i=0}^{p}(1 - \alpha_i z^{-1})}{\prod_{j=0}^{p}(1 - \beta_j z^{-1})} = C\frac{\prod_{i=0}^{p}(z - \alpha_i)}{\prod_{j=0}^{p}(z - \beta_j)}$$

where α_i and β_j represent the zeros of the numerator and denominator polynomials, respectively. The zeros of the denominator correspond to the poles of $H(z)$.

Stability is a crucial property of a filter. A filter is considered *stable* if, for any input with bounded amplitude, the output also remains bounded. According to a theorem, which is widely accepted in control theory and signal processing, a filter is stable if and only if *all* poles of $H(z)$ have a modulus less than 1. This condition ensures that the system will not diverge and will produce a finite response to bounded inputs.

EXAMPLE 7.11.– The transfer function is given by:

$$H(z) = \frac{1+z^{-1}}{2} = \frac{z+1}{2z}$$

– Zero: there is a zero at $z = -1$ since the numerator $z + 1 = 0$ gives $z = -1$.

– Pole: there is a pole at $z = 0$ because the denominator $2z = 0$ leads to $z = 0$.

– Stability: for stability, all poles must have a modulus less than 1. The pole at $z = 0$ has a modulus of $|0| = 0$, which is less than 1. However, a pole at the origin indicates marginal stability, as it may lead to unbounded output for certain inputs.

EXAMPLE 7.12.– The transfer function is:

$$H(z) = \frac{1}{2-z^{-1}} = \frac{z}{2z-1}$$

Zero: $z = 0$.
Pole: $z = \frac{1}{2}$.
Stability: the filter is stable, as the pole's modulus is less than 1.

EXAMPLE 7.13.– The transfer function is:

$$H(z) = \frac{1}{1-4z^{-1}} = \frac{z}{z-4}$$

We have a zero for $z = 0$ and a pole for $z = 4$. The filter is not stable.

EXAMPLE 7.14.– The transfer function is:

$$H(z) = \frac{1+z^{-2}}{1-4z^{-1}+z^{-2}} = \frac{z^2+1}{z^2-4z+1}$$

– Zeros: the zeros are found by setting the numerator equal to zero,

$$z^2 + 1 = 0 \quad \Rightarrow \quad z^2 = -1 \quad \Rightarrow \quad z = \mp i$$

– Poles: the poles are found by setting the denominator equal to zero,

$$z^2 - 4z + 1 = 0$$

Solving the quadratic equation,

$$z = 2 \pm \sqrt{3}$$

– Modulus of poles: the modulus of the poles $z = 2 \pm \sqrt{3}$,

$$|2 + \sqrt{3}| = 2 + \sqrt{3} \text{ and } |2 - \sqrt{3}| = 2 - \sqrt{3}$$

Since $\sqrt{3} \approx 1.732$, then:

$$|2+\sqrt{3}| \approx 3.732 > 1 \text{ and } |2-\sqrt{3}| \approx 0.268 < 1,$$

the filter is not stable since at least one pole has a modulus greater than 1.

7.3.5. *Frequency behavior*

The expression for the Fourier transform of a sampled signal can be summarized as follows.

If $x_e(t)$ is a signal sampled from a continuous signal $x(t)$ at a sampling frequency f_e with a period $T = 1/f_e$, the sampled signal expression of $x_e(t)$ can be expressed as:

$$x_e(t) = \sum_{n=-\infty}^{+\infty} x(nT)\delta(t-nT)$$

The Fourier transform of the sampled signal of $x_e(t)$ is given by:

$$X_e(f) = \int_{-\infty}^{+\infty} x_e(t)e^{-2i\pi ft}\,dt$$

$$= \sum_{n=-\infty}^{+\infty} x(nT) \int_{-\infty}^{+\infty} \delta(t-nT)e^{-2i\pi ft}dt$$

$$= \sum_{n=-\infty}^{+\infty} x(nT)e^{-2i\pi fnT}$$

By redefining $x(nT)$ as $x(n)$, the expression can be rewritten as:

$$X_e(f) = \sum_{n=-\infty}^{+\infty} x(n)\,(e^{-2i\pi fT})^n = \sum_{n=-\infty}^{+\infty} x(n)\,(e^{2i\pi fT})^{-n}$$

By setting $z = e^{2i\pi fT}$, the z-transform is obtained. This establishes the link between the Fourier transform and the z-transform. Applying this result to the transfer function:

$$z \quad \rightarrow \quad e^{2i\pi fT}$$

$$H(z) \quad \rightarrow \quad H(f)$$

EXAMPLE 7.15.– Let the transfer function be given by:

$$H(z) = \frac{1}{2}(1 + z^{-1})$$

Substituting $z = e^{2i\pi fT}$ for the frequency domain:

$$H(f) = \frac{1}{2}(1 + e^{-2i\pi fT})$$

This can be simplified as:

$$H(f) = e^{-i\pi fT}\frac{e^{i\pi fT}+e^{-i\pi fT}}{2} = e^{-i\pi fT}\cos(\pi fT)$$

Consequently, the magnitude of the transfer function is $|H(f)| = |cos(\pi fT)|$ (see Figure 7.21).

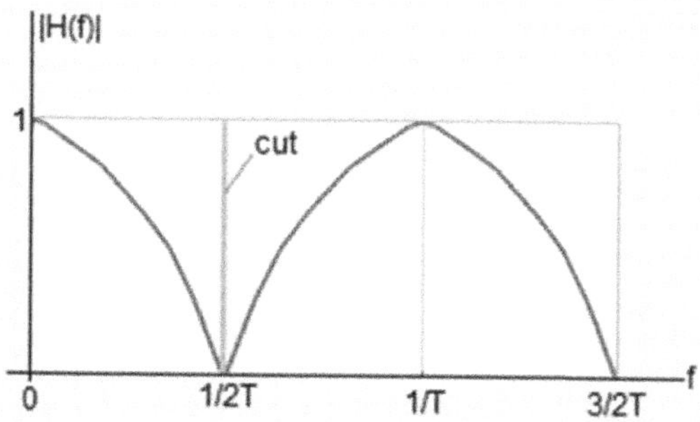

Figure 7.21. *Representation of |H(f)|. For a color version of this figure, see www.iste.co.uk/cochard/mathematics2.zip*

In accordance with Shannon's sampling theorem, the sampling frequency $\left(\frac{1}{T}\right)$ must be greater than twice the maximum frequency f_{max} to avoid aliasing. This condition can be expressed as:

$$f_{max} < \frac{1}{2T}$$

This constraint limits the frequencies in the signals to the range $\left[0, \frac{1}{2T}\right]$. Given this, the transfer function operates within this frequency range, effectively acting as a low-pass filter, allowing only frequencies below $\left(\frac{1}{2T}\right)$ to pass while attenuating higher frequencies.

EXAMPLE 7.16.– Given the transfer function:

$$H(z) = \frac{1}{2-z^{-1}}.$$

Substituting $z = e^{2i\pi fT}$ gives the frequency response:

$$H(f) = \frac{1}{2-e^{-2i\pi fT}} = \frac{1}{2-\cos(2\pi fT)+i\sin(2\pi fT)}$$

To find $|H(f)|^2 = \frac{1}{5-4\cos(2\pi fT)}$, multiply $H(f)$ by its complex conjugate:

$$|H(f)|^2 = \frac{1}{(2-\cos(2\pi fT))^2+\sin^2(2\pi fT)}$$

Simplifying:

$$|H(f)|^2 = \frac{1}{5-4\cos(2\pi fT)}$$

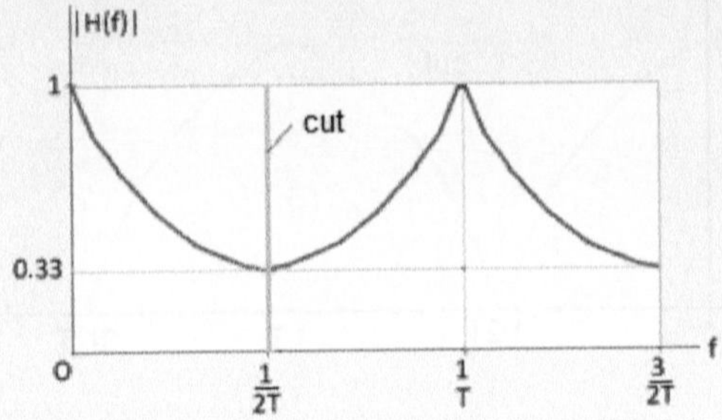

Figure 7.22. *Representation of $|H(f)|$. For a color version of this figure, see www.iste.co.uk/cochard/mathematics2.zip*

The shape of $|H(f)|$ is shown in Figure 7.22.

Considering Shannon's rule, the sampling frequency $\left(\frac{1}{T}\right)$ must be greater than twice the maximum signal frequency, which implies $f_{max} < \frac{1}{2T}$. Consequently, the filter restricts frequencies to the range $\left[0, \frac{1}{2T}\right]$, making it a low-pass filter.

7.3.6. *FIR filters*

An FIR filter is a *non-recursive* filter: the output signal depends solely on the input signals. In the general expression for the output signal:

$$s(n) = \sum_{i=0}^{M} a_i e(n-i) - \sum_{j=1}^{N} b_j \, s(n-j)$$

The coefficients b_j are zero, so the equation simplifies to:

$$s(n) = \sum_{i=0}^{M} a_i e(n-i)$$

If the input signal is a pulse represented by the function δ, the output signal represents the impulse response:

$$h(n) = \sum_{i=0}^{M} a_i \delta(n-i) = a_0\delta(n) + a_1\delta(n-1) + \cdots + a_M\delta(n-M)$$

This shows that $h(0) = a_0, \dots, h(M) = a_M$. The $M+1$ coefficients a_i are clearly identified, and their number $(M+1)$ is finite.

An FIR filter can be represented schematically using a "delay" function (see Figure 7.23).

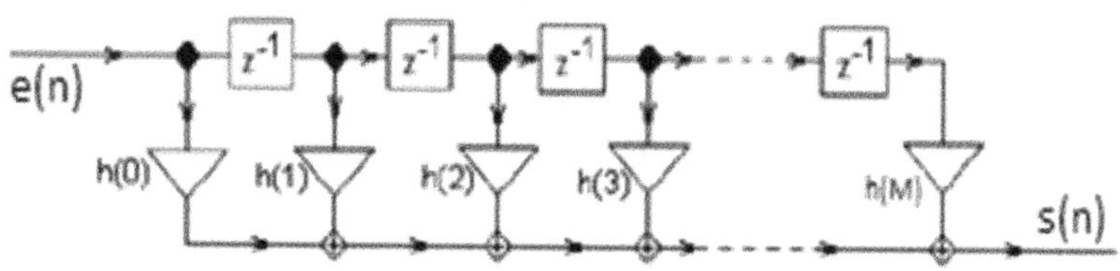

Figure 7.23. *Diagram of an FIR filter*

FIR filters are inherently stable because:

$$\sum_{k} |h(k)| < \infty$$

The transfer function for an FIR filter is given by:

$$H(z) = \sum_{k=0}^{M} h(z) z^{-k}$$

This can be expressed as:

$$H(z) = \frac{h(0)z^M + h(1)z^{M-1} + \cdots + h(M)}{z^M}$$

In the case of FIR filters, there is only one (multiple) pole at $z = 0$, and since the modulus is less than 1, it confirms that all FIR filters are stable.

In the frequency domain, by setting $z = e^{2i\pi fT}$, the transfer function can be expressed as:

$$H(f) = \sum_{k=0}^{M} h(k)e^{-2i\pi kfT}$$

This shows that $H(f)$ is a periodic function with period $f_e = \frac{1}{T}$. As a result, $H(f)$ can therefore be developed into a Fourier series:

$$H(f) = \sum_{p=-\infty}^{+\infty} c_p e^{2i\pi pfT}$$

where the Fourier coefficients c_p are defined as:

$$c_p = \frac{1}{f_e} \int_{-f_e/2}^{+f_e/2} H(f)\, e^{-2i\pi pfT} df = \frac{1}{f_e} \int_{-\infty}^{+\infty} H(f)\, e^{-2i\pi pfT} df$$

$$= \frac{1}{f_e} \int_{-\infty}^{+\infty} H(f)\, e^{+2i\pi(-p)fT} df = Th(-p)$$

Given that $H(f) = 0$ for f outside the interval $[-\frac{f_e}{2}, +\frac{f_e}{2}]$ (according to Shannon's theorem). As a result:

$$H(f) = \sum_{p=-\infty}^{+\infty} Th(-p)e^{2i\pi pfT} = T \sum_{p=-\infty}^{+\infty} h(p)\, e^{-2i\pi pfT}$$

Thus, the Fourier coefficients c_p are proportional to the components of the impulse response $h(n)$, with the proportionality constant being the sampling period T. This establishes a clear link between the time-domain impulse response and the frequency domain representation of the filter.

EXAMPLE 7.17.– The filter given by:

$$s(n) = \frac{1}{2}[e(n) + e(n-1))$$

is an FIR filter. The impulse response $h(n)$ can be determined from the equation. In this case, the coefficients of the impulse response are:

– $h(0) = \frac{1}{2}$ (the coefficient for the current input $e(n)$);

– $h(1) = \frac{1}{2}$ (the coefficient for the delayed input $e(n-1)$);

– $h(n) = 0$ for all other n (since there are no other terms in the equation).

This means the impulse response can be explicitly written as:

$$h(n) = \begin{cases} \frac{1}{2} & \text{If } \ n = 0 \\ \frac{1}{2} & \text{If } \ n = 1 \\ 0 & \text{otherwise} \end{cases}$$

This confirms that the filter is an FIR filter with a finite number of non-zero coefficients (specifically, two in this case). The finite nature of the impulse response ensures the stability and bounded output characteristics of the FIR filter.

7.3.7. *IIR filters*

An IIR filter is a recursive filter, where the output signal depends not only on previous input signals but also on previous output signals. The general form of the output signal for an IIR filter can be expressed as:

$$s(n) = \sum_{i=0}^{M} a_i e(n-i) - \sum_{j=1}^{N} b_j s(n-j)$$

In this equation, the coefficients b_j are not zero, indicating the feedback mechanism that characterizes recursive filters.

This difference equation can be separated into two components:

$$s_1(n) = \sum_{i=0}^{M} a_i e(n-i) \quad \text{and} \quad s(n) = s_1(n) - \sum_{j=1}^{N} b_j s(n-j)$$

The first component represents the response of the filter to the input signal $e(n)$ and can be viewed as an FIR filter (see Figure 7.24) since it relies only on current and past input samples. The second component incorporates previous output samples, establishing the filter's recursive nature.

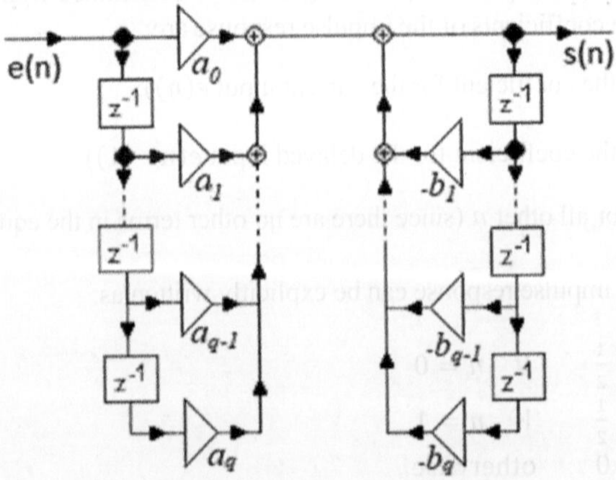

Figure 7.24. *Diagram of an IIR filter*

EXAMPLE 7.18.– The filter defined by

$$s(n) = \frac{1}{2}[e(n) + s(n-1)]$$

is indeed an IIR filter that can be represented by the diagram in Figure 7.25.

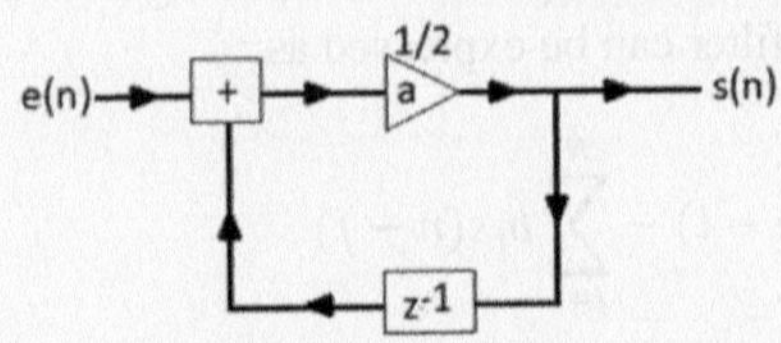

Figure 7.25. *Diagram of the IIR filter*

The pole of the transfer function for the IIR filter given by $s(n)$ can be found from its transfer function. The transfer function can be derived as follows:

$$S(z) = \frac{1}{2}[E(z) + z^{-1}S(z)]$$

Rearranging gives:

$$S(z)\left(1-\frac{1}{2}z^{-1}\right)=\frac{1}{2}E(z)$$

Thus, the transfer function $H(z)$ becomes:

$$H(z)=\frac{S(z)}{E(z)}=\frac{\frac{1}{2}}{1-\frac{1}{2}z^{-1}}=\frac{1}{2}\cdot\frac{1}{1-\frac{1}{2}z^{-1}}$$

From the transfer function, it is clear that the pole occurs at:

$$z=\frac{1}{2}.$$

The filter is stable if all poles of the transfer function lie within the unit circle (i.e. their moduli are less than 1). In this case:

$$\left|\frac{1}{2}\right|=0.5\ <1$$

Since the only pole of the transfer function is $\frac{1}{2}$, which is indeed less than 1, the filter is considered stable.

EXAMPLE 7.19.– The given filter described by the equation:

$$s(n)=e(n)+4s(n-1)$$

is indeed an IIR filter, which is represented by the diagram shown in Figure 7.26.

Deriving the transfer function:

– Apply the z-transform to the difference equation:

$$S(z)=E(z)+4z^{-1}S(z)$$

– Rearranging the equation gives:

$$S(z)(1-4z^{-1})=E(z)$$

– Thus, the transfer function $H(z)$ becomes:

$$H(z)=\frac{S(z)}{E(z)}=\frac{1}{1-4z^{-1}}=\frac{z}{z-4}$$

Finally, to determine stability, finding the poles of the transfer function is essential. The pole occurs when the denominator is equal to zero:

$$z - 4 = 0 \qquad \Rightarrow \qquad z = 4$$

Since the only pole of the transfer function is at $z = 4$, which is outside the unit circle, the filter is not stable.

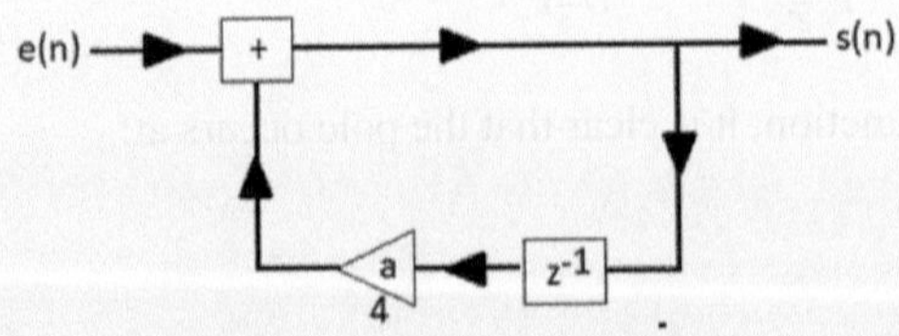

Figure 7.26. *Diagram of the IIR filter*

8

The Digital Image

CONCEPTS COVERED IN THIS CHAPTER.–

Computers did not make it possible to visualize, analyze and process digital images until the 1960s. A digital image can be thought of as a sequence of discrete signals, each coded on a certain number of bits.

Various definitions and concepts are specific to digital images, and the aim of this chapter is to introduce them.

After explaining the difference between raster and vector images, key concepts of colorimetry essential for the following sections will be introduced.

The chapter concludes with examples of image coding, an overview of image compression (to be covered in detail later) and a list of the main formats used in computer systems.

References: [COU 92, FOL 91, GON 93, HEA 97, LEP 91, RIM 93].

8.1. Raster and vector images

Digital images intended for viewing on computer screens can be categorized into two main types: raster (or bitmap) images and vector images.

8.1.1. *Raster images*

A raster image, also known as a bitmap image, is represented as a matrix of points (pixels). In the case of a binary image, pixels can either be "on", indicating

Mathematics for Digital Science 2,
by Gérard-Michel COCHARD and Mhand HIFI.

the presence of a dot, or "off", indicating its absence. For a color image, each pixel is associated with a specific color. These coded dots are arranged in rows and columns, following a simple correspondence: an element of the matrix (coded dot) represents a dot on the computer screen.

The main advantages of raster images are as follows:

– Simple display mechanisms: the dots are directly coded, and display devices, such as screens, use a grid of dots to reproduce the image easily.

– Ease of representing complex images: complex images can be easily represented in raster form, as the arrangement of dots does not need to follow a specific logical order.

– Efficient image processing: raster structures facilitate the application of image processing algorithms, allowing for operations such as pixel manipulation, filtering and transformations.

– Straightforward format conversion: converting between different image formats is relatively simple, as it mainly involves transforming the dot matrix.

– Independence of image parts: each part of a raster image is independent of others, meaning degradation in one part does not necessarily affect the integrity of the whole image.

However, there are several disadvantages associated with the use of raster images:

– Fixed resolution: raster images have a fixed resolution determined by the number of dots per row and column. Displaying the image on a different playback device may require resolution adaptation, which can result in a loss of quality.

– Issues with spatial modifications: modifications to a raster image, such as resizing or rotation, can lead to problems such as loss of information and aliasing.

– High processing requirements: raster images contain a large number of dots that need to be processed, requiring powerful processors for efficient operations.

– Large file sizes: Bitmap files are generally large and consume significant memory space. A document containing high-resolution images can take up several megabytes per image.

Despite these drawbacks, raster images are still widely used, as they allow for the representation of realistic, complex and detailed images, while also offering flexibility in pixel processing and manipulation.

8.1.2. *Vector images*

A vector image is a graphical representation composed of basic geometric shapes, such as lines, circles, rectangles, splines, Bézier curves, etc. Each shape is described by geometric attributes, as well as appearance-related attributes such as thickness, color and style. Displaying or printing a vector image requires conversion to raster format (pixel mode) since peripherals typically operate in this mode.

The advantages of vector images include the following:

– Device independence: a vector image is a symbolic representation that can be interpreted by various types of display or printing device.

– Flexible spatial modifications: transformations such as resizing, rotating or moving objects can be performed without loss of information or image quality, as they involve geometric operations.

– Suitability for CAD work: vector images are particularly well-suited for computer-aided design (CAD) applications, where precision and flexibility are essential.

– Small file size: the size of a vector image primarily depends on its geometric complexity, resulting in relatively small file sizes. A complex vector image typically occupies only 1–2 MB.

However, vector images also have their drawbacks:

– Challenges with analog images: transforming images of analog origin, such as photographs, into vector format is difficult due to their complexity and the need to represent them using simple geometric shapes.

– Conversion requirements: displaying or printing a vector image necessitates calculations to convert it into a raster image, as most display devices operate in raster mode. A vector file essentially serves as a program that describes the operations needed to display or print the image. If part of the file becomes corrupted, it can lead to significant issues. However, vector files in “text” mode, such as PostScript files, provide some resilience against this problem.

8.2. Notions of colorimetry

The perception of an object’s color is influenced by several factors, including its geometry, the light source illuminating it, the surrounding environment and the human visual system itself. As a result, color appreciation involves both subjective and objective components.

8.2.1. *Grayscale*

An image is characterized by the amount of light attributed to each point, which is represented by a gray level measured in terms of intensity. To simplify the representation of intensity levels, a quantitative scale is commonly used. On this scale, intensity level 0 corresponds to absolute black, while intensity level 1 represents absolute white, with a full range of gray levels in between. When defining n different intensity levels from black to white, the lowest intensity, I_0, represents black, and the highest intensity, 1, represents white. However, it is important to note that achieving an absolute intensity of 0 is technically not feasible, so the lowest possible intensity typically ranges between 0.005 and 0.025.

Studies on physiological vision have shown that intensity levels should be spaced logarithmically rather than linearly. To achieve this, a geometric sequence with a common ratio r can be defined. For n levels, the intensity levels are given by:

$$I_0, I_1 = rI_0, I_2 = rI_1 = r^2 I_0, \dots, I_{n-1} = r^{(n-1)} I_0 = 1$$

To determine the value of r, the following relation is used:

$$r = \left(\frac{1}{I_0}\right)^{\frac{1}{n-1}}$$

This formula calculates the factor of the geometric sequence r as a function of the number of levels n and lowest intensity I_0. Additionally, the intensity values I_j for each level can be calculated using the relation:

$$I_j = r^j I_0 = I_0^{\frac{(n-1-j)}{(n-1)}}.$$

This allows for the specific intensity for each level j to be determined based on the lowest intensity I_0, the total number of levels n and the position of the level j in the sequence.

Thus, for four levels, the common ratio would be $r = \left(\frac{1}{I_0}\right)^{\left(\frac{1}{3}\right)}$ and for 256 levels, $r = \left(\frac{1}{I_0}\right)^{\left(\frac{1}{255}\right)}$.

In the case of 256 levels, starting with intensity $I_0 = 0.010$, the common ratio would be $r = 1.018$, and the sequence of intensity levels would be: $0.01000 \quad 0.01018 \quad 0.01036 \quad ... \quad 0.9823 \quad 1.0000$.

Here, $\left(\frac{1}{I_0}\right)$ represents the ratio between the highest and lowest intensities, known as the dynamic range. Studies indicate that distinguishing between two intensity levels requires a value of r of at least "1.01"; below this threshold, the human eye cannot effectively differentiate between two successive levels.

Using this value of r and the dynamic range, the number of distinguishable levels can be determined with the relation:

$$r = \left(\frac{1}{I_0}\right)^{\frac{1}{n-1}}.$$

For example, with a dynamic range between 50 and 100 (typical for cathode ray tubes), it is possible to achieve between 400 and 530 gray levels. In contrast, for a newspaper printed in black and white, where the dynamic range is approximately 10, the number of gray levels would be approximately 234.

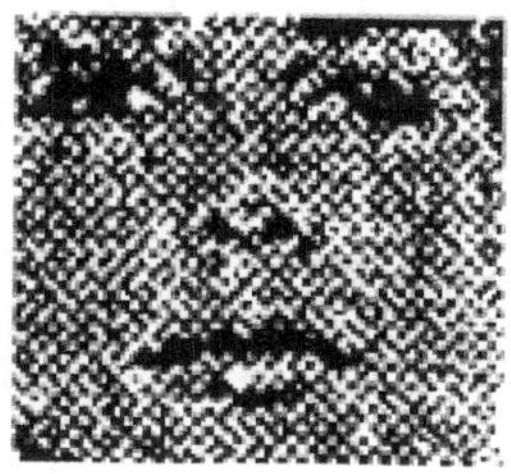

Figure 8.1. *Grayscale image*

However, elementary displays are binary, meaning they can only represent two states: on or off. This raises the question of how halftones can be represented on such screens. This issue is particularly relevant when reproducing black-and-white photographs in newspapers. One solution is to use "circular" black dots, with the area of each dot proportional to the gray value $(1 - I)$, as shown in Figure 8.1.

Graphic displays use a pointillism technique in which small disks of variable radius are replaced by square blocks that group together several pixels. Each of these blocks represents a different shade of gray.

For example, using 2×2 blocks allows for the distinction of five different levels of gray, as shown in Figure 8.2. Each block consists of four pixels, and the combination of their light intensities creates the impression of a specific shade of gray. By varying the size of the blocks or using different pixel arrangements, a greater number of gray levels can be achieved.

Figure 8.2. *Grayscale with 2x2 blocks*

When 3×3 blocks are used, it becomes possible to distinguish 10 different intensity levels, as shown in Figure 8.3. Each block consists of nine pixels, and the combination of their light intensities creates the impression of a specific shade of gray. By varying the size of the blocks or using different pixel arrangements, a greater number of gray levels can be provided.

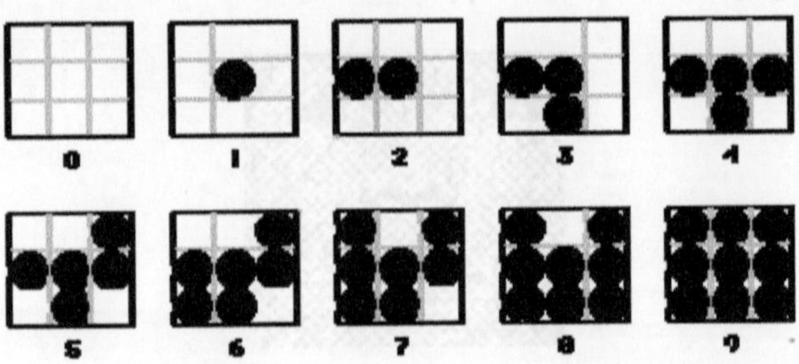

Figure 8.3. *Grayscale with 3x3 blocks*

More generally, when block sizes of $(n \times n)$ are used, it is possible to represent $n^2 + 1$ intensity levels. This method can be synthesized using a threshold matrix, also known as a dithering matrix

For example, consider the threshold matrix $\boldsymbol{D}(3) = \begin{bmatrix} 6 & 8 & 4 \\ 1 & 0 & 3 \\ 5 & 2 & 7 \end{bmatrix}$.

The interpretation of this matrix is as follows: for each intensity pixel, if its value is less than the corresponding intensity I in the threshold matrix, that pixel is turned on; otherwise, it is turned off.

For $n = 2$, the threshold matrix is $\boldsymbol{D}(2) = \begin{bmatrix} 0 & 2 \\ 3 & 1 \end{bmatrix}$.

For values $n = 2$ or $n = 3$, the threshold matrices and associated levels are selected to adhere to certain rules:

1) Pixels must not be aligned horizontally or vertically, as this would create visible horizontal or vertical stripes in the image.

2) The black pixels of one level must remain black for subsequent darker gray levels.

3) Gray levels should increase progressively as we transition to darker levels.

4) Black pixels must be contiguous, meaning they grouped together to form uninterrupted areas of black.

By adhering to these rules, threshold matrices can be used to optimally distribute lit pixels, simulating additional intensity levels and achieving smoother grayscales that are more faithful to the original image. Threshold matrices can be defined using a recurrence relation that calculates $D(n)$ as a function of $\boldsymbol{D}\left(\frac{n}{2}\right)$ and $\boldsymbol{U}\left(\frac{n}{2}\right)$, where $\boldsymbol{U}$ is a square matrix of dimensions $\left(\frac{n}{2}\right) \times \left(\frac{n}{2}\right)$ containing only elements equal to 1

$$\boldsymbol{D}(n) = \begin{bmatrix} 4\boldsymbol{D}\left(\frac{n}{2}\right) + D(2)_{00}\boldsymbol{U}(\frac{n}{2}) & 4\boldsymbol{D}\left(\frac{n}{2}\right) + D(2)_{01}\boldsymbol{U}(\frac{n}{2}) \\ 4\boldsymbol{D}\left(\frac{n}{2}\right) + D(2)_{10}\boldsymbol{U}(\frac{n}{2}) & 4\boldsymbol{D}\left(\frac{n}{2}\right) + D(2)_{11}\boldsymbol{U}(\frac{n}{2}) \end{bmatrix}$$

Applying this relationship, the matrix $\boldsymbol{D}(4)$ is obtained, resulting in four rows and four columns, as follows:

$$\boldsymbol{D}(4) = \begin{bmatrix} 0 & 8 & 2 & 10 \\ 12 & 4 & 14 & 6 \\ 3 & 11 & 1 & 9 \\ 15 & 7 & 13 & 5 \end{bmatrix}$$

To obtain the elements of the matrix $\boldsymbol{D}(4)$ (which has four rows and four columns, representing 17 shades), the recurrence relation is expressed as follows:

$$\boldsymbol{D}(4) = \begin{bmatrix} 4\boldsymbol{D}(2) + 0 \times \boldsymbol{U}(2) & 4\boldsymbol{D}(2) + 2 \times \boldsymbol{U}(2) \\ 4\boldsymbol{D}(2) + 3 \times \boldsymbol{U}(2) & 4\boldsymbol{D}(2) + 1 \times \boldsymbol{U}(2) \end{bmatrix}$$

where

$$\boldsymbol{U}(2) = \begin{bmatrix} 1 & 1 \\ 1 & 1 \end{bmatrix} \text{ and } \boldsymbol{D}(2) = \begin{bmatrix} 0 & 2 \\ 3 & 1 \end{bmatrix}$$

The element (which is actually a group of four elements) at the bottom left of the matrix $\boldsymbol{D}(4)$ is calculated as follows:

$$\boldsymbol{D}(2) + 3 \times \boldsymbol{U}(2) = 4\begin{bmatrix} 0 & 2 \\ 3 & 1 \end{bmatrix} + 3\begin{bmatrix} 1 & 1 \\ 1 & 1 \end{bmatrix} = \begin{bmatrix} 3 & 11 \\ 15 & 7 \end{bmatrix}.$$

Performing similar calculations for the other groups of elements yields the matrix $\boldsymbol{D}(4)$ shown above.

8.2.2. *Colors*

Producing images with transparency effects, textures and shading requires a wide range of colors that must be appropriately coded. To understand this, it is essential to recall that light is an electromagnetic wave characterized by its spectral distribution in terms of frequency or wavelength. Visible light, which the human eye can perceive, lies within the spectrum between 380 and 770 nanometers, spanning from ultraviolet to infrared. Experimental evidence has demonstrated that all visible light can be reconstituted from three carefully selected monochromatic lights, called primary colors (as established by Thomas Young in 1802). This principle is referred to as trichromacy. The primary colors typically used are red, green and blue, which form the basis of various color-coding systems, such as the RGB (Red, Green, Blue) model. In this model, each color is represented by a combination of the three primary components in specific intensities.

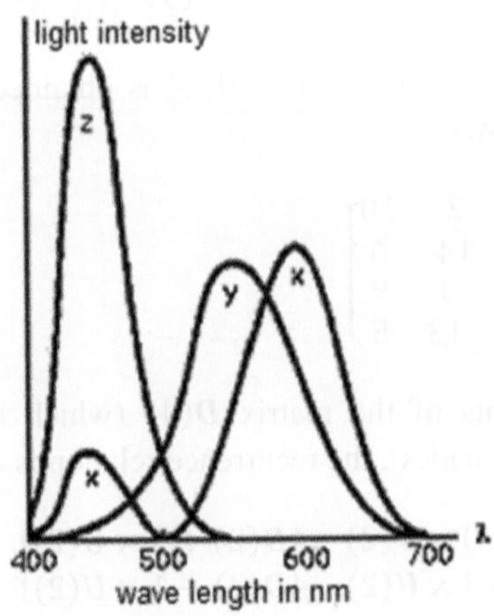

Figure 8.4. *Spectral curves (from Wikimedia Commons)*

The retina of the human eye comprises three types of cone cells, each sensitive to a specific frequency range corresponding to different wavelength: the blue range, the green range and the red range. In 1931, the International Commission on Illumination (ICI) established a mathematical representation of the human eye's sensitivity to various wavelengths λ of light. This is described by three spectral curves, denoted as $x(\lambda), y(\lambda)$ and $z(\lambda)$ (see Figure 8.4).

These curves quantify the relative response of the human eye to different wavelengths of visible light. In particular, the $y(\lambda)$ curve corresponds to the eye's maximum sensitivity, which occurs in the green region of the light spectrum. The $x(\lambda)$ and $z(\lambda)$ curves represent sensitivity to shorter (blue) and longer (red) wavelengths, respectively.

To characterize the color of light with spectral distribution $P(\lambda)$, trichromaticity values X, Y and Z are used. These values are calculated by integrating the spectral distribution $P(\lambda)$ weighted by the spectral curves $x(\lambda), y(\lambda)$ and $z(\lambda)$, respectively. The equations are as follows:

$$X = \int P(\lambda)\, x(\lambda)\, d\lambda, \quad Y = \int P(\lambda) y(\lambda) d\lambda \quad \text{and} \quad Z = \int P(\lambda) z(\lambda) d\lambda.$$

These trichromaticity values characterize the relative contribution of the primary colors (represented by the spectral curves $x(\lambda), y(\lambda)$ and $z(\lambda)$) to the overall color of the light. Using these trichromaticity values, each color can be represented in a three-dimensional space, typically using X, Y and Z associated with each primary color. These three values correspond to base vectors in three-dimensional space. Thus, each color corresponds to a vector $\boldsymbol{C}$ in this space, defined by:

$$\boldsymbol{C} = x'\boldsymbol{X} + y'\boldsymbol{Y} + z'\boldsymbol{Z},$$

considering $\boldsymbol{X}$, $\boldsymbol{Y}$ and $\boldsymbol{Z}$ as three-unit vectors in orthonormal space. The color $\boldsymbol{C}$ is then considered as a vector with components x', y' and z'.

To normalize color vectors, each component is divided by the sum of the components x', y' and z', which gives

$$x = \frac{x'}{(x' + y' + z')}, \quad y = \frac{y'}{(x' + y' + z')} \quad \text{and} \quad z = \frac{z'}{(x' + y' + z')}$$

The normalization relation $x + y + z = 1$ allows calculation of the z component as a function of x and y. The values x, y and z are known as chromaticity values.

The primary colors X, Y and Z are theoretical constructs, and not all vectors in XYZ space correspond to visible colors. Visible colors exist only within a specific cone in the first octant of XYZ space. However, the primary colors X, Y and Z are linearly related to the more familiar primary colors Red (R), Green (G) and Blue (B) *by* the following relationship:

$$\begin{bmatrix} X \\ Y \\ Z \end{bmatrix} = \begin{bmatrix} 2.7690 & 1.7518 & 1.1300 \\ 1.0000 & 4.5907 & 0.0601 \\ 0.0000 & 0.0565 & 5.5943 \end{bmatrix} \begin{bmatrix} R \\ G \\ B \end{bmatrix}$$

This matrix equation expresses the components X, Y and Z in terms of the components R, G and B, allowing conversion between the two color spaces. The base colors R, G and B were defined before primary colors X, Y and Z. The ICI formalized the use of X, Y and Z as base colors in 1931.

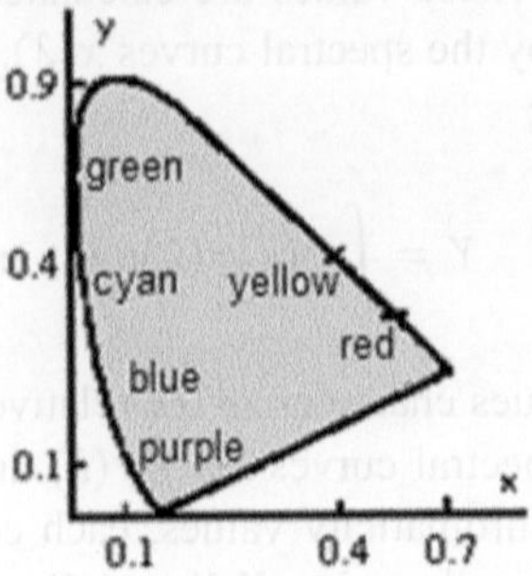

Figure 8.5. *ICI chromaticity diagram*

Since $x + y + z = 1$, the visible colors can be represented in a two-dimensional plane using the coordinates (x, y). This representation is known as the ICI chromaticity diagram (see Figure 8.5). This diagram visually maps the chromaticity values of the various visible colors, allowing a graphical depiction of color relationships in ICI trichromatic space.

Several models are commonly used in computer graphics to represent colors. These models serve different purposes based on the application and provide various methods for encoding and manipulating color information:

– *The RGB model (Red, Green, Blue):*

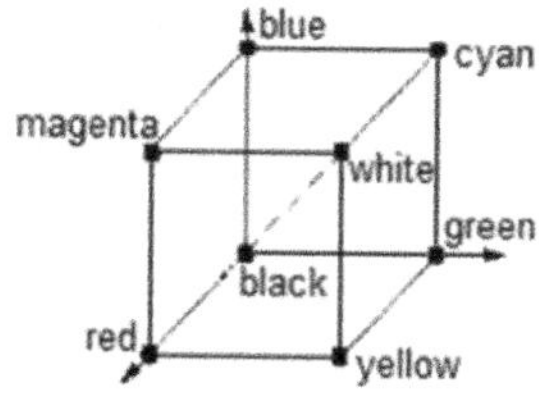

Figure 8.6. *The RGB model*

This model (see Figure 8.6) is widely used for displays and lighting, where colors are produced by mixing red, green and blue lights in different intensities. It is an additive color model.

– *The CMY model (Cyan, Magenta, Yellow).* The primary colors in this model are complementary to the RGB primary colors. This system is used by some color-imaging devices and inkjet printers. When used in the CMYK model (with K representing black), a black ink cartridge is added to achieve a true black, as mixing the other colors does not always produce a satisfactory result.

– *The YIQ model.* This model is a recoding of the RGB model, used by the National Television Standards Committee (NTSC). It is primarily used for video signal transmission.

$$\begin{bmatrix} Y \\ I \\ Q \end{bmatrix} = \begin{bmatrix} 0.30 & 0.59 & 0.11 \\ 0.60 & -0.27 & -0.32 \\ 0.21 & -0.52 & 0.31 \end{bmatrix} \begin{bmatrix} R \\ G \\ B \end{bmatrix}$$

– *The HSL (Hue, Saturation, Luminance) or HSV (Hue, Saturation, Value) model:*

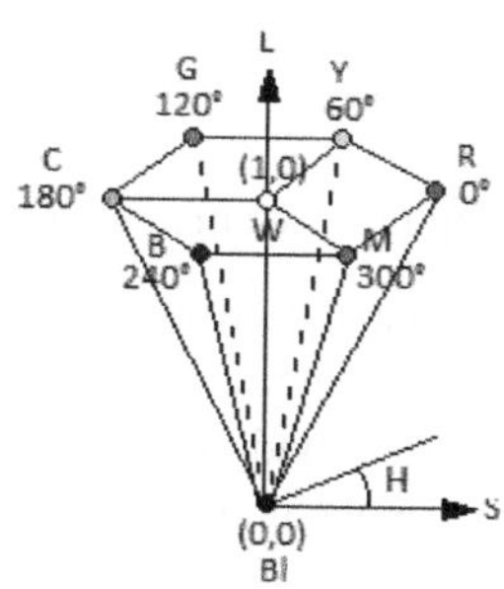

Figure 8.7. *The HSL model. For a color version of this figure, see www.iste.co.uk/cochard/mathematics2.zip*

This model is more aligned with human color perception. This model uses a hexagon-shaped space, with the luminance axis L (see Figure 8.7). When $L = 1$, maximum intensity colors are achieved. The hue H is determined by the angle between the red axis and a point within the hexagon, while saturation S is defined by the distance from the luminance axis to a point on the hexagon.

– *The HLS (Hue, Lightness, Saturation) model:*

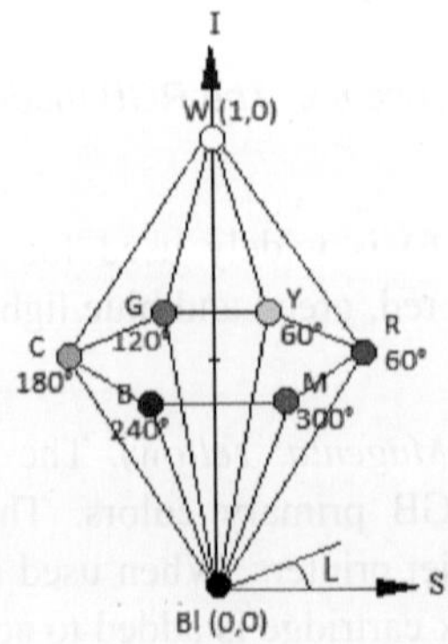

Figure 8.8. *The HLS model. For a color version of this figure, see www.iste.co.uk/cochard/mathematics2.zip*

This model, similar to the HLS model, is a color representation system that describes colors in terms of their hue, lightness and saturation. It uses a double cone-shaped space (see Figure 8.8). The axis running from black to white represents intensity, hue is determined by the polar angle and saturation is defined by the polar radius.

8.2.3. *True color and indexed color*

Graphic systems can be divided into two main categories:

1) True color systems: these systems encode a pixel using 16, 24 or 32 bits, allowing for a wide range of color nuances. For instance, with 24 bits, up to 16.7 million different colors can be represented. This color depth is commonly used in applications such as photographic retouching, where high color accuracy is required for achieving realistic results.

2) Indexed color systems: these systems encode a pixel with 8 bits or less, resulting in a limited number of colors. For example, with 8 bits, up to 256 different shades can be represented.

However, thanks to the color palette technique, it is possible to achieve a similar number of colors as with true color systems. In this mechanism, each pixel is associated with an index in a color table (palette) that contains a predefined set of colors.

The index for each pixel is used to determine the actual color displayed, which saves memory space and reduces computing power requirements.

If a pixel is coded on p bits, up to 2^p different colors can be displayed. For example, with $p = 8$ bits, 256 different colors are available. This range is often sufficient for simplified graphics and illustrations but is inadequate for representing realistic images with high color accuracy.

To overcome this limitation, the color palette technique is used. Instead of directly assigning a color to each pixel with the p bits, each pixel is associated with an index (address) in a color table, known as a palette. Each entry in the palette, coded on b bits, represents a specific color.

For example, with $p = 12$ bits, 2^{12} (4096) addresses are available in the palette. Each address points to an entry in the color table, encoded with b bits. If each palette entry is encoded with 3×8 bits (24 bits), 4096 different colors can be represented from a range of 16 million colors (see Figure 8.9).

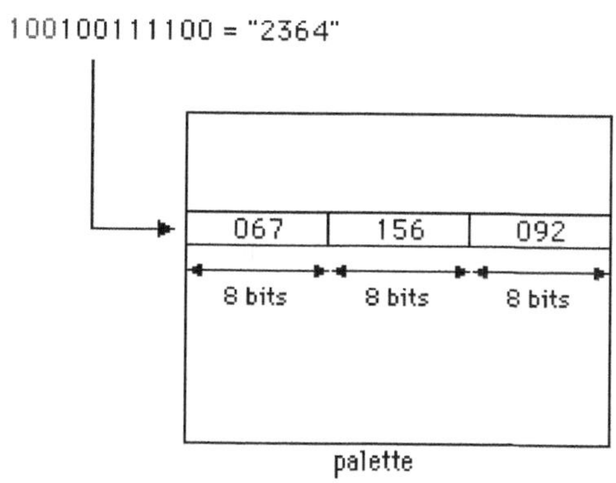

Figure 8.9. *Color palette*

8.4. Image display modes

8.4.1. *Matrix coding*

8.4.1.1. *Raw point coding*

A screen is a matrix of pixels, and to represent an image, a matrix is used where each element encodes a pixel. If an image has m rows and n columns, it contains $(m \times n)$ points or pixels:

– For a binary image (black or white pixels): each pixel is coded using one bit (0 or 1), resulting in a bit matrix $(m \times n)$.

– For a grayscale image (with 2^l shades): each pixel is coded on l bits, resulting in a matrix of $(m \times n \times l)$ bits.

– For a color image (with 2^l levels of Red, Green and Blue): each pixel is encoded using $(3 \times l)$ bits, requiring 3-bit matrices $(m \times n \times l)$ (or a matrix of $m \times n \times l \times (3 \times l)$ bits) to represent the color image.

Consider an image of size 512×512 pixels:

– For a binary image, the corresponding file (excluding format information) would be 32 KB (512×512 bits).

– For a 256-grayscale image, with each pixel coded on 8 bits, the file size would be 256 KB ($512 \times 512 \times 8$ bits).

– For a color image with 256 shades of R (Red), G (Green), B (Blue), the file size would be three times that the grayscale image, i.e. 768 KB ($3 \times 512 \times 512 \times 8$ bits).

High-quality image files, containing many colors and high resolution, can be very large, requiring compression for efficient storage and transmission.

8.4.1.2. *Range coding*

The compression principle involves avoiding the repetition of identical pixel sequences on the same line. Based on Figure 8.10, the following coding scheme is used, where B represents "white" and N represents "black"

line *	(B,21)
line **	(B,6) (N,7) (B,8)
line ***	(B,5) (N,5) (B,3) (N,5) (B,3)
line ****	(N,3) (B,2) (N,4) (B,2) (N,3) (B,1) (N,3) (B,3)

line ***** (B,1) (N,1) (B,1) (N,1) (B,1) (N,1) (B,1) (N,1) (B,1) (N,1) (B,1) (N,1) (B,1) (N,1) (B,1) (N,1) (B,1) (N,1) (B,3)

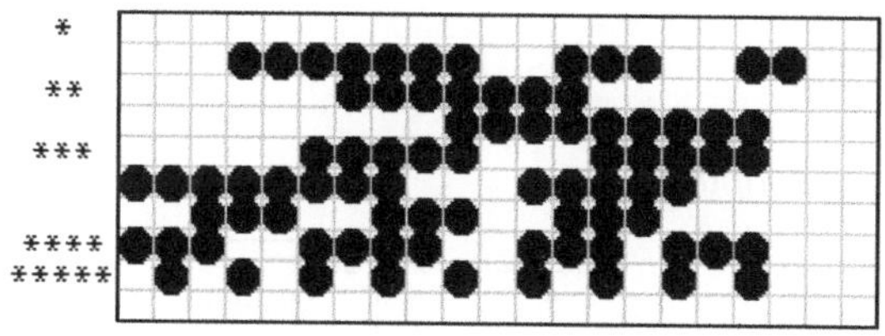

Figure 8.10. *Range coding*

For line **, there are six white pixels followed by seven black pixels, and then eight more white pixels.

8.4.1.3. *Quad trees*

A quaternary tree is created by dividing a square image into four quadrants. Each quadrant is further divided into four sub-quadrants, and this process continues recursively. Figure 8.11 shows the principle of quaternary tree coding.

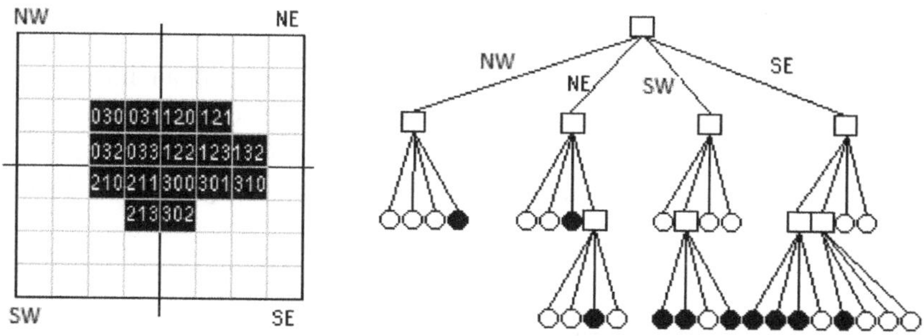

Figure 8.11. *Quaternary tree*

The lists can be identified as follows (with NW: 0, NE: 1, SW: 2, SE: 3):

– Ordered list of black pixels: 030, 031, 032, 033, 120, 121, 122, 123, 132, 210, 211, 213, 300, 301, 302, 310.

– Compacted list: for four consecutive codes that differ only in the last digits, the last digit is replaced with “x”. Thus, the compacted list is 03x, 12x, 132, 210, 211,

213, 300, 301, 302, 310. For example, 12x indicates NW/SE, while 213 represents SW/NE/SE.

8.4.2. *Vector coding*

8.4.2.1. *Dynamic description*

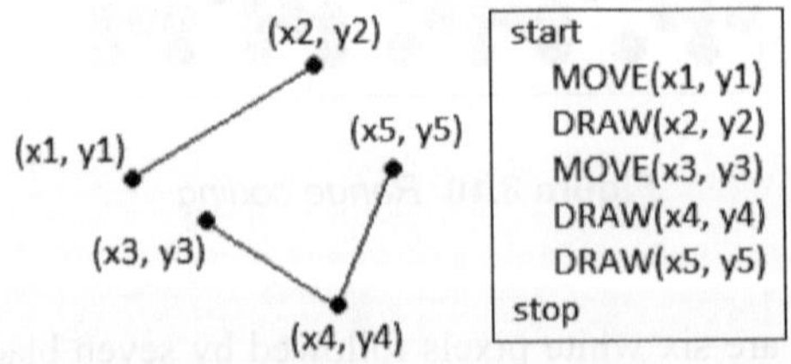

Figure 8.12. *Dynamic coding*

The dynamic description of the image is expressed as plot programs (see Figure 8.12). These programs outline the actions needed to generate the image from an abstract representation.

8.4.2.2. *Freeman coding*

Freeman coding is a method of representing the edges of an image. In its simplest form, it describes each pixel in relation to the previous pixel, using a plot direction. This method defines eight basic directions (see Figure 8.13). Each direction is represented by a number from 0 to 7, indicating the plot direction between neighboring pixels.

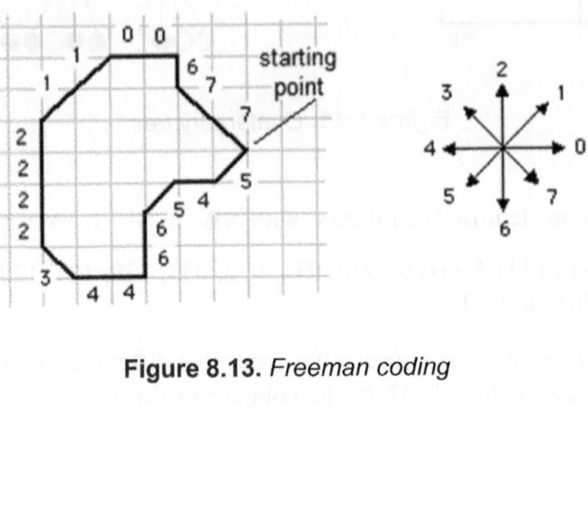

Figure 8.13. *Freeman coding*

The Freeman coding codes can be identified as follows:

– Simple coding: 5 4 5 6 6 4 4 3 2 2 2 1 1 0 0 6 7 7.

– Repeat coding: (5.1) (4.1) (5.1) (6.2) (4.2) (3.1) (2.4) (1.2) (0.2) (6.1) (7.2).

8.4.3. *Fractal curves*

Fractal curves, introduced by B. Mandelbrot in 1975, involve repeating the same pattern thousands of times, often incorporating a random factor into the generation process. They are widely used in realistic rendering to create texture effects and complex structures.

In contrast, when precise rules are used instead of a random factor, the curves are referred to as geometric curves. Geometric curves adhere to well-defined rules for their construction. Figure 8.14 shows an example of how to construct a geometric curve.

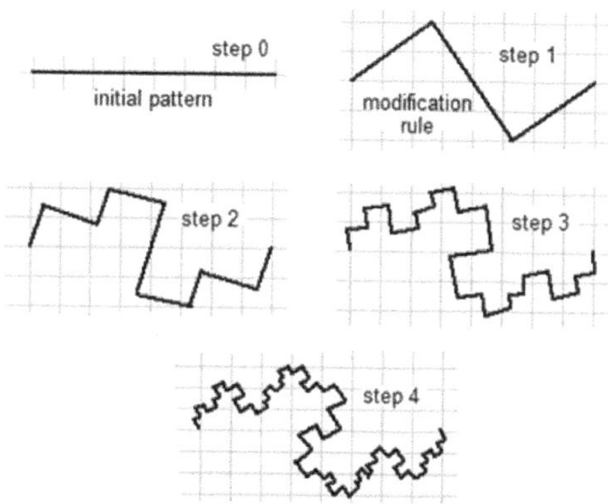

Figure 8.14. *Fractal curve*

8.5. Compression and compaction

A raster image with a resolution of 640×480 pixels and 256 colors (8-bit coding) would occupy approximately 300 KB of space (1 K = 1024 bytes). Four of these images would fill an old 1.4 MB (1 MB = 1024 KB) floppy disk.

However, digitizing moving images poses greater challenges. For such images at a video frequency of 25 frames per second, a one-second sequence would require

7,500 KB, or approximately 7.32 MB of space, while an hour-long sequence would occupy 25.75 GB (1 GB = 1024 MB). Additionally, the transfer rate of magneto-optical media such as CD-ROMs is approximately 300 KB/s. Therefore, it is essential to "compress" the image size by a factor of approximately 25 ($\frac{7500}{25} = 300$) to achieve an acceptable transfer rate. For a photographic-quality image with 16 million colors coded on 24 bits, the compression factor must be even higher, at approximately 75.

Moreover, the capacity of an old CD-ROM is limited to 650 MB, which can store only 2,218 uncompressed images of 640×480 pixels at 8-bit coding, equivalent to approximately 1 minute and 28 seconds of video. Therefore, reducing the image size is imperative for both storage capacity and transfer rate.

Reducing image size can be accomplished using two classes of methods: *compaction*, which reduces volume without losing information, and *compression*, which reduces volume by sacrificing some of the information.

The techniques used for reducing image size are based on the following methods:

– Eliminating redundancy: using a repetition factor.

– Elimination of certain information: removing less perceptible details.

– For moving images: recording only the differences between two successive frames (a method used by the MPEG standard).

Here is a brief description of some common image compression techniques:

– RLC coding (Run Length Coding): this method replaces a sequence of identical bytes with the byte followed by the number of occurrences. For example, the sequence 666666666 would be encoded as 6/9, where 6 represents the byte and 9 indicates the number of consecutive occurrences.

– Huffman coding: this method is based on the statistical coding of bit blocks according to their frequency of occurrence. Infrequent blocks are coded with a "higher" (longer) number of bits, while very frequent blocks are coded with a "lower" (shorter) number of bits. An initial pass is necessary to calculate frequencies before proceeding with coding.

– LZW coding (Lempel, Ziv, Welch): this method involves identifying and listing repeating blocks of bits in a table. The blocks are then coded using their addresses in this table, thus reducing the overall size of the image.

In recent years, particularly since 1978, a significant standardization process has taken place, resulting in the establishment of standards such as JPEG for still images and MPEG for moving images. These standards have facilitated the adoption of various compression processes.

When it comes to digital image formats, there is unfortunately a wide variety of encoding modes. However, some formats have emerged as de facto "standards", while standardization processes are still underway. The following section reviews the most commonly used digital image formats.

8.6. Image formats

8.6.1. *Raster image formats*

Below is a list of some of the most commonly used digital image formats:

– BMP: the original Microsoft format for Windows 3.x, widely used in the PC environment.

– PCX: Z-soft's original Paintbrush format, used in the PC environment. It can process 8-bit images but is not suitable for 16-, 24- or 32-bit images. Its compression algorithm is RLC (Run Length Coding).

– GIF: a widely used format of CompuServe origin, used in the PC environment. It uses 8-bit coding (256 colors) and the LZW compression algorithm. GIF is gradually being replaced by PNG.

– PNG: a format that is increasingly replacing GIF. It uses lossless compression thanks to the deflate algorithm (RFC 1951).

– TIFF: the original Aldus and Microsoft format, mainly used for scanned images. It is compatible with both PC and Mac environments and supports several compression algorithms, including RLC and LZW.

– PICT: the basic format for QuickDraw on the Mac, which can also process vector images. It is primarily used in the Mac environment.

– TGA: the original true vision format, which uses the RLC compression algorithm. It is mainly used in the PC environment.

– FAX: a format used for document transmission, especially faxes. It uses binary coding with a combination of RLC and Huffman coding.

– JPG (or JPEG): a commonly used format for complex images such as photographs. The classic format uses lossy compression based on the discrete cosine transform (DCT).

8.6.2. *Vector image formats*

Below is a list of other commonly used digital image formats:

– PICT: already mentioned above.

– DXF: the original AutoCAD format, widely used in CAD. Suitable for 2D work, it is well known and widely adopted.

– HPGL: the original Hewlett-Packard format, used for plotter peripherals. It is extensively used in this field.

– EPS: the original Adobe format, based on the PostScript language. It is used in both PC and Mac environments.

– CGM: derived from GKS (Graphical Kernel System), this is a universal, standardized format.

– IGES: a format used for CAD data exchange, primarily in workstation environments.

– EDIGEO: a format adapted to cartography, standardized by AFNOR (Association française de normalisation). It is specifically designed for this field of application.

9

2D Computer Graphics

CONCEPTS COVERED IN THIS CHAPTER.–

2D computer graphics involves drawing plane figures on a discrete space composed of pixels. The chapter begins with an examination of basic plots, such as line segments and circle arcs.

The following sections deal with the transition from user representation to screen representation (windowing), and surface coloring.

Homogeneous coordinates are introduced to represent common geometric transformations in matrix form, simplifying computational processing.

The chapter concludes with an introduction to parametric curves (Hermite, Bézier, B-splines), which are frequently used in industrial (CAD) or artistic environments.

References: [FOL 91, GAR 84, GON 93, HEA 97, PLA 87].

9.1. Basic graphics processing

9.1.1. *Drawing a segment*

Since a computer screen operates as a 2D plane with discrete coordinates, it inherently limits the exact drawing of line segments. This limitation arises because pixels on the screen are discrete, indivisible units, making it impossible to draw a perfectly straight line.

To visually represent a line segment on a computer screen, several techniques have been developed to obtain a visually close approximation to a straight line (some of these techniques are described in Figure 9.1).

Mathematics for Digital Science 2,
by Gérard-Michel COCHARD and Mhand HIFI.

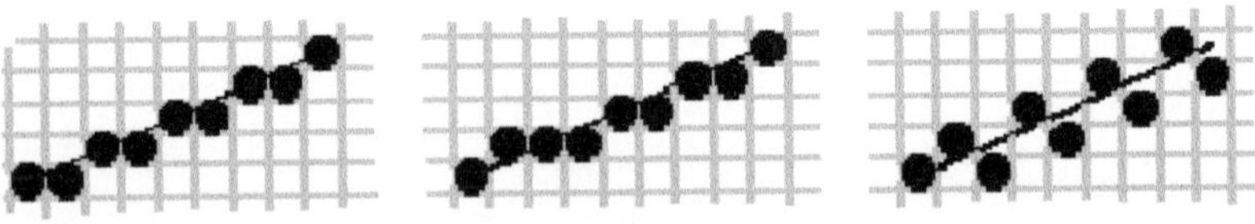

Figure 9.1. *Line drawing techniques*

Among various line drawing techniques, Bresenham's algorithm (1965) is widely recognized for minimizing deviation from a true line segment. This algorithm uses integer calculations to optimally select which pixels to illuminate, resulting in an accurate approximation of the straight line. Its goal is to draw a straight line segment using only the points of a discrete grid, as shown in Figure 9.2.

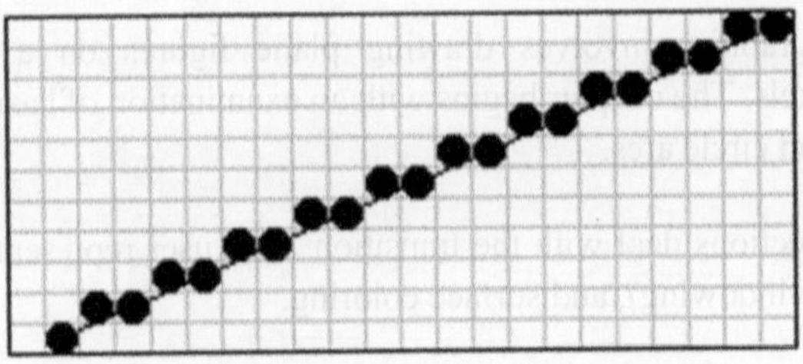

Figure 9.2. *Bresenham algorithm*

To achieve this, Bresenham's algorithm distinguishes between eight cases, depending on the slope of the line and the direction in which the segment is drawn. These eight cases divide the Euclidean plane into eight regions known as octants, as shown in Figure 9.3.

Octant 1: consider a straight line with slope m where $0 \leq m \leq 1$ on a grid between points (x_1, y_1) and (x_2, y_2), such that $x_1 < x_2$. At each step of the plotting algorithm, x is incremented after plotting a point at (x, y). The algorithm must determine the next point to be displayed, with two possibilities: it can either plot the point $(x+1, y)$ or at $(x+1, y+1)$. Therefore, within *the first octant* of the plane, tracing a line becomes a decision problem with two options at each stage.

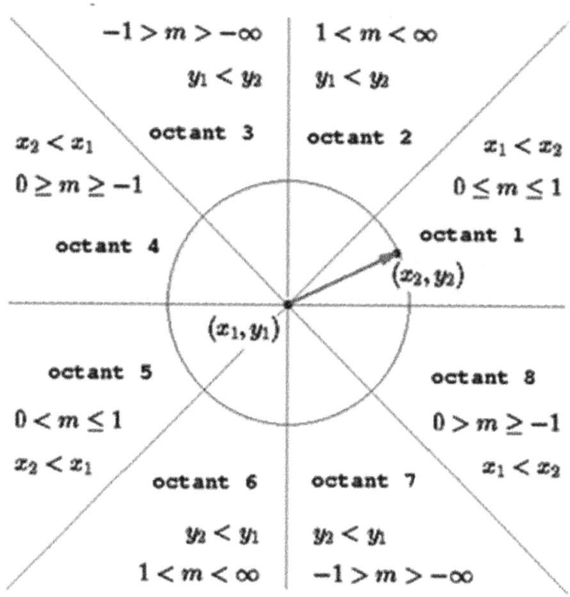

Figure 9.3. *Eight applications of Bresenham's algorithm*

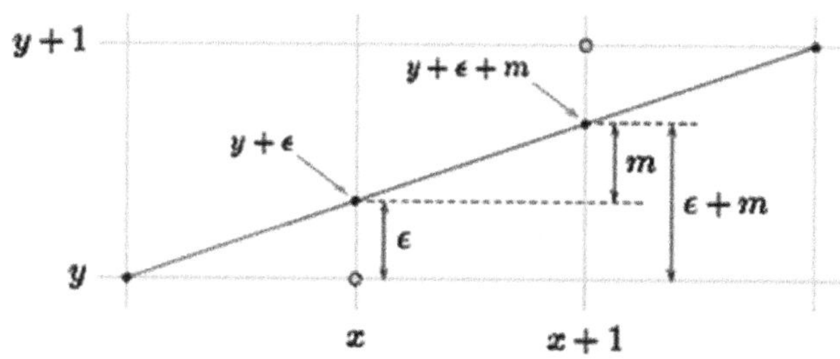

Figure 9.4. *Geometry for the first octant*

A diagram of the situation can be drawn after the algorithm displays the point (x, y) (see Figure 9.4). After plotting the point (x, y), the algorithm must make a compromise between the intended drawing and the actual resolution of the grid. Usually, the point (x, y) represents an error, as the actual point on the line may not be accessible on the grid. Therefore, an ordinate error ε is associated with each abscissa x, since the real ordinate should be $y+\varepsilon$. This error lies in the interval $[-\frac{1}{2}, +\frac{1}{2}[$. Moving from x to $x+1$, the value of the (mathematical) ordinate increases by an amount equal to the slope of the line, m. The point $(x+1, y)$ is chosen if the difference between this new ordinate and y is less than 0.5, specifically if:

$$y + \varepsilon + m < y + \frac{1}{2}.$$

Otherwise, the point $(x+1, y+1)$ will be plotted. This approach minimizes the overall error between the mathematical line segment and the plotted segment. The error resulting from this new point can now be regarded as the new ε. This process can be repeated for the next point on the line, where the abscissa is $x+2$. The new error value can take one of two possible values, depending on the chosen point. If $(x+1, y)$ is selected, the new error value is given by:

$$\varepsilon = (y + \varepsilon + m) - y.$$

Alternatively, if $(x+1, y+1)$ is chosen, it becomes:

$$\varepsilon = (y + \varepsilon + m) - (y + 1).$$

This leads to the following algorithm:

```
ε = 0; y = y1 ; x = x1;
While x ≤ x2
     draw (x,y)
     If (ε+m < 0.5)
               ε = ε + m
     Otherwise
               y = y+1; ε = ε + m -1
     EndIf
     x = x+1
EndWhile
```

This algorithm for octant 1 can be optimized to eliminate the use of floating-point variables. Instead, by defining $\Delta x = x_2 - x_1$ and $\Delta y = y_2 - y_1$ with $m = \frac{\Delta y}{\Delta x}$, the test expression can be manipulated. By multiplying by 2 and Δx, the inequality can be expressed as:

$$2\varepsilon\Delta x + 2\Delta y < \Delta x.$$

All quantities in this inequality are now integers. By setting $\varepsilon' = \varepsilon\Delta x$, the test becomes:

$$2(\varepsilon' + \Delta y) < \Delta x.$$

The update rules for the error on each step can also be adapted to ε':

1) $\varepsilon = \varepsilon + m \Leftrightarrow \varepsilon' = \varepsilon' + \Delta y$

2) $\varepsilon = \varepsilon + m - 1 \Leftrightarrow \varepsilon' = \varepsilon' + \Delta y - \Delta x$

The algorithm, which uses only integer calculations, is as follows:

```
Δx = x2 -x1 ; Δy = y2 -y1;
ε' = 0; y = y1 ; x = x1;
While x ≤ x2 do
      draw(x,y);
      If (2(ε'+Δy) < Δx) then
            ε' = ε' + Δy;
      Otherwise
            y = y+1;
            ε' = ε' + Δy - Δx;
      EndIf
      x = x+1;
EndWhile
```

REMARKS.– The algorithm has been improved in the following areas:

– The test is performed using only whole numbers, resulting in greater efficiency.

– Multiplication by 2 can be achieved through left-shifting in binary.

– This version is limited to slopes in the first octant.

9.1.2. *Drawing a circle*

An extension of Bresenham's algorithm, Michener's algorithm addresses the problem of circle drawing. The principle remains the same: illuminate the pixels closest to the circle, with the algorithm also dependent on the position within the octant. For example, in octant 1 (indicating movement upwards and to the left, with the slope of the tangent between -1 and 0), x is decreased at each step of the algorithm. The algorithm checks whether the distance between the center of the circle and the point $(x - 1, y)$ is less than the distance between the center of the circle and $(x - 1, y + 1)$. If it is, the point $(x - 1, y))$ is selected (see Figure 9.5).

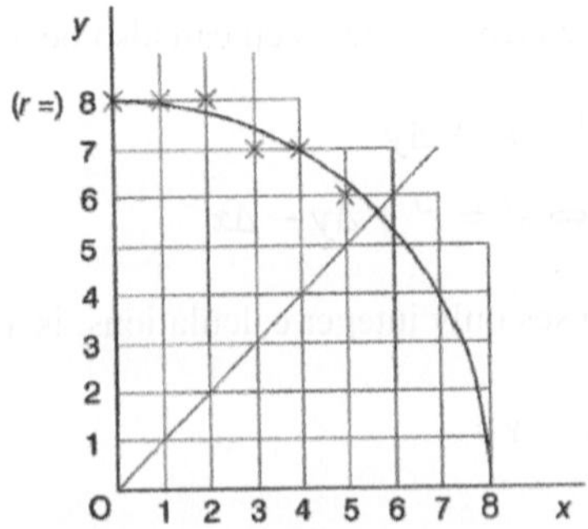

Figure 9.5. *Michener's algorithm*

Michener's algorithm for octant 1, for a circle centered at (0,0) with radius R, is given below:

```
x = R; y = 0;
While x ≥ y do
        draw(x,y);
        y = y + 1;
        d1 = |R*R-(x*x+y*y)|;
        d2 = |R*R-((x-1)*(x-1)+y*y)|;
        If d1 > d2 then
                x = x - 1;
        EndIf
EndWhile
```

9.1.3. *Windowing*

It is useful to distinguish between two types of space: user space, which represents the user's work plan and is independent of the visualization system, and screen space, which is obviously dependent on the visualization system in use. A correspondence is established between a rectangle in user space, called a window, and a rectangle in screen space, referred to as a clip window (see Figure 9.6).

If the window and clip window are not homothetic, this can lead to visual distortion, causing objects to appear distorted from their true shape. For example, straight lines in user space may appear curved or slanted in screen space if the correspondence between the two rectangles is not linear.

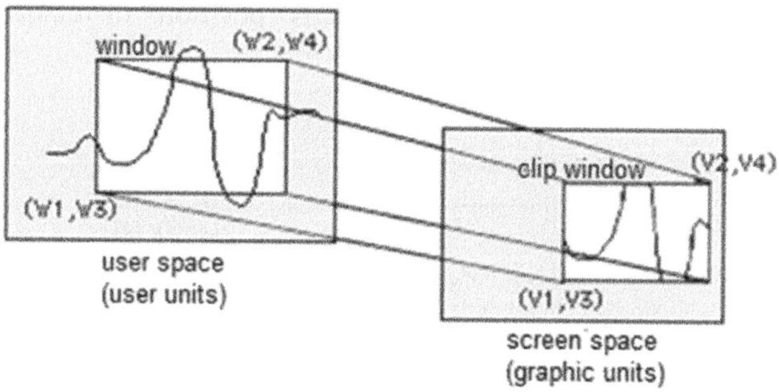

Figure 9.6. *User space and screen space*

The correspondence between user space and screen space involves an operation called clipping, which eliminates portions of graphics that fall outside the window. To illustrate this, consider the example of a line segment. This segment may be entirely inside the window, entirely outside the window or it may intersect an edge of the window. To determine which case the segment is in, a preliminary approach might involve comparing the coordinates of each point on the segment with the coordinates of the window's vertices. However, this method would be computationally inefficient.

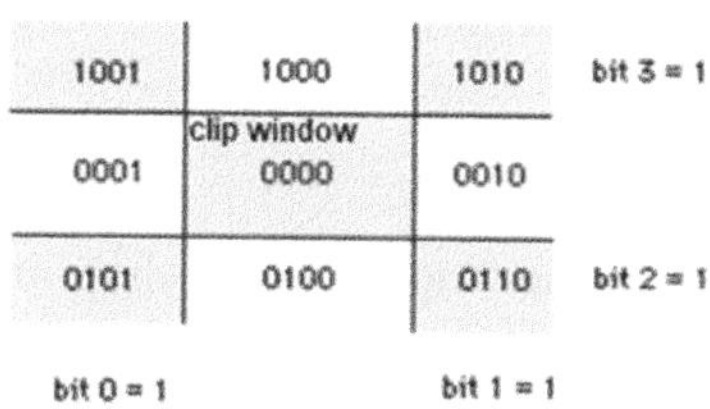

Figure 9.7. *Sutherland algorithm*

To optimize this operation, the Sutherland algorithm is used. It divides the screen space into nine zones, numbered with four bits, using the fence as the basis for this division. The fence itself is numbered 0000 (see Figure 9.7). This division of screen space simplifies the clipping process by quickly identifying the zones in which the segment points are located.

Figure 9.8 shows several scenarios where segment ends are coded by the zone numbers corresponding to their locations. These zone numbers facilitate the rapid identification of the relative positions of the segment ends concerning the window, with each zone number determined by the ends' positions in relation to the clip window edges.

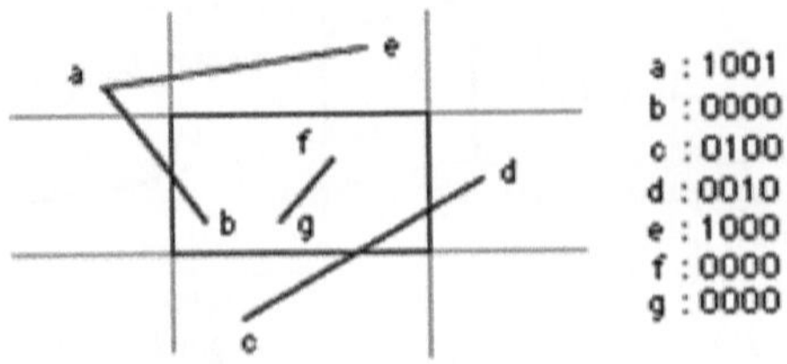

Figure 9.8. *Several situations*

The Cohen–Sutherland algorithm can be used to differentiate among various scenarios:

```
If code(a) = 0000 and code(b) = 0000 then
    ab is fully visible;
Otherwise
    If code(a)∧code(b) = 0000 then
       ab is partially visible or invisible (see note below);
    Otherwise
       ab is totally invisible;
    EndIf
EndIf
```

NOTE 9.1.– The symbol ∧ is the "logical and" whose truth table is $1 \wedge 1 = 1$, $1 \wedge 0 = 0$, $0 \wedge 1 = 0$, $0 \wedge 0 = 0$.

Note that the symbol ∧ can also be interpreted as AND, representing an equivalent notation. The previous algorithm eliminates some invisible segments, thereby saving computation time by avoiding their display. However, there may be situations where the algorithm fails to detect all invisible segments, as shown in Figure 9.8, where points c and d are slightly lower. In these cases, the clipping algorithm based on zone numbers would not be sufficient to identify these invisible segments.

9.1.4. *Filling and coloring*

Filling a surface bounded by a closed contour with points of a defined color can be achieved using a simple method: vertical translation of a horizontal segment.

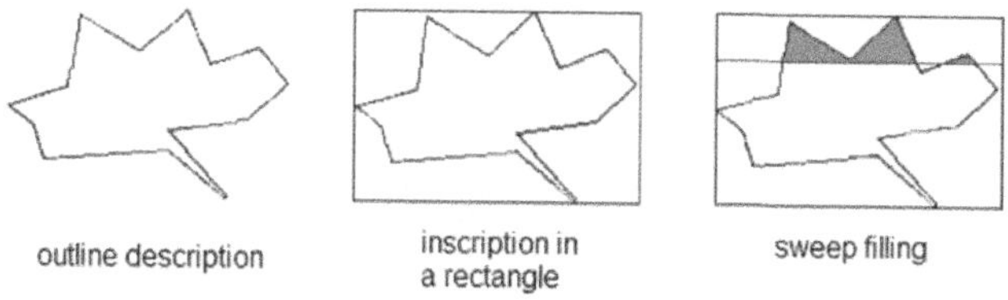

Figure 9.9. *Sweep filling. For a color version of this figure, see www.iste.co.uk/cochard/mathematics2.zip*

Assuming that the contour can be described analytically and the figure is inscribed in a rectangle, a horizontal segment is translated line-by-line, where the length of the segment corresponds to the length of the rectangle. At each line, the color of the points on the segment changes every time the contour of the surface to be filled is crossed: the background color is used on the outside, and a defined color is applied on the inside (see Figure 9.9).

However, this method encounters several problems, particularly when passing through a summit. This situation and the rules to be applied are shown in Figure 9.10.

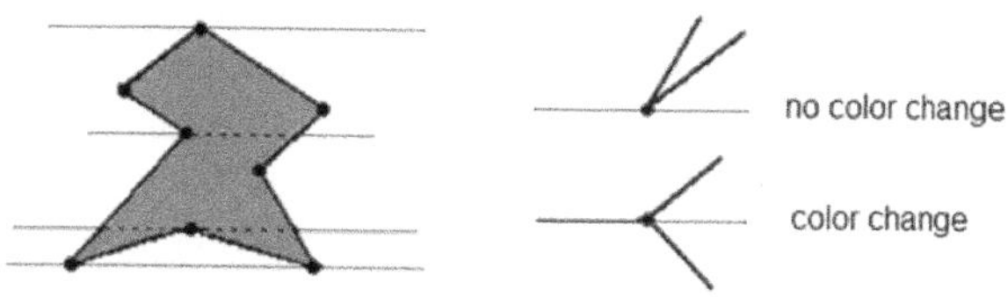

Figure 9.10. *Rules to apply. For a color version of this figure, see www.iste.co.uk/cochard/mathematics2.zip*

9.2. 2D geometric transformations

9.2.1. *Homogeneous coordinates*

Using the usual Cartesian coordinates, a point (x, y) on the screen can be represented by a line matrix $[x\ y]$ (see Figure 9.11).

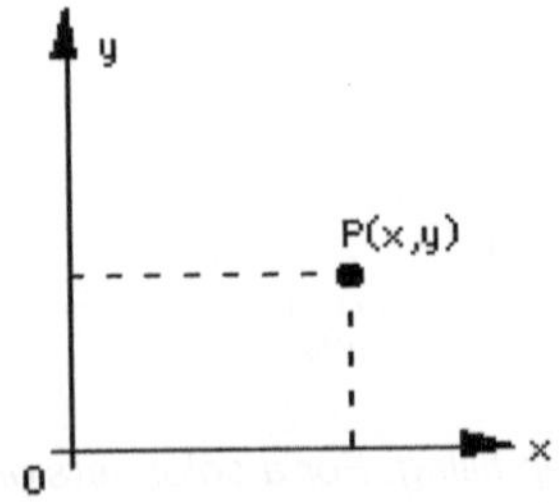

Figure 9.11. *Cartesian representation*

In homogeneous coordinates, a point is represented by a line matrix that includes an additional coordinate, the scale factor s: $[x\ y\ s\ ]$.

The advantages of using homogeneous coordinates for representation are threefold:

1) *Integer manipulation*: this representation allows only integers to be manipulated, facilitating mathematical operations and calculations. For example, a point with decimal coordinates such as [0.6, 0.1] can be represented by integers such as [6, 1, 10]. This simplifies calculations and avoids precision problems associated with floating-point numbers.

2) *Overflow resolution*: homogeneous coordinates address overflow issues when representing numbers. For instance, if coordinates exceed the available integer range, such as [70000, 30000], they can be represented by smaller integers using an appropriate scale. For example, [70000, 30000] could be represented by [7000, 3000, 0.1]. This allows for working with larger values without exceeding the limits of integer representation.

3) *Matrix representation of translations*: homogeneous coordinates allow for matrix representation of translations, rotations and scaling. This unification simplifies the process of applying geometric transformations in computer graphics.

9.2.2. *Translation*

A translation is defined by a vector with components (M, N). Under the effect of this translation, a point (x, y) is transformed into a point (x', y') (see Figure 9.12).

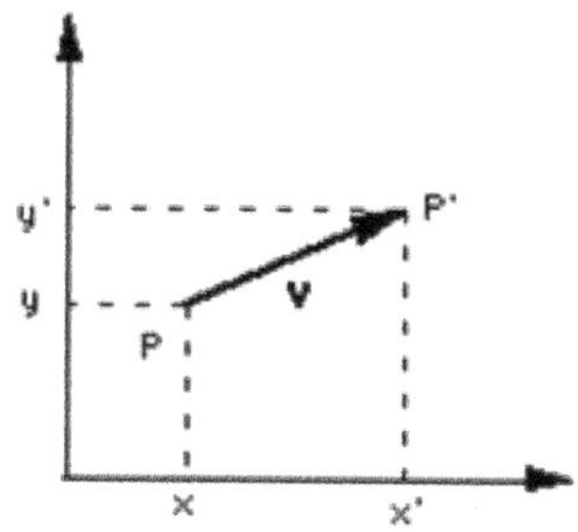

Figure 9.12. *Translation of the direction vector* ***V***

The equation system linking (x, y) to (x′, y′) is:

$$x' = x + M$$

$$y' = y + N$$

This system cannot be represented in matrix form as:

$$[x' \;\; y'] = [x \;\; y]\boldsymbol{T},$$

where $\boldsymbol{T}$ is a 2×2 matrix. However, in homogeneous coordinates, it can be expressed as

$$[x' \;\; y' \;\; 1] = [x \;\; y \;\; 1]\boldsymbol{T}(M, N),$$

where $\boldsymbol{T}(M, N)$ is a 3×3 matrix defined as follows:

$$\boldsymbol{T}(M, N) = \begin{bmatrix} 1 & 0 & 0 \\ 0 & 1 & 0 \\ M & N & 1 \end{bmatrix}$$

9.2.3. *Rotation around the origin*

An angle rotation θ around the origin O of the Cartesian coordinate system maps a point (x, y) to the point (x', y') obtained by rotation around O (see Figure 9.13).

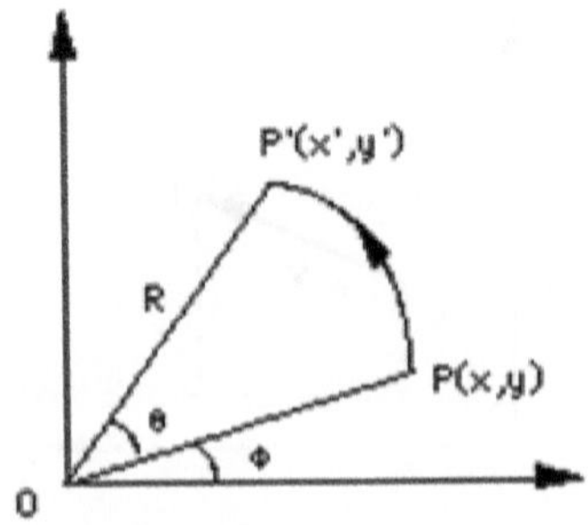

Figure 9.13. *Rotation around O by an angle θ*

The equation system linking (x, y) to (x', y') is:

$$x' = x\cos\theta - y\sin\theta$$

and

$$y' = x\sin\theta + y\cos\theta.$$

These equations can be written in matrix form in homogeneous coordinates as:

$$[x' \ \ y' \ \ 1] = [x \ \ y \ \ 1]\mathbf{R}(\theta)$$

with

$$\boldsymbol{R}(\theta) = \begin{bmatrix} cos\theta & sin\theta & 0 \\ -sin\theta & cos\theta & 0 \\ 0 & 0 & 1 \end{bmatrix}.$$

9.2.4. *Dilation*

A dilation is simply a change of scale; under a dilation defined by two numbers E and F, a point (x, y) is transformed into a point (x', y') (see Figure 9.14).

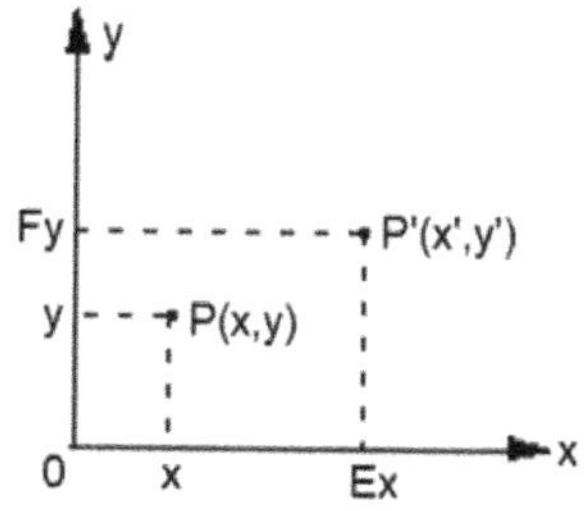

Figure 9.14. *Dilation (E, F)*

The equation system linking (x, y) to (x', y') is given by $x' = Ex$ and $y' = Fy$. In homogeneous coordinates, the transformation can be expressed as:

$$[x' \ y' \ 1] = [x \ y \ 1]\boldsymbol{D}(E, F),$$

where

$$\boldsymbol{D}(E, F) = \begin{bmatrix} E & 0 & 0 \\ 0 & F & 0 \\ 0 & 0 & 1 \end{bmatrix}$$

9.2.5. *Symmetries*

A symmetry is a geometric transformation that preserves distances and angles but reverses the orientation of objects. It is a special case of dilation. There are different types of symmetry:

1) *Symmetry with respect to the y-axis* $(E = -1, F = 1)$: when a point is symmetrical about the y-axis, if the coordinates of the point are (x, y), then the coordinates of the symmetrical point are $(-x, y)$.

2) *Symmetry with respect to the x-axis* $(E = 1, F = -1)$: when a point is symmetrical about the x-axis, if the coordinates of the point are (x, y), then the coordinates of the symmetrical point are $(x, -y)$.

3) *Symmetry with respect to the origin* $(E = -1, F = -1)$: when a point is symmetrical about the origin, if the coordinates of the point are (x, y), then the coordinates of the symmetrical point are $(-x, -y)$.

9.2.6. *Composition of transformations*

Figure 9.15 shows a decomposition of a rotation by an angle α around the point $\Omega = (a, b)$ using elementary transformations. The decomposition of the rotation into elementary transformations is as follows:

1) *Translation* ($\boldsymbol{\Omega O}$): translate the point Ω to the origin by performing a vector translation $(-a, -b)$. This involves moving all the points in the plane so that the point Ω is at the origin (0, 0).

2) *Rotation* (with angle α): rotate by an angle α around the origin (0, 0) using the appropriate rotation formula. This rotates all points in the plane around the origin.

3) *Reverse translation* ($\boldsymbol{O\Omega}$): translate the point Ω back to its initial position by performing a translation of vector (a, b). This returns all points in the plane to their original positions.

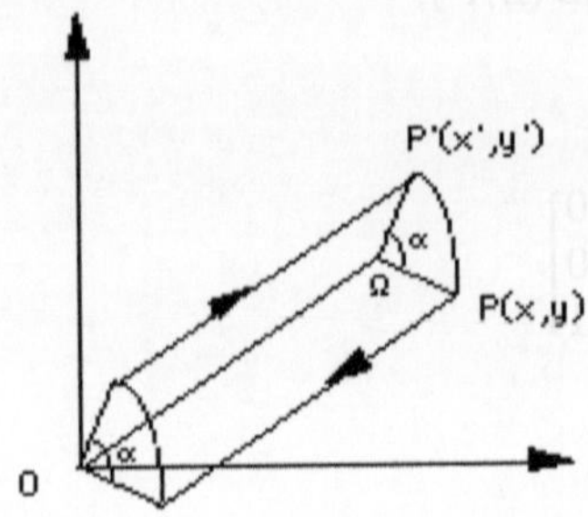

Figure 9.15. *Composition of transformations*

Let $P_1 = (x_1, y_1)$ be the image of (x, y) after the vector translation $\boldsymbol{\Omega O}$. This can be expressed as:

$$[x_1 \quad y_1 \quad 1] = [x \quad y \quad 1]\begin{bmatrix} 1 & 0 & 0 \\ 0 & 1 & 0 \\ -a & -b & 1 \end{bmatrix}$$

Let $P_2 = (x_2, y_2)$ be the image of P_1 after the rotation by an angle α around the origin O. This can be expressed as:

$$[x_2 \quad y_2 \quad 1] = [x_1 \quad y_1 \quad 1]\begin{bmatrix} \cos\alpha & \sin\alpha & 0 \\ -\sin\alpha & \cos\alpha & 0 \\ 0 & 0 & 1 \end{bmatrix}$$

Finally, let $P' = (x', y')$ be the image of P_2 after the vector translation $\boldsymbol{O\Omega}$. This can be expressed as:

$$[x' \quad y' \quad 1] = [x_2 \quad y_2 \quad 1]\begin{bmatrix} 1 & 0 & 0 \\ 0 & 1 & 0 \\ a & b & 1 \end{bmatrix}$$

This leads to the following relation:

$$[x' \quad y' \quad 1] = [x \quad y \quad 1]\begin{bmatrix} 1 & 0 & 0 \\ 0 & 1 & 0 \\ -a & -b & 1 \end{bmatrix}\begin{bmatrix} \cos\alpha & \sin\alpha & 0 \\ -\sin\alpha & \cos\alpha & 0 \\ 0 & 0 & 1 \end{bmatrix}\begin{bmatrix} 1 & 0 & 0 \\ 0 & 1 & 0 \\ a & b & 1 \end{bmatrix}$$

$$= [x \quad y \quad 1]\begin{bmatrix} \cos\alpha & \sin\alpha & 0 \\ -\sin\alpha & \cos\alpha & 0 \\ -a\cos\alpha + b\sin\alpha + a & -a\sin\alpha - b\cos\alpha + b & 1 \end{bmatrix}$$

NOTE 9.1.– There is no need to remember the result of this matrix; it is sufficient to understand the elementary transformations to derive it through matrix multiplication.

NOTE 9.2.– The product of matrices is not commutative; thus, swapping the matrices in the product changes the result. Care must be taken with the order of transformations.

9.2.7. *Object representation*

A plane image is a collection of points $(x_1, y_1), (x_2, y_2), \ldots, (x_n, y_n)$. To apply a transformation to the image, simply apply the transformation to each of the points. These calculations can be performed in a single operation by arranging the points in a matrix:

$$\begin{bmatrix} x_1 & y_1 & 1 \\ x_2 & y_2 & 1 \\ x_3 & y_3 & 1 \\ \vdots & \vdots & \vdots \\ x_n & y_n & 1 \end{bmatrix}$$

and multiplying this matrix by the transformation matrix $\boldsymbol{M}$:

$$\begin{bmatrix} x'_1 & y'_1 & 1 \\ x'_2 & y'_2 & 1 \\ x'_3 & y'_3 & 1 \\ \vdots & \vdots & \vdots \\ x'_n & y'_n & 1 \end{bmatrix} = \begin{bmatrix} x_1 & y_1 & 1 \\ x_2 & y_2 & 1 \\ x_3 & y_3 & 1 \\ \vdots & \vdots & \vdots \\ x_n & y_n & 1 \end{bmatrix} \boldsymbol{M}$$

9.3. 2D parametric curves

9.3.1. *Using cubic curves*

Cubic curves offer great flexibility in representing "linear" shapes, making them a popular tool for font representation. A cubic curve is defined by a parameter t that varies from 0 (origin-point A) to 1 (end-point B), using parametric equations. The parametric equations of cubic curves (see Figure 9.16) are as follows:

$$x(t) = a_x t^3 + b_x t^2 + c_x t + d_x$$

$$y(t) = a_y t^3 + b_y t^2 + c_y t + d_y$$

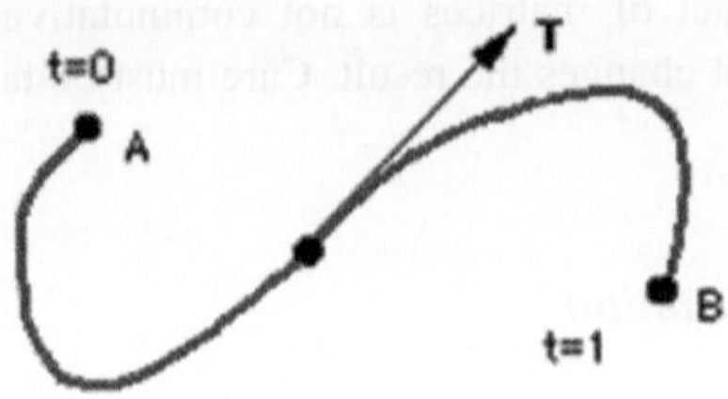

Figure 9.16. *Cubic curve. For a color version of this figure, see www.iste.co.uk/cochard/mathematics2.zip*

Note that eight constants are required to completely determine a plane cubic curve. The preceding expressions can be formulated in a matrix form as follows:

$$x(\boldsymbol{t}) = \boldsymbol{t} \times \boldsymbol{C_x}$$

$$y(\boldsymbol{t}) = \boldsymbol{t} \times \boldsymbol{C_y}$$

where

$$\mathbf{t} = [t^3 \quad t^2 \quad t \quad 1], \quad \boldsymbol{C}_x = \begin{bmatrix} a_x \\ b_x \\ c_x \\ d_x \end{bmatrix}, \quad \boldsymbol{C}_y = \begin{bmatrix} a_y \\ b_y \\ c_y \\ d_y \end{bmatrix}$$

The vector $\boldsymbol{T}$, tangent to the curve at a point defined by the parameter t, is determined by its components:

$$T_x = \frac{\partial x}{\partial t} = 3a_x t^2 + 2b_x t + c_x$$

$$T_y = \frac{\partial y}{\partial t} = 3a_y t^2 + 2b_y t + c_y.$$

In computer graphics, the most commonly used cubic curves are Hermite curves, Bézier curves and B-splines. Each of these curves has its own properties and advantages, making them suitable for different applications in computer graphics or industrial design.

9.3.2. *Hermite curves*

A Hermite curve (see Figure 9.17) is defined by its origin P_1 (with two constants X_1 and Y_1), its extremity P_4 (with two constants X_4 and Y_4), as well as the tangent vectors at the origin and endpoints $\boldsymbol{T_1}$ and $\boldsymbol{T_2}$ (with four constants $T_{1x}, T_{1y}, T_{2x}, T_{2y}$). This definition implies the following conditions:

$$x(0) = X_1 \qquad y(0) = Y_1 \qquad x(1) = X_4 \qquad y(1) = Y_4$$

$$x'(0) = T_{1x} \qquad y'(0) = T_{1y} \qquad x'(1) = T_{2x} \qquad y'(1) = T_{2y}$$

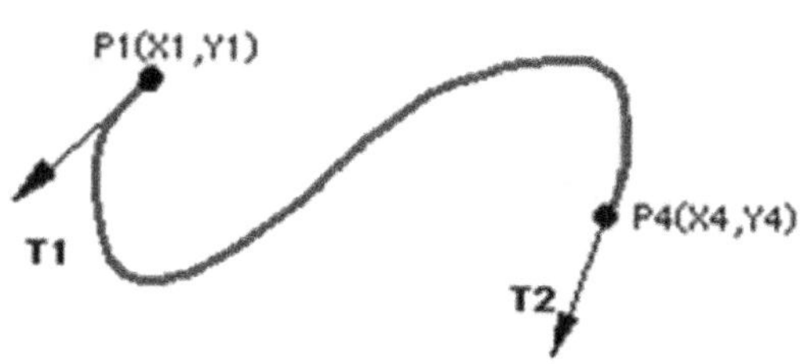

Figure 9.17. *Hermite curve. For a color version of this figure, see www.iste.co.uk/cochard/mathematics2.zip*

The following relationships are deduced:

$$\begin{bmatrix} X_1 \\ X_4 \\ T_{1x} \\ T_{2x} \end{bmatrix} = \begin{bmatrix} 0 & 0 & 0 & 1 \\ 1 & 1 & 1 & 1 \\ 0 & 0 & 1 & 0 \\ 3 & 2 & 1 & 0 \end{bmatrix} \boldsymbol{C_x}$$

which leads to the following relationship:

$$\boldsymbol{C_x} = \begin{bmatrix} 2 & -2 & 1 & 1 \\ -3 & 3 & -2 & -1 \\ 0 & 0 & 1 & 0 \\ 1 & 0 & 0 & 0 \end{bmatrix} \cdot \begin{bmatrix} X_1 \\ X_4 \\ T_{1x} \\ T_{2x} \end{bmatrix} = \boldsymbol{M_h G_{hx}}$$

Thus, $x(t) = \boldsymbol{t\, M_h\, G_{hx}}$ and $y(t) = \boldsymbol{tM_hG_{hy}}$.

Therefore, the cubic Hermite curve equations become:

$$x(t) = X_1(2t^3 - 3t^2 + 1) + X_4(-2t^3 + 3t^2) + T_{1x}(t^3 - 2t^2 + t) + T_{2x}(t^3 - t^2)$$

$$y(t) = Y_1(2t^3 - 3t^2 + 1) + Y_4(-2t^3 + 3t^2) + T_{1y}(t^3 - 2t^2 + t) + T_{2y}(t^3 - t^2)$$

9.3.3. *Bézier curves*

A Bézier curve is defined by four points: two points serve as the endpoints $P1$ and $P4$, while the other two points, known as "attractors", $P2$ and $P3$, are positioned along the tangents at the origin (start) and the end of the curve (see Figure 9.18). The influence of the attractors determines the shape of the curve, allowing for smooth and flexible designs.

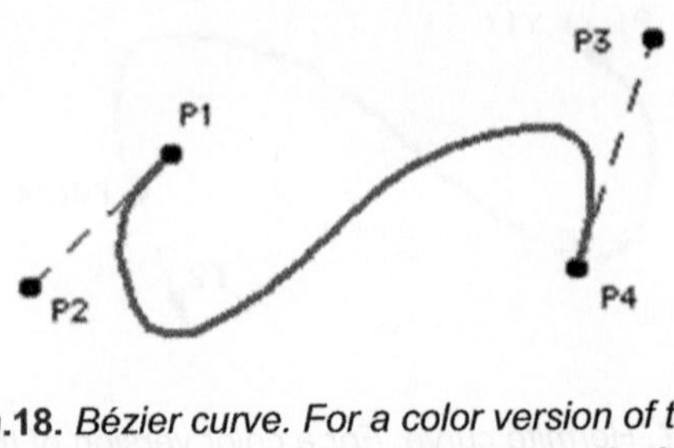

Figure 9.18. *Bézier curve. For a color version of this figure, see www.iste.co.uk/cochard/mathematics2.zip*

Tangent vector at the origin $\boldsymbol{T_1}$:

$$\boldsymbol{T_1} = 3\,\boldsymbol{P_1P_2} = \begin{bmatrix} T_{1x} \\ T_{1y} \end{bmatrix} = \begin{bmatrix} 3X_2 - 3X_1 \\ 3Y_2 - 3Y_1 \end{bmatrix}$$

This indicates that $\boldsymbol{T_1}$ is three times the vector from point $\boldsymbol{P_1}$ to point $\boldsymbol{P_1}$.

Tangent vector at the origin $\boldsymbol{T_2}$:

$$\boldsymbol{T}_2 = 3\,\boldsymbol{P_3P_4} = \begin{bmatrix} T_{2x} \\ T_{2y} \end{bmatrix} = \begin{bmatrix} 3X_4 - 3X_3 \\ 3Y_4 - 3Y_3 \end{bmatrix}$$

This indicates that $\boldsymbol{T_2}$ is three times the vector from point $\boldsymbol{P_3}$ to point $\boldsymbol{P_4}$.

Hence, the following relationships are deduced:

$$x(t) = \boldsymbol{t\,M_h\,G_{hx}} = \boldsymbol{t\,M_h\,M_{hb}\,G_{bx}} = \boldsymbol{t\,M_b\,G_{bx}}$$

where

$$\boldsymbol{G_{hx}} = \begin{bmatrix} X_1 \\ X_4 \\ T_{1x} \\ T_{2x} \end{bmatrix} = \begin{bmatrix} 1 & 0 & 0 & 0 \\ 0 & 0 & 0 & 1 \\ -3 & 3 & 0 & 0 \\ 0 & 0 & -3 & 3 \end{bmatrix} \begin{bmatrix} X_1 \\ X_2 \\ X_3 \\ X_4 \end{bmatrix} = \boldsymbol{M_{hb}G_{bx}}$$

By expanding, the explicit expressions are:

$$x(t) = (X_4 - X_1 + 3X_2 - 3X_3)t^3 + 3(X_1 - 2X_2 + X_3)t^2 + 3(X_2 - X_1)t + X_1$$

and

$$y(t) = (Y_4 - Y_1 + 3Y_2 - 3Y_3)t^3 + 3(Y_1 - 2Y_2 + Y_3)t^2 + 3(Y_2 - Y_1)t + Y_1$$

Thus, the final equations are:

$$x(t) = (1-t)^3X_1 + 3t(1-t)^2X_2 + 3t^2(1-t)X_3 + t^3X_4$$

$$y(t) = (1-t)^3Y_1 + 3t(1-t)^2Y_2 + 3t^2(1-t)Y_3 + t^3Y_4$$

9.3.4. *B-spline curves*

B-spline curves are a powerful and flexible way to represent complex shapes in computer graphics and geometric modeling. They are defined recursively and provide an optimal interpolation of points $P_1, P_2, P_3, \dots, P_n$. The process begins by defining the curve section that best passes through the points P_1, P_2, P_3 and P_4. Subsequently, the next section is constructed to best pass through P_2, P_3, P_4 and P_5 superimposing it on the previous section, etc. (see Figure 9.19).

This interpolation method generates smooth, continuous curves that optimally pass through a given set of control points.

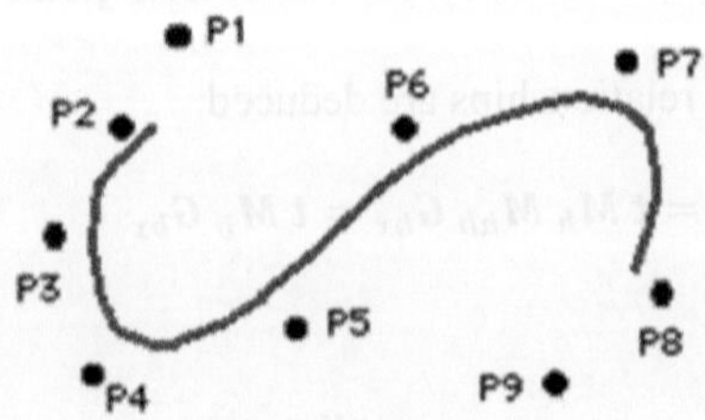

Figure 9.19. *B-spline curve. For a color version of this figure, see www.iste.co.uk/cochard/mathematics2.zip*

It offers greater flexibility and accuracy than other interpolation methods. These curves are used in 3D modeling, animation and computer-aided design to represent complex curves and free-form shapes. As before, let us assume

$$x(\mathbf{t}) = \boldsymbol{t}\, \boldsymbol{M}_s\, \boldsymbol{G}_{isx} \qquad \text{and} \qquad y(t) = \boldsymbol{t}\, \boldsymbol{M}_s\, \boldsymbol{G}_{isy}$$

where $\boldsymbol{M}_s$, $\boldsymbol{G}_{isx}$ and $\boldsymbol{G}_{isy}$ are defined as follows:

$$\boldsymbol{M}_s = \frac{1}{6}\begin{bmatrix} -1 & 3 & -3 & 1 \\ 3 & -6 & 3 & 0 \\ -3 & 0 & 3 & 0 \\ 1 & 4 & 1 & 0 \end{bmatrix} \qquad \boldsymbol{G}_{isx} = \begin{bmatrix} X_{i-1} \\ X_i \\ X_{i+1} \\ X_{i+2} \end{bmatrix} \qquad \boldsymbol{G}_{isy} = \begin{bmatrix} Y_{i-1} \\ Y_i \\ Y_{i+1} \\ Y_{i+2} \end{bmatrix}$$

These expressions are derived from the continuity of the first and second derivatives at a point:

$$\frac{dx_i}{dt}(t=1) = \frac{dx_{i+1}}{dt}(t=0) \quad \text{and} \quad \frac{d^2x_i}{dt^2}(t=1) = \frac{d^2x_{i+1}}{dt^2}(t=1)$$

10

Concepts in Image Processing and Analysis

CONCEPTS COVERED IN THIS CHAPTER.–

This chapter covers the fundamental concepts and methods of image processing and analysis.

It begins with image display methods, including simple matching, random threshold display and threshold matrix display.

Next, basic tools for image analysis are introduced: histograms, profiles and level searches. The concept of entropy, a key element in information theory, is presented along with the method for calculating the entropy of an image.

Basic image processing tools are then explored, such as look-up table (LUT) histogram transformations, subtraction and image averaging.

Image filtering is discussed in its mathematical formulation, both in the spatial and frequency domains.

The final section of the chapter deals with binary images, including dilation, erosion, skeletonization, contour extraction and segmentation.

References: [COS 89, HAY 99, KUN 93, LEL 00, TIS 08, TOU 87].

For simplicity, throughout the following sections, unless otherwise specified, initial raster images will be considered to have 256 levels of grayscale (0 = black to 255 = white). If more levels are used, the algorithms will need to be adapted, but this can be done without significant difficulty.

Mathematics for Digital Science 2,
by Gérard-Michel COCHARD and Mhand HIFI.

10.1. Image display

Since an image is encoded with 256 grayscale levels, it may be necessary to display it on a device that supports fewer levels, denoted as N (where $N < 256$). This situation often arises when dealing with highly detailed images while using a display device with limited capacities. Several algorithms are available to handle this reduction in grayscale levels effectively.

10.1.1. *Simple correspondence*

Consider the case $N = 4$, which corresponds to a four-level display (denoted by 0, 1, 2, 3). The mapping between the 256 levels and the 4 display levels is established by dividing the range $0 - 255$ into four intervals (of equal length): $0 - 63, 64 - 127, 128 - 191$ and $192 - 255$. Grayscale levels from 0 to 63 (inclusive) are mapped to display level 0, levels 64–127 (inclusive) to display level 1, etc.

Let $I(x, y)$ represent the pixel level at position (x, y). The following algorithm describes the 256-to-4 level color mapping:

```
For each pixel (x,y) with grayscale level I(x,y)
    If (I(x,y) < 64)
        display with level 0
    Else If (I(x,y) < 128)
            display with level 1
        Else If (I(x,y) < 92)
                display with level 2
            Else
                display with level 3
            EndIf
        EndIf
    EndIf
EndFor
```

10.1.2. *Random threshold display*

The previous display method is rather crude and does not consider the density of lit point on the screen. A large group of points of the same level should be distinguishable on the display. The random threshold method, which operates on a probabilistic principle, offers a solution to this issue.

First, consider a two-level representation ($N = 2$), where the image is represented with two levels (e.g. 0 = black and 1 = white). For each point (x, y), a value is drawn from a uniform random variable X (ranging between 0 and 254). The decision to display the point is based on the following rules:

– If $I(x, y) < X$, the point at (x, y) is turned on (new level =1).

– If $I(x, y) > X$, the point at (x, y) is turned off (new level =0).

The two-level random threshold algorithm is outlined below:

```
For each pixel (x,y) of level I(x,y)
   Generate a random threshold S between 0 and 254
   If (I(x,y) > S)
      turn on the pixel (set to level 1)
   Else
       turn off the pixel (set to level 0)
   EndIf
EndFor
```

Statistically, for a population of p pixels with a color level I_0, approximately half of the random thresholds will be higher than I_0, and half will be lower. As a result, at least half of the points in this population will be visible on the display. This ensures a more balanced representation of pixels, even when using a reduced number of display levels.

Now, consider a four-level representation ($N = 4$). The range $0 - 255$ is divided into three zones, with the available display levels marked as 0, 1, 2 and 3.

The algorithm applies the following logic in each zone:

– In the first zone, a two-level representation is applied with colors 0 and 1: "on" means the pixel is assigned color 1, and "off" means it is assigned color 0.

– In the second zone, a two-level representation is applied with colors 1 and 2: "on" means the pixel is assigned color 2, and "off" means it is assigned color 1.

– Etc.

This process ensures a smooth transition across the different levels while applying the random threshold method for each zone. The algorithm corresponding to the random threshold at four levels is outlined as follows:

```
For each pixel (x,y) with intensity level I(x,y)
   If (I(x,y) < 85)
      Generate a random threshold S between 0 and 84
      If (I(x,y) > S)
         display with level 1
      Else
         display with level 0
      EndIf
   Else
      If (I(x,y) < 170)
         Generate a random threshold S between 85 and 169
         If (I(x,y) > S)
            display with level 2
         Else
            display with level 1
         EndIf
      Else
         Generate a random threshold S between 170 and 254
         If (I(x,y) > S)
            display with level 3
         Else
            display with level 2
         EndIf
      EndIf
   EndIf
EndFor
```

This algorithm divides the intensity range into three zones (0–84, 85–169 and 170–254) and applies the random threshold method for each zone to determine the final display level for the pixel.

10.1.3. *Threshold matrix display*

Recall that threshold matrices are defined by the following relationships:

$$\boldsymbol{D}(2) = \begin{bmatrix} D_{00}(2) & D_{01}(2) \\ D_{10}(2) & D_{11}(2) \end{bmatrix} = \begin{bmatrix} 0 & 2 \\ 3 & 1 \end{bmatrix} \qquad \boldsymbol{D}(4) = \begin{bmatrix} 0 & 8 & 2 & 10 \\ 12 & 4 & 14 & 6 \\ 3 & 11 & 1 & 9 \\ 15 & 7 & 13 & 5 \end{bmatrix}$$

and for any n:

$$\boldsymbol{D}(2n) = \begin{bmatrix} 4D(n) + D_{00}(2)U(n) & 4D(n) + D_{01}(2)U(n) \\ 4D(n) + D_{10}(2)U(n) & 4D(n) + D_{11}(2)U(n) \end{bmatrix}$$

where

$$\boldsymbol{U}(n) = \begin{bmatrix} 1 & 1 & \dots & 1 \\ 1 & 1 & \dots & 1 \\ \vdots & \vdots & \ddots & \vdots \\ 1 & 1 & \dots & 1 \end{bmatrix}$$

To use the threshold matrix-based thresholding method $\boldsymbol{D}(2n)$, it is necessary to select an appropriate matrix. The process for choosing the correct matrix will be discussed in more detail later. For now, here is how the display method operates.

Let a pixel be (x, y). Calculate $x_0 \equiv x\,[2n]$ and $y_0 \equiv y\,[2n]$, where $a\,[b]$ expresses (modulo operation) the remainder of the integer division of a by b. Therefore, x_0 and y_0 are between 0 and $2n - 1$. In the threshold matrix, take the threshold $S = D_{x_0 y_0}(2n)$. If $I(x, y) > S$, the pixel is lit at a given level. Otherwise, it is lit at another level, determined similarly to the random threshold method. Applying this method to the case where $N = 4$, the range of values from 0 to 255 is divided into three zones, each comprising 85 gray levels. The threshold matrix $\boldsymbol{D}(8)$ contains 64 thresholds, while $\boldsymbol{D}(10)$ contains 100. Therefore, $\boldsymbol{D}(8)$ is chosen, and normalization is done by the factor

$$\frac{63}{84} = \frac{3}{4} = 0.75.$$

For each zone, light intensity is normalized according to the following formulas:

– Zone 1: $P = 0.75 \times I(x, y)$.

– Zone 2: $P = 0.75 \times [I(x, y) - 85]$.

– Zone 3: $P = 0.75 \times [I(x, y) - 170]$.

Using these normalization formulas, adjusted values for each pixel in each zone are obtained. The resulting P values will be used to determine the pixel's lighting level in the four-level thresholding method.

The algorithm for the four-level threshold matrix, described below, is quite similar to that of the random threshold method:

```
For each pixel (x,y) of intensity level I(x,y)
   Compute x0 =x mod 8  and  y0 =y mod 8
   Set the threshold S=Dx y00(8) from the threshold matrix D(8)
   If (I(x,y) < 85)
      Compute P=0.75*I(x,y)
      If (P < S)
         display with level 1
      Else
         display with level 0
      EndIf
   Else
      If (I(x,y) < 170)
         Compute P=0.75*(I(x,y)-85)
         If (P > S)
            display with level 2
         Else
            display with level 1
         EndIf
      Else
         Compute P=0.75*(I(x,y)-170)
         If (P > S)
            display with level 3
         Else
            display with level 2
         EndIf
      EndIf
   EndIf
EndFor
```

10.2. Basic image analysis tools

10.2.1. *Histogram*

Given an image with n rows and p columns, i.e. $n \times p$ pixels, each pixel is encoded using q bits (e.g. if $q = 8$, there are 256 levels). To perform statistical analysis on the pixel levels, the number of pixels that possess each level can be counted. The graphical representation of this statistic is a histogram, which displays the frequency of each level across the image (see Figure 10.1).

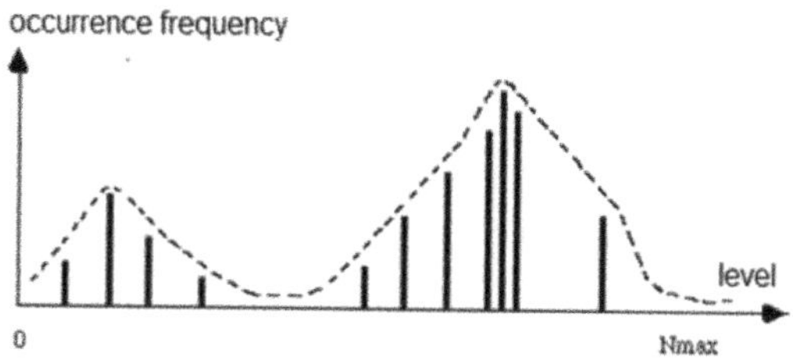

Figure 10.1. *Histogram*

If the image contains many different levels, the histogram tends to form of a curve; otherwise, it appears as a series of "sticks". The histogram can be interpreted probabilistically: if n_i is the number of pixels at level i, then $f_i = \frac{n_i}{n \times p}$ represents the occurrence frequency of level i. The (approximate) probability of this level appearing in the image is given by $p_i \approx f_i$.

The histogram provides quick information about the image's characteristics. For instance, it can help distinguish between an image that is too dark (with levels mostly close to 0, as shown in Figure 10.2) and one that is too light (with levels mostly near 255, as shown in Figure 10.3).

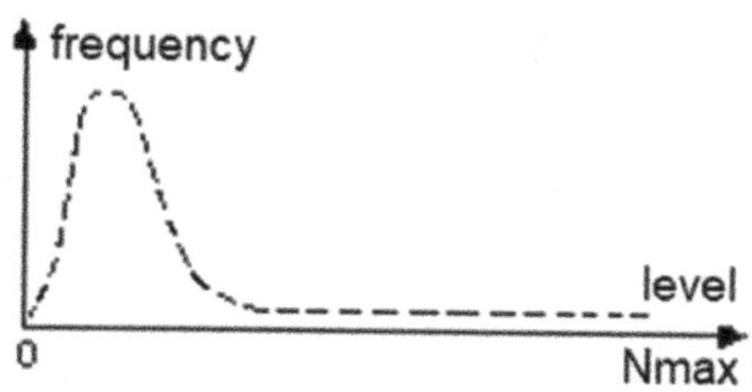

Figure 10.2. *Image too dark*

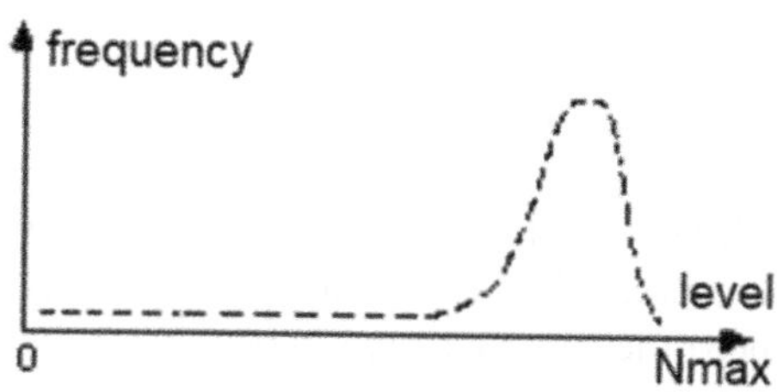

Figure 10.3. *Image too bright*

It is also possible to distinguish between the image background and the objects of interest in the image (foreground) (see Figure 10.4).

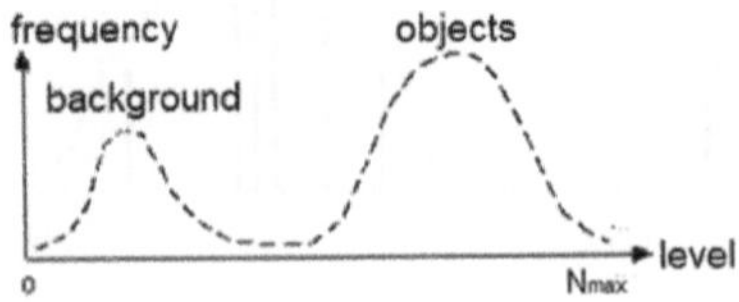

Figure 10.4. *Background/foreground difference*

10.2.2. *Profiles*

A profile is a representation of the variation in pixel intensity $I(x,y)$ along a specified row or column.

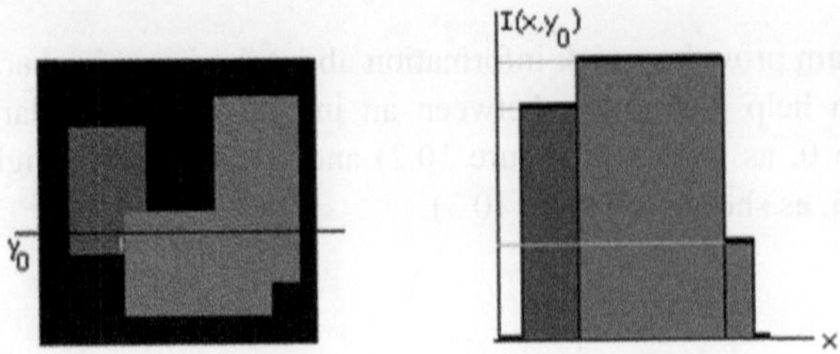

Figure 10.5. *Horizontal line profile. For a color version of this figure, see www.iste.co.uk/cochard/mathematics2.zip*

10.2.3. *Level search*

This operation involves displaying all pixels that share the same level as a given pixel, which helps highlight similar details in the image.

10.2.4. *Information contained in an image*

Given an image coded on N color levels, the amount of information provided by a particular level i is represented by the quantity $Q_i = -\log_2(p_i)$, where p_i is the probability of occurrence of level i. Since p_i lies between 0 and 1, it follows that the quantity of information Q_i varies from 0 to infinity: $p_i = 0 \Rightarrow Q_i = +\infty$ and $p_i = 1 \Rightarrow Q_i = 0$.

NOTE 10.1.– The rarer a level (i.e. when p_i is low), the greater the quantity of information it provides. Conversely, a common or ubiquitous level offers very little information.

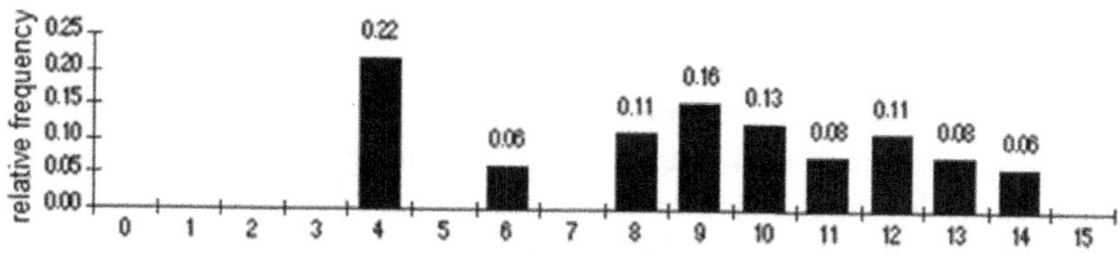

levels	frequency	pi	Qi	Ei
0	0	0.00	infinite	0.00
1	0	0.00	infinite	0.00
2	0	0.00	infinite	0.00
3	0	0.00	infinite	0.00
4	28	0.22	2.19	0.48
5	0	0.00	infinite	0.00
6	8	0.06	4.00	0.25
7	0	0.00	infinite	0.00
8	14	0.11	3.19	0.35
9	20	0.16	2.68	0.42
10	16	0.13	3.00	0.38
11	10	0.08	3.68	0.29
12	14	0.11	3.19	0.35
13	10	0.08	3.68	0.29
14	8	0.06	4.00	0.25
15	0	0.00	infinite	0.00
sums	128	1.00		3.05 ← entropy

Figure 10.6. *First-order entropy calculation*

In practice, the probability p_i is approximated through statistical counting, following the law of large numbers, which naturally leads to approximations in the calculation of the amount of information. For an image, the average quantity of information calculated by taking a weighted arithmetic mean of the information quantities contributed by each level (with the coefficients being the probabilities p_i). The result is known as the *entropy* (specifically, first-order entropy, as pixel correlations are not considered here) of the image:

$$E = \sum_{i=0}^{N-1} E_i = \sum_{i=0}^{N-1} p_i Q_i = -\sum_{i=0}^{N-1} p_i \log_2(p_i)$$

Figure 10.6 shows how entropy is calculated for a 16-level image with a known histogram.

10.3. Basic processing

Basic processing, also known as pre-processing, aims to enhance an image for a specific application. A common enhancement is the restoration of a degraded image which typically occurs in two distinct phases:

Phase 1: modeling degradation. In this phase, the goal is to understand and model the degradation process that has affected the image. This may involve various factors such as blurring, noise or loss of contrast. Modeling degradation helps quantify and characterize these undesirable effects.

Phase 2: applying inverse modeling. Once the degradation model is established, inverse modeling techniques can be applied to reverse the degradation. These restoration techniques aim to compensate for the effects of degradation and restore the image to its original or enhanced state. Such techniques may include operations such as blur reduction, noise reduction and contrast enhancement.

Working directly on the pixels of an image is known as spatial domain processing. In contrast, performing operations on the image's Fourier transform is referred to as frequency domain processing.

In this section, as before, the gray level of a pixel at (x, y) is denoted by $I(x, y)$, with values typically ranging from 0 (black) and $N - 1$ (white), representing the grayscale. Three basic types of processing can be distinguished: histogram-based methods, spatial-domain filtering and frequency domain filtering.

10.3.1. *Histogram transformation*

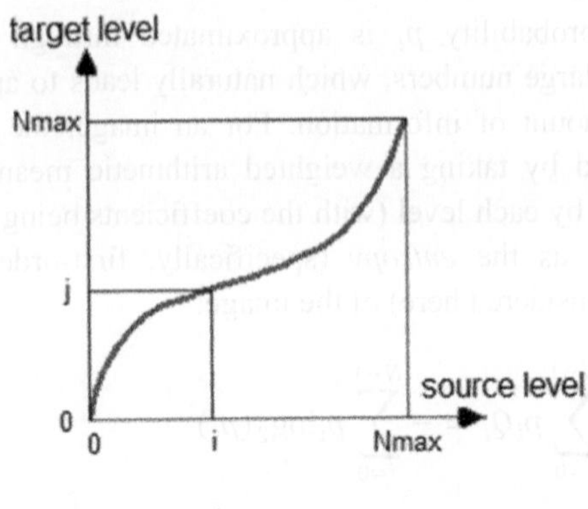

Figure 10.7. *Defining a LUT. For a color version of this figure, see www.iste.co.uk/cochard/mathematics2.zip*

Also known as a *LUT*, a look-up table is a tool that maps any level (from 0 to 255) to another level within the same range: $j = Transfo(i)$ (see Figure 10.7). This operation performs a level transformation, effectively modifying the histogram of the image.

Figure 10.8 shows two simple examples of LUTs: an identity LUT, where each level maps to itself, and an inversion LUT, where the levels are reversed.

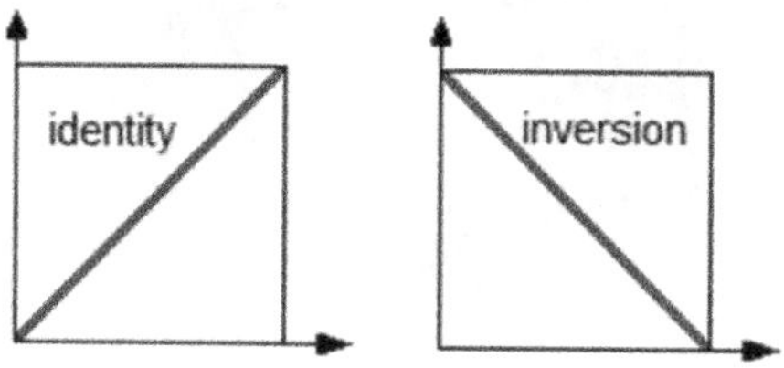

Figure 10.8. *Identity LUT and inversion LUT. For a color version of this figure, see www.iste.co.uk/cochard/mathematics2.zip*

Application 1: dynamic range adjustment. For images that are either too light, or have pixel levels over a narrow range, the pixel values are redistributed across the full range $(0 - 255)$.

Application 2: histogram-based binarization. This method highlights specific areas of the histogram. All pixels with gray levels between a and b are set to the maximum level N_{max}, while all other pixels are set to level 0 (see Figure 10.9).

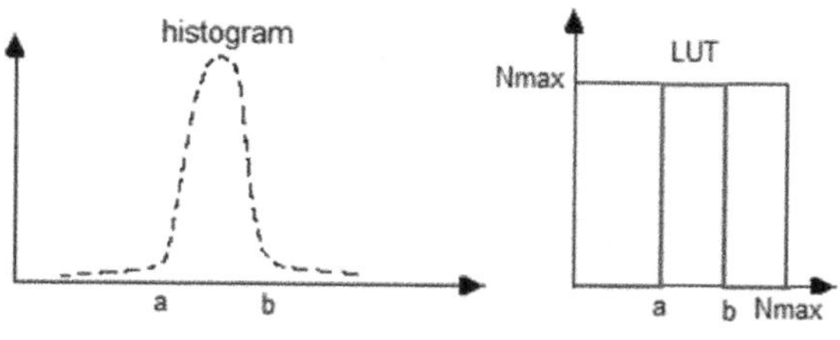

Figure 10.9. *Highlighting specific areas. For a color version of this figure, see www.iste.co.uk/cochard/mathematics2.zip*

EXAMPLE 10.1.– Consider the image I1 shown in Figure 10.10. The aim is to modify the pixel values so that points with a gray value between 0 and 127 are converted to white (255), while points with a gray value between 128 and 255 have become too black (0). By applying the LUT shown in Figure 10.12, the transformed image I2 is produced, as shown in Figure 10.11.

Figure 10.10. *Image I1*

Figure 10.11. *Image I2*

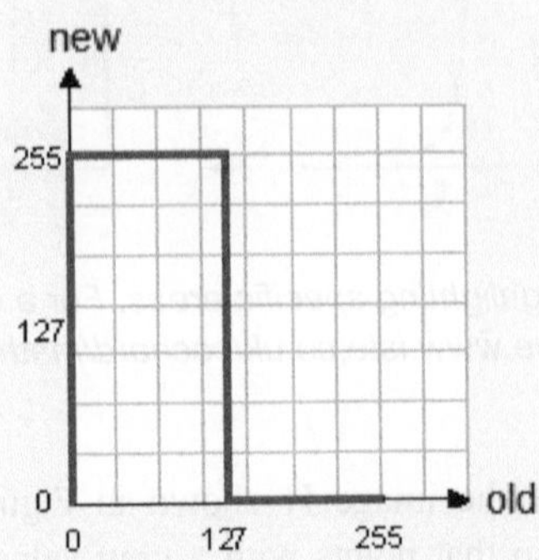

Figure 10.12. *LUT used. For a color version of this figure, see www.iste.co.uk/cochard/mathematics2.zip*

EXAMPLE 10.2.– The I1 image shown in Figure 10.13 contains unwanted gray traces. To eliminate these traces, a look-up table (LUT) can be applied to replace the pale gray shades with white. The results of this transformation are shown in Figures 10.14 and 10.15.

Figure 10.13. *Image I1*

Figure 10.14. *Image I2*

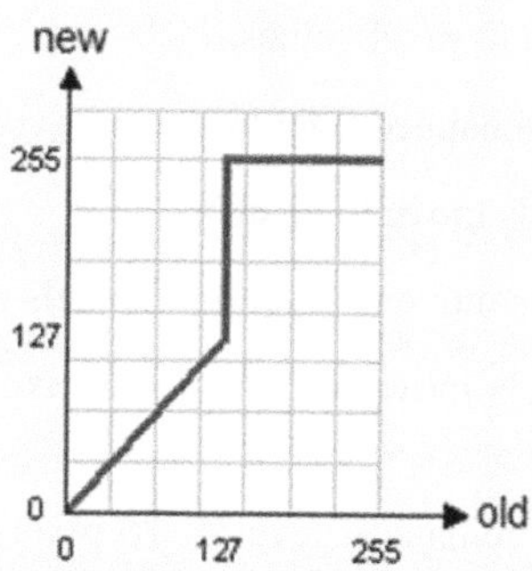

Figure 10.15. *LUT used. For a color version of this figure, see www.iste.co.uk/cochard/mathematics2.zip*

10.3.2. *Changing the shape of the histogram: equalization*

This process rebalances the histogram towards a predetermined shape. When the target shape corresponds to a uniform distribution (as is often the case), the technique is called histogram equalization. The aim is to generate a new image from an original, where all gray levels occur with the same frequency, either exactly or approximately. If the histogram of the new image follows a uniform distribution, it is described as "flat". If the distribution is nearly uniform, the histogram is said to be "equalized".

a) Flat histogram

Consider a simple example with an I1 image of eight rows and 16 columns. For simplicity, only eight gray levels are available. The average number of pixels per level is $p_0 = \frac{8\times16}{8} = 16$ pixels. The histogram of image I1 is represented by the diagram in Figure 10.16.

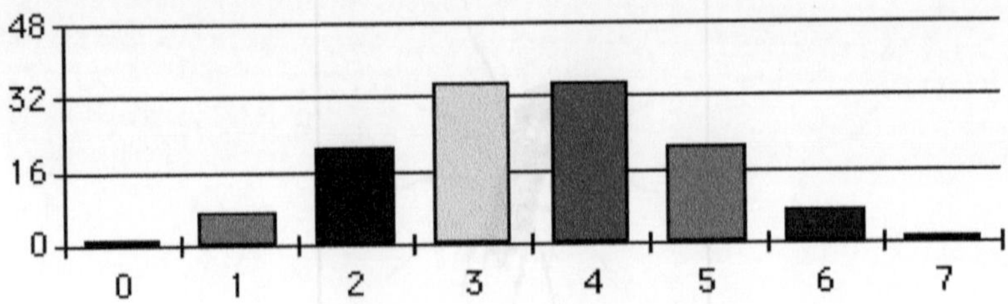

Figure 10.16. *A histogram*

To achieve a flat histogram, each color level from 0 to 7 must contain 16 pixels. This requires a redistribution of pixel values:

– Old level 0: remains unchanged (identical).

– Old level 1: seven pixels move to level 0.

– Old level 2: eight pixels move to level 0, 13 pixels move to level 1.

– Old level 3: three pixels move to level 1, 16 pixels move to level 2, the rest remain unchanged.

– Old level 4: 16 pixels remain at level 4, 16 pixels move to level 5 and three pixels move to level 6.

– Old level 5: 13 pixels move to level 6, eight pixels move to level 7.

– Old level 6: seven pixels move to level 7.

– Old level 7: remains unchanged.

By modifying the pixel values in left-right, top-to-bottom order, the resulting image I2 has a histogram as shown in Figure 10.17.

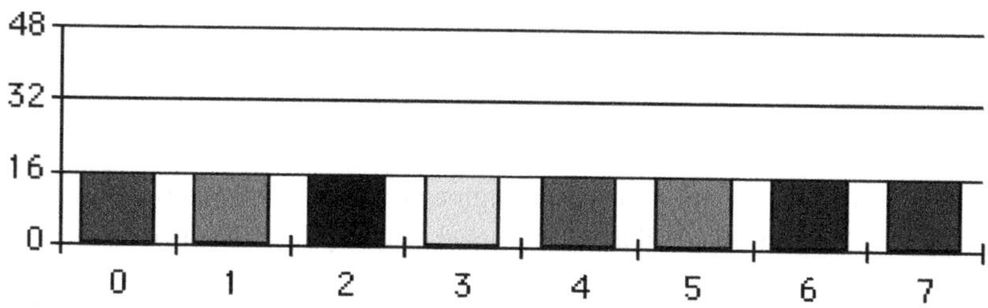

Figure 10.17. *Flat histogram*

b) Equalized histogram

In the previous method, there was an arbitrary element: the selection of which pixels change levels. A more effective approach is to treat all pixels of the same level uniformly and modify the levels globally. This is done by reducing the number of levels, replacing several levels with a single one. Let N_r be the number of grouping levels ($N_r < N$). In this example, $N_r = 4$, and the histogram is then divided into bands, each with a width of N/N_r. The average number of pixels per band is np/N_r, which, in this case, is $128/4 = 32$.

The pixel levels are distributed as follows:

– Level 0, 1 and 2 → 29 pixels assigned to level 0.

– Level 3 → 35 pixels assigned to level 1.

– Level 4 → 35 pixels assigned to level 2.

– Level 5, 6 and 7 → 29 pixels assigned to level 3.

This redistribution results in image I3, with its histogram displayed in Figure 10.18.

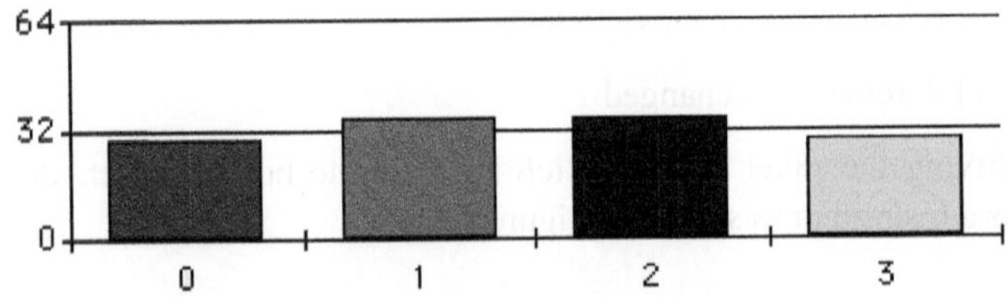

Figure 10.18. *Equalized histogram*

The choice of N_r is generally made through a process of dichotomy: the value of N is divided by 2 repeatedly until a suitable level of grouping is reached.

The algorithm can be formalized as follows:

```
//Initialization:
average = number of pixels / N
band = N/Nr
center = band/2
startid = 0
ifin = 0
//Loop until all levels are processed:
While (ifin < Nr)
   Accumulation = 0
   Accumulate pixels:
   While (accumulation not close to average) and (ifin < Nr)
      Cumul = cumul + histogram(ifin)
      increment ifin
   EndWhile
   //Assign transform values:
   For (i=start to ifin-1)
      Transfo(i) = center
   EndFor
   Idbut = ifin
   Center = center+strip
EndWhile
//Apply the transformation to the image:
For all points I(I,j)
   I(i,j) = Transfo(I(i,j))
EndFor
```

10.3.3. *Image subtraction and averaging*

Image subtraction is used in certain applications to highlight specific characteristic details. In radiography, for example, the goal may be to emphasize particular organs through the use of contrast agents. A radiographic image of a subject is first captured, denoted as $I1(x,y)$. After injecting the contrast agent, TV images of the same subject are taken, denoted as $I2(x,y)$. The difference $g(x,y) = I1(x,y) - I2(x,y)$ provides a "dynamic" visualization of the details of interest.

Additionally, image summations can also be performed. Consider an image $I(x,y)$ tainted by noise $b(x,y)$. The result is a noisy image represented as $g(x,y) = I(x,y) + b(x,y)$. Taking M images of the same subject results in $g_k(x,y) = I(x,y) + b_k(x,y)$ for each image k.

The average value (noted here $<.>$) of the results is calculated as follows:

$$< g(x,y) > = \frac{1}{M}\sum_{k=1}^{M} g_k(x,y) = I(x,y) + \frac{1}{M}\sum_{k=1}^{M} b_k(x,y)$$

This operation can be repeated multiple times to consider the statistical distribution of the obtained values $< g(x,y) >$, characterized by its mathematical expectation $E(< g >)$ and standard deviation $\sigma(< g >)$. Assuming that the noise has a mean value of zero leads to the following results:

$$E(< g >) = I(x,y)$$

These results indicate that the average of noisy images $g_k(x,y)$ tends towards the "clean" image $I(x,y)$ when M is sufficiently large.

10.4. Filtering

10.4.1. *Filtering in the spatial domain*

There are two classes of filters: linear and nonlinear, which are categorized based on whether the applied operator is linear or nonlinear.

10.4.1.1. *Linear filtering*

Linear filters fall into three categories, depending on their frequency domain properties. The names used will be explained later:

1) *Low-pass filters:* these filters attenuate image details that present sharp transitions from the rest of the image, such as edges or specific features. Applying a "low-pass" filter results in an image that appears blurred or loses sharpness.

2) *High-pass filters:* unlike low-pass filters, high-pass filters attenuate "neutral" image components to highlight details that stand out.

3) *Bandpass filters*: these filters are less commonly used for image enhancement but are useful in specific cases of image restoration.

The principle of linear filtering involves replacing the gray level of a pixel with a linear combination of the gray levels of the surrounding pixels. This linear combination is generally represented by a mask or convolution matrix. Consider, for example, a pixel (x, y) with a gray level of $I(x, y)$. The gray value of this pixel can be replaced by a linear combination of the gray values of all immediate neighboring pixels, including the pixel itself (see Figure 10.19).

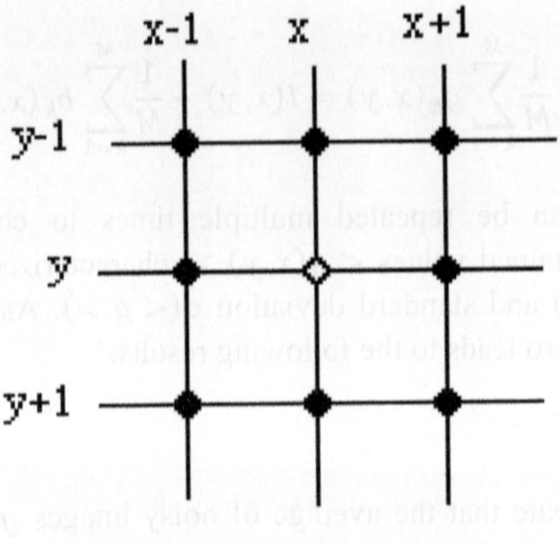

Figure 10.19. *Pixels to consider*

The function $g(x, y)$ is expressed as follows:

$$g(x,y) = \frac{1}{9}[I(x-1,y-1) + I(x-1,y) + I(x-1,y+1) + I(x,y-1) \\ + I(x,y) + I(x,y+1) + I(x+1,y-1) + I(x+1,y) \\ + I(x+1,y+1)]$$

Alternatively, it can be represented using the summation notation:

$$g(x,y) = \sum_{i=-1}^{+1} \sum_{j=-1}^{+1} h_1(i,j) I(x+i, y+j)$$

where the mask h_1 is defined by the matrix:

$$\boldsymbol{h_1} = \frac{1}{9}\begin{bmatrix} 1 & 1 & 1 \\ 1 & 1 & 1 \\ 1 & 1 & 1 \end{bmatrix}$$

The process can be generalized for a filter h_p of size $(2n+1) \times (2n+1)$:

$$g(x,y) = \sum_{i=-n}^{+n} \sum_{j=-n}^{+n} h_p(i,j) I(x+i, y+i)$$

The matrices below provide examples of linear filters where the central pixel is given greater weight than the peripheral pixels

$$\boldsymbol{h_2} = \frac{1}{10}\begin{bmatrix} 1 & 1 & 1 \\ 1 & 2 & 1 \\ 1 & 1 & 1 \end{bmatrix} \qquad \mathbf{h_3} = \frac{1}{16}\begin{bmatrix} 1 & 2 & 1 \\ 2 & 4 & 2 \\ 1 & 2 & 1 \end{bmatrix}$$

EXAMPLE 10.3.– Applying a linear filter to an image involves processing each pixel based on its surrounding pixel values. Suppose that the image I in Figure 10.20 is composed of four gray levels, coded as follows: 0 for black, 1 for dark gray, 2 for light gray and 3 for white.

Assume that point A is a "stray" point that has been incorrectly coded, possibly due to transmission errors over an unreliable network. In such cases, filtering can partially correct the error, although this is not always possible.

Let us consider a linear filter defined by a 3×3 mask:

$$\boldsymbol{H} = \frac{1}{15}\begin{bmatrix} 1 & 2 & 1 \\ 2 & 3 & 2 \\ 1 & 2 & 1 \end{bmatrix}$$

Now, apply this filter to point A, whose coding is assumed to be 1. The center of the mask is superimposed on point A (see Figure 10.21), and the following calculation is performed:

$$g(A) = \left(\begin{matrix} 1 \times 0 + 2 \times 0 + 1 \times 0 + 2 \times 0 + 3 \times 1 + 2 \times 0 + \\ 1 \times 0 + 2 \times 0 + 1 \times 3 \end{matrix} \right) / 15 = 0.4.$$

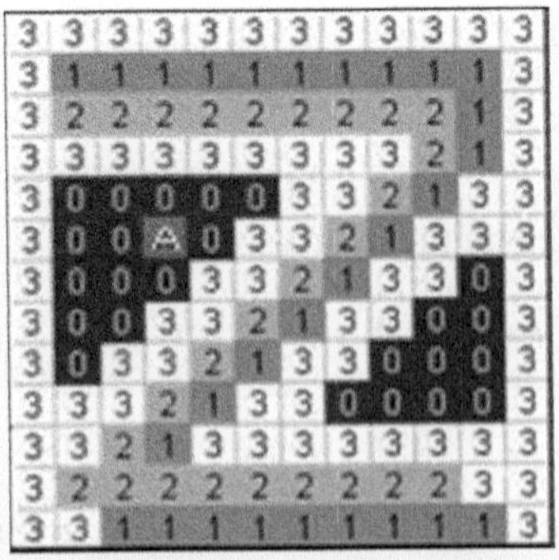

Figure 10.20. *Image to be processed*

Rounding 0.4 to the nearest value gives 0. Therefore, point A, now "filtered", is black.

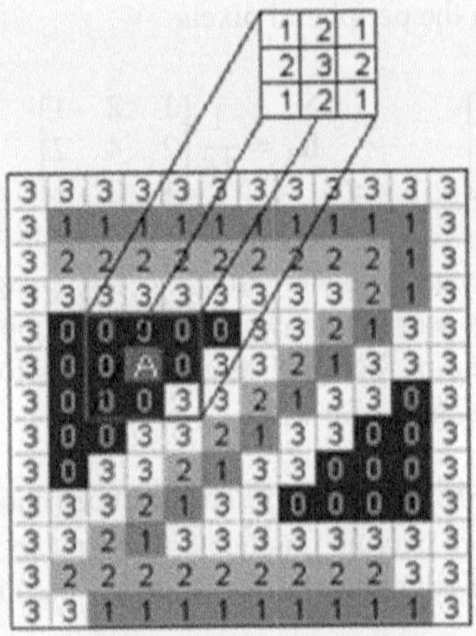

Figure 10.21. *After filtering*

The coefficients of all these filters are positive, helping to "soften" the image by replacing the level of each pixel with a weighted arithmetic mean of the levels of neighboring pixels. These are known as "low-pass" filters. Conversely, linear filters can have positive coefficients at the center and negative coefficients for neighboring pixels. These are referred to as "high-pass" filters, which highlight details that "stand out" in the image. An example of such a filter is defined by the following mask:

$$h_4 = \begin{bmatrix} -1 & -1 & -1 \\ -1 & 8 & -1 \\ -1 & -1 & -1 \end{bmatrix}$$

To "harden" the image, more elaborate methods can be used. First, a "low-pass" filter is applied to obtain an intermediate image $g_b(x,y)$. Subsequently, this image can be subtracted from the original image $l(x,y)$ to achieve a "high-pass" filtered image.

The resulting image is expressed as: $g_h(x,y) = l(x,y) - g_b(x,y)$.

High boost filtering can also be used by introducing an amplification coefficient A into the previous expression. The high boost filtered image $g_{hb}(x,y)$ is calculated as follows:

$$\begin{aligned} g_{hb}(x,y) = A.I(x,y) - g_b(x,y) &= (A-1).I(x,y) + I(x,y) - g_b(x,y) \\ &= (A-1).I(x,y) + g_h(x,y) \end{aligned}$$

Another method for "hardening" the image involves the use of derivative filters, mainly used for edge detection, which will be discussed later.

The linear filtering operation described earlier is essentially the application of a convolution product. In the discrete case, a convolution product for functions of two variables is defined as follows:

$$g(x,y) = h_g(x,y) * I(x,y) = \sum_{k=-n}^{+n} \sum_{l=-n}^{+n} h_g(k,l).I(x+k, y+l)$$

This equation represents the "composition" of two images: the original image I and the mask h_g.

In the filtering operation, a significant problem arises with pixels located at the edge of the image, as the mask may extend beyond the image boundaries. Various strategies can be adopted to address this issue:

– Image reduction: one approach is to "reduce" the image. If the mask is of dimension $(2n+1) \times (2n+1)$, the border of width n can be removed from the final image, resulting in a smaller output image.

– Zero padding: another strategy involves extrapolating the pixels outside the image with zeros. In this case, these out-of-bound pixels are assigned an arbitrary level 0.

– Mirror padding: alternatively, a "mirror" extrapolation can be used. For a mask of dimension $(2n + 1) \times (2n + 1)$, a border of width n can be added, filled by mirroring the edge pixels of the original image.

10.4.1.2. *Gaussian filtering*

Recall that the Gauss function, which represents the probability density of the normal distribution, is defined as:

$$G(x) = \frac{1}{\sigma\sqrt{2\pi}} e^{-\frac{x^2}{2\sigma^2}}$$

This function is widely used in statistics and is often referred to as the Laplace–Gauss or normal law. Its characteristic curve is commonly known as the "bell" curve, as shown in Figure 10.22.

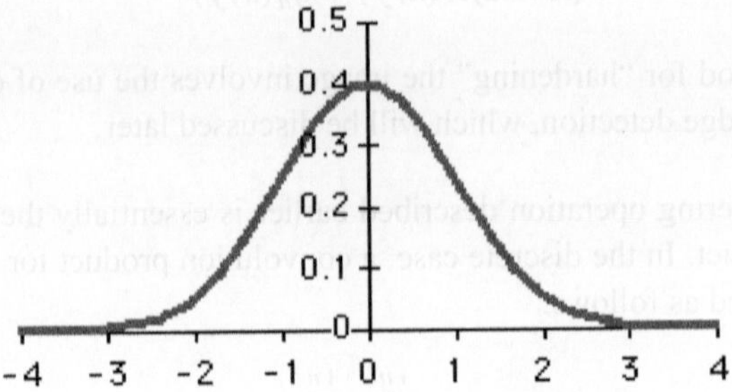

Figure 10.22. *Gaussian curve. For a color version of this figure, see www.iste.co.uk/cochard/mathematics2.zip*

The parameter σ represents the standard deviation and indicates the dispersion of the points on the curve (with $\sigma = 1$ in Figure 10.22).

The following useful results are established:

$$\int_{-\infty}^{+\infty} G(x)\,dx = 1 \qquad \int_{-2\sigma}^{+2\sigma} G(x)\,dx = 0.95 \qquad \int_{-3\sigma}^{+3\sigma} G(x)\,dx = 0.99$$

These results indicate that it is often sufficient to consider intervals $[-2\sigma, +2\sigma]$ or $[-3\sigma, +3\sigma]$.

In classical linear filtering, mask elements act as statistical weights. The value of the two-dimensional Gaussian function is given by

$$G(x,y) = \frac{1}{2\pi\sigma^2} e^{-\frac{x^2+y^2}{2\sigma^2}}$$

For practical purposes, x and y are limited to the range $[-3\sigma, +3\sigma]$, as significant values of the Gaussian fall within this range (covering approximately 99.9% of the Gaussian distribution). With a mask size of $(2n + 1) = 5$ and a standard deviation of $\sigma = 5/6$, the mask of the Gaussian filter, one of the most widely used nonlinear filters, is defined as follows (with the sum of all coefficients being close to 1):

$$\boldsymbol{G} = \begin{bmatrix} 0.0007 & 0.0063 & 0.0129 & 0.0063 & 0.0007 \\ 0.0063 & 0.0543 & 0.1116 & 0.0543 & 0.0063 \\ 0.0129 & 0.1116 & 0.2292 & 0.1116 & 0.0129 \\ 0.0063 & 0.0543 & 0.1116 & 0.0543 & 0.0063 \\ 0.0007 & 0.0063 & 0.0129 & 0.0063 & 0.0007 \end{bmatrix}$$

Gaussian filtering therefore consists of the discrete convolution of the filter with the image, which is expressed as either $g = G * I$ or:

$$g(x,y) = \frac{1}{\left(\sigma\sqrt{2\pi}\right)^2} \sum_{k=-2}^{+2} \sum_{l=-2}^{+2} e^{-\frac{k^2+l^2}{2\sigma^2}} \,.\, I(x+k; y+l)$$

EXAMPLE 10.4.– Gaussian filtering with a 3 × 3 mask. Consider the image from the previous example to observe how the shade of point A (initial coding of 1) is transformed. A 3 × 3 Gaussian filter is selected (see Figure 10.23).

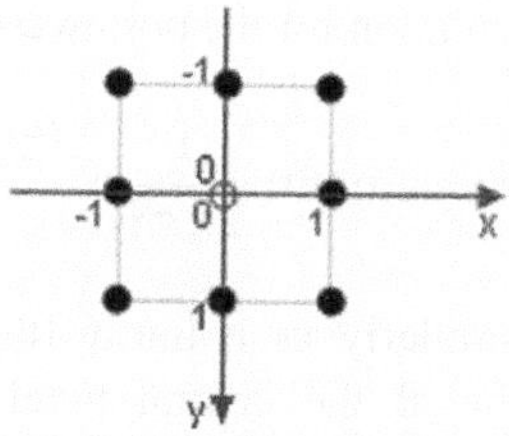

Figure 10.23. *3x3 filter*

The filter width is set to three pixels, which corresponds to 6σ (i.e. $2 \times 3\sigma$) to contain almost the entire Gaussian function. This leads to a calculated standard deviation of $\sigma = 0.5$. With this standard deviation, the values of the filter terms can be computed, resulting in the following 3 × 3 Gaussian filter matrix:

$$\sigma = 0.5 \qquad \boldsymbol{G} = \begin{bmatrix} 0.0117 & 0.0862 & 0.0117 \\ 0.0862 & 0.6366 & 0.0862 \\ 0.0117 & 0.0862 & 0.0117 \end{bmatrix}$$

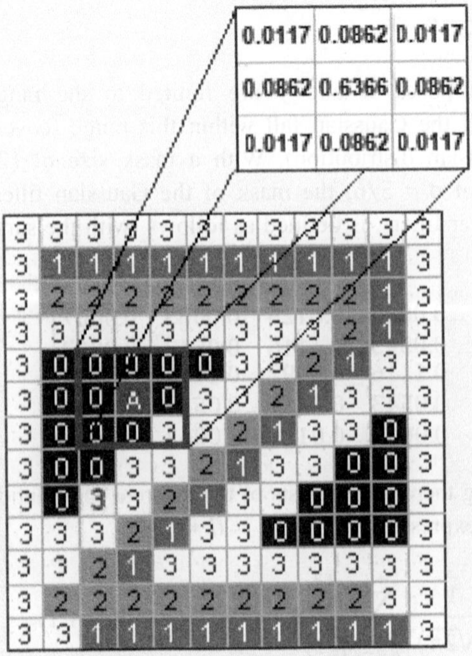

Figure 10.24. *Application of the Gaussian filter. For a color version of this figure, see www.iste.co.uk/cochard/mathematics2.zip*

The sum of the coefficients in the Gaussian filter is approximately equal to 1. To illustrate this, consider the calculation for the northwest element of the filter:

$$\text{Value of the northwest element: } \frac{1}{2\pi\sigma^2} e^{-\frac{(-1)\times(-1)+1\times 1}{2\sigma^2}} = 0.0117$$

Next, the mask is applied similarly to a linear filter (see Figure 10.24). The calculation for the new value at the central pixel involves multiplying each surrounding pixel's value by its corresponding filter coefficient:

$$0 \times 0.0117 + 0 \times 0.0862 + 0 \times 0.0117 + 0 \times 0.0862 + 1 \times 0.6366 + 0 \times 0.0862 + 0 \times 0.0117 + 0 \times 0.0862 + 3 \times 0.0117 = 0.6717.$$

Thus, the resulting value after applying the Gaussian filter to point A (initially coded as 1) is approximately 0.6717.

For each point, assuming that the mask values have already been calculated, the operation requires $(2n+1)\times(2n+1)$ multiplications and $(2n+1)\times(2n+1)-1$ additions.

For an image of 200×200 points and a mask size of 5, this amounts to 1,000,000 multiplications and 960,000 additions! Note that computation time is roughly proportional to the mask size, so smaller masks are preferable when possible. However, a Gaussian filter can be decomposed into smaller Gaussian filters to reduce computational load. The mask can be written in separable form as follows:

$$G(k,l) = G_1(k).G_2(l) \quad \text{with} \quad G_1(k) = \frac{1}{\sigma\sqrt{2\pi}} e^{-\frac{k^2}{2\sigma^2}} \quad \text{and} \quad G_2(k) = \frac{1}{\sigma\sqrt{2\pi}} e^{-\frac{k^2}{2\sigma^2}}$$

In this form, G_1 corresponds to a $(2n+1)\times 1$ rectangular mask, while G_2 corresponds to a $1\times(2n+1)$ rectangular mask. The result can thus be expressed through successive convolutions:

$$g = G_2 * G_1 * I = G_1 * G_2 * I$$

10.4.2. *Frequency domain filtering*

The principle of this approach is straightforward and relies on Fourier transforms: filtering is applied to the Fourier transforms of the image and mask, and the result is then transformed back using the inverse Fourier transform. Letting FT denote the Fourier transform:

$$\begin{pmatrix} mask\ h \rightarrow TF(h) = H \\ image\ I \rightarrow TF(I) = J \end{pmatrix} \rightarrow H.J = TF(h * I)$$

$$\rightarrow filtered\ image\ h * I = TF^{-1}(H.J)$$

10.4.2.1. *Low-pass filters*

These filters are designed to attenuate high frequencies, as illustrated by the ideal filter in Figure 10.25, where the filter function $H(u,v)$ is defined by:

$$H(u,v) = \begin{cases} 1 & \text{if } r < R \\ 0 & \text{if } r > R \end{cases} \quad \text{where} \quad r^2 = u^2 + v^2.$$

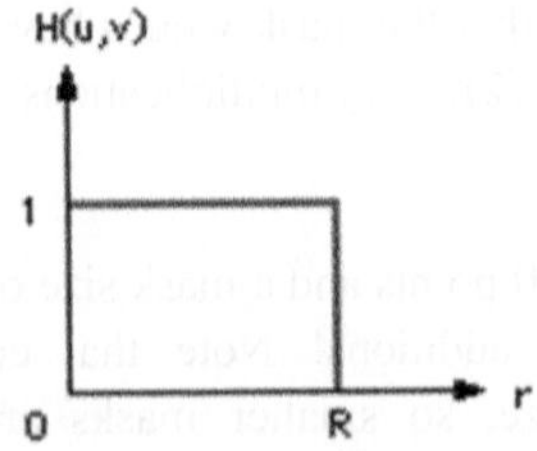

Figure 10.25. *Low-pass filter. For a color version of this figure, see www.iste.co.uk/cochard/mathematics2.zip*

The Butterworth filter, shown in Figure 10.26, serves a similar purpose to the ideal filter but with a smoother transition between passing and attenuating frequencies, avoiding the "brutal" cutoff of the ideal filter. It is defined by the function:

$$H(u,v) = \frac{1}{1+\alpha\left(\frac{r}{R}\right)^{2n}}$$

where:

– $\alpha = 1$ or $\alpha = \sqrt{2} - 1$;

– R represents the cutoff frequency radius;

– $r = \sqrt{u^2 + v^2}$ denotes the radial distance from the origin in the frequency domain;

– n is an integer ≥ 1 and controls the sharpness of the filter's roll-off.

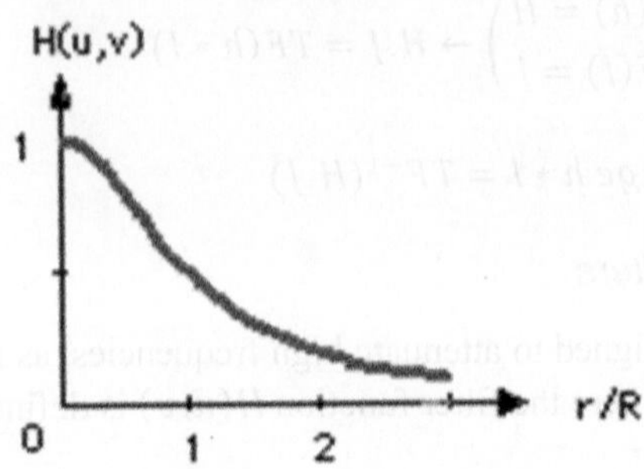

Figure 10.26. *Butterworth filter. For a color version of this figure, see www.iste.co.uk/cochard/mathematics2.zip*

Another example of a linear filter is the Gaussian filter, which also belongs to this category. An interesting property of the Gaussian filter is that it can be replaced by a cascade of simpler Gaussian filters, each with its own standard deviation. To illustrate this, consider a set of K Gaussian filters defined as:

$$g_i(x) = \frac{1}{\sigma_i\sqrt{2\pi}} e^{-\frac{x^2}{2\sigma_i^2}}$$

where each filter $g_i(x)$ has a standard deviation σ_i.

To apply these filters in succession, the combined filter g results from sequentially convolving each elementary filter g_i:

$$g = g_1 * g_2 * g_3 * \dots * g_K$$

Using the fundamental property of the Fourier transform of a convolution product, the resulting function $g(x)$ takes a Gaussian form:

$$g(x) = \frac{1}{\sigma\sqrt{2\pi}} e^{-\frac{x^2}{2\sigma^2}}$$

where the effective standard deviation σ of the combined filter is:

$$\sigma = \sqrt{\sum_{i=1}^{k} \sigma_i^2}$$

This expression shows that a Gaussian filter G with standard deviation σ can be interpreted as a cascade of simpler Gaussian filters g_i with standard deviations σ_i, achieving the same overall smoothing effect.

If a size $(-3\sigma, +3\sigma)$ is chosen for the Gaussian filter mask G, the mask width in pixels can be expressed as $2n + 1 = 6\sigma$; therefore, $n = 3\sigma - \frac{1}{2}$.

For an elementary Gaussian filter mask, a similar size $(-3\sigma_0, +3\sigma_0)$ is chosen, resulting in the mask width of $2n_0 + 1 = 6\sigma_0$; therefore, $n_0 = 3\sigma_0 - \frac{1}{2}$.

Assuming identical elementary filters, the standard deviation of the combined filter can be represented as $\sigma = \sqrt{\sum_{i=1}^{k} \sigma_0} = \sigma_0\sqrt{k}$. Substituting this into the expression for n ($n = 3\sigma - \frac{1}{2}$) gives:

$$n = 3\sigma - \frac{1}{2} = 3\sigma_0\sqrt{k} - \frac{1}{2}$$

Using $n_0 = 3\sigma_0 - \frac{1}{2}$ leads to:

$$n = \left(n_0 + \frac{1}{2}\right)\sqrt{k} - \frac{1}{2}$$

For example, when using nine times a filter size of 5 (where $k = 9$ and $n_0 = 2$), the operation can be interpreted as equivalent to applying a single filter of size:

$$2n + 1 = 2\left(n_0 + \frac{1}{2}\right)\sqrt{k} = 2\left(2 + \frac{1}{2}\right)\sqrt{9} = 15 = 2 \times 7 + 1$$

Thus, applying nine elementary filters of size 5 effectively creates a larger filter, resulting in enhanced smoothing. This combined approach can be considered equivalent to using a single filter of size 15 pixels.

10.4.2.2. *High-pass filters*

High-pass filters are designed to attenuate low frequencies while allowing high frequencies to pass through. The characteristics of the ideal high-pass filter are shown in Figure 10.27.

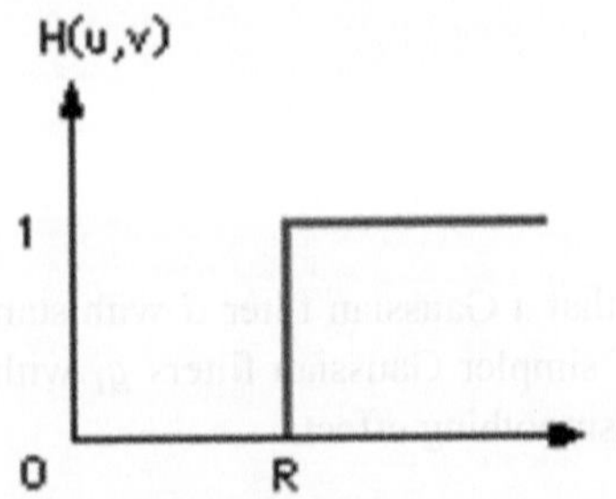

Figure 10.27. *High-pass filter. For a color version of this figure, see www.iste.co.uk/cochard/mathematics2.zip*

The Butterworth filter provides a gentler transition between passing and attenuating frequencies compared to the ideal filter, as shown in Figure 10.28.

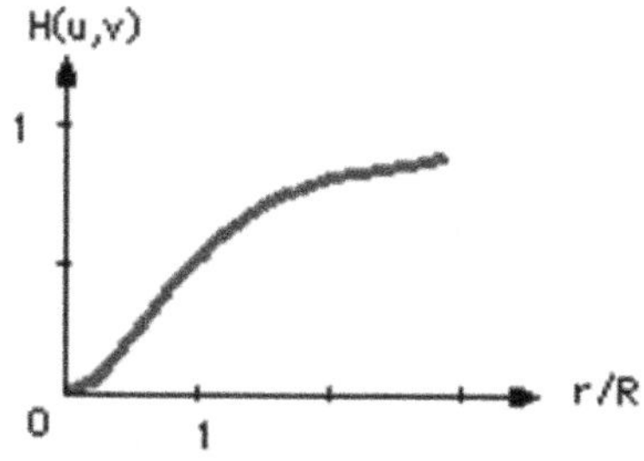

Figure 10.28. *Butterworth filter. For a color version of this figure, see www.iste.co.uk/cochard/mathematics2.zip*

$$H(u,v) = \frac{1}{1+\alpha\left(\frac{R}{r}\right)^{2n}} \quad \text{with} \quad \alpha = 1 \quad \text{or} \quad \alpha = \sqrt{2} - 1, \quad n \quad \text{integer} \quad \geq 1.$$

10.5. Binary images

Binarization is the process of converting a multi-gray image into a binary black-and-white image, resulting in only two levels.

This method is commonly used for isolating objects in an image (see Figure 10.28).

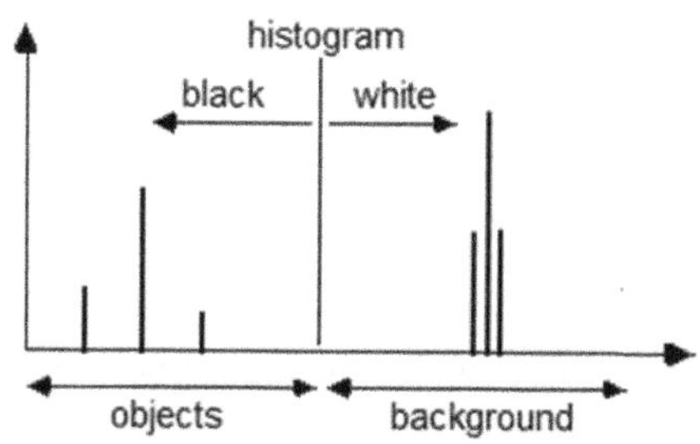

Figure 10.29. *Object isolation*

As a result, a binary image can be represented as a Boolean matrix, where each element indicates true (1 = white) or false (0 = black).

10.5.1. *Morphological operators*

10.5.1.1. *Dilation*

When an image is noisy, its binarization may reveal isolated, spurious black or white dots that need to be removed using techniques such as dilation and erosion.

Dilation involves eliminating isolated "black" dots by expanding the "white" areas, which "consumes" the "black" dots. The method scans the image using a window of size $(2n + 1) \times (2n + 1)$ with a logical "OR" operation. The center of the window is placed on the current pixel, and a logical "OR" is performed on the surrounding $(2n + 1)^2 - 1$ pixels. If the result is 1, the pixel is set to 1; if the result is 0, the pixel is retained.

10.5.1.2. *Erosion*

The aim here is to eliminate isolated "white" spots, and the method is quite similar to that of dilatation:

– The image is scanned using a window of size $(2n + 1) \times (2n + 1)$, with the center of the window on the current pixel.

– A logical "AND" is performed on the surrounding pixels.

– If the result is 1, the pixel is retained; if the result is 0, the pixel is set to 0.

In general, the process involves a succession of erosion and dilation operations.

10.5.1.3. *Skeletonization*

The aim of this method is to reduce the redundancy of information present in an image, thereby decreasing the amount of data to be analyzed.

The method involves isolating the main lines of the image through successive thinning operations until only lines with a thickness of one pixel remain.

To achieve this, eight different masks are iteratively applied to the image. Each mask corresponds to a specific transformation: if a particular configuration is detected in the vicinity of the pixel being processed, that pixel is replaced by 0.

During each pass, each mask is applied successively to the entire image, with the central point of the mask used to process each pixel. This process is repeated until the results obtained between two successive passes are identical, indicating that the image has reached a stabilized state

$$M_1 = \begin{bmatrix} 0 & X & 1 \\ 0 & 1 & 1 \\ 0 & X & 1 \end{bmatrix} \rightarrow \begin{bmatrix} 0 & X & 1 \\ 0 & 0 & 1 \\ 0 & X & 1 \end{bmatrix} \quad M_2 = \begin{bmatrix} 0 & 0 & X \\ 0 & 1 & 1 \\ X & 1 & 1 \end{bmatrix} \rightarrow \begin{bmatrix} 0 & 0 & X \\ 0 & 0 & 1 \\ X & 1 & 1 \end{bmatrix}$$

$$M_3 = \begin{bmatrix} 0 & 0 & 0 \\ X & 1 & X \\ 1 & 1 & 1 \end{bmatrix} \rightarrow \begin{bmatrix} 0 & 0 & 0 \\ X & 0 & X \\ 1 & 1 & 1 \end{bmatrix} \quad M_4 = \begin{bmatrix} X & 0 & 0 \\ 1 & 1 & 0 \\ 1 & 1 & X \end{bmatrix} \rightarrow \begin{bmatrix} X & 0 & 0 \\ 1 & 0 & 0 \\ 1 & 1 & X \end{bmatrix}$$

$$M_5 = \begin{bmatrix} 1 & X & 0 \\ 1 & 1 & 0 \\ 1 & X & 0 \end{bmatrix} \rightarrow \begin{bmatrix} 1 & X & 0 \\ 1 & 0 & 0 \\ 1 & X & 0 \end{bmatrix} \quad M_6 = \begin{bmatrix} 1 & 1 & X \\ 1 & 1 & 0 \\ X & 0 & 0 \end{bmatrix} \rightarrow \begin{bmatrix} 1 & 1 & X \\ 1 & 0 & 0 \\ X & 0 & 0 \end{bmatrix}$$

$$M_7 = \begin{bmatrix} 1 & 1 & 1 \\ X & 1 & X \\ 0 & 0 & 0 \end{bmatrix} \rightarrow \begin{bmatrix} 1 & 1 & 1 \\ X & 0 & X \\ 0 & 0 & 0 \end{bmatrix} \quad M_8 = \begin{bmatrix} X & 1 & 1 \\ 0 & 1 & 1 \\ 0 & 0 & X \end{bmatrix} \rightarrow \begin{bmatrix} X & 1 & 1 \\ 0 & 0 & 1 \\ 0 & 0 & X \end{bmatrix}$$

Figure 10.30 shows the result of skeletonization.

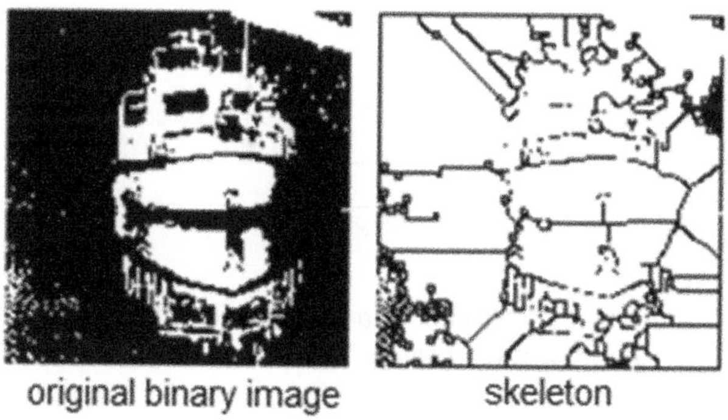

Figure 10.30. *Skeletonization*

10.6. Segmentation

Segmentation is the process of dividing an image into distinct regions. There are two common approaches to characterizing these regions: by their edges and by their homogeneity. The latter often relies on criteria such as similarity in color, texture or other visual properties.

10.6.1. *Outline extraction*

An outline is defined as a sudden variation in color level (see Figure 10.31). It represents a complex case of discontinuity. To detect a discontinuity, a linear mask can be used, which, when applied to a point, yields a result R. For a mask of order 3, the general form can be expressed as follows:

$$\begin{bmatrix} a_1 & a_2 & a_3 \\ a_4 & a_5 & a_6 \\ a_7 & a_8 & a_9 \end{bmatrix}$$

The result is given by:

$$R = \sum_{i=1}^{9} a_i z_i$$

where z_i represents the pixel level at position i.

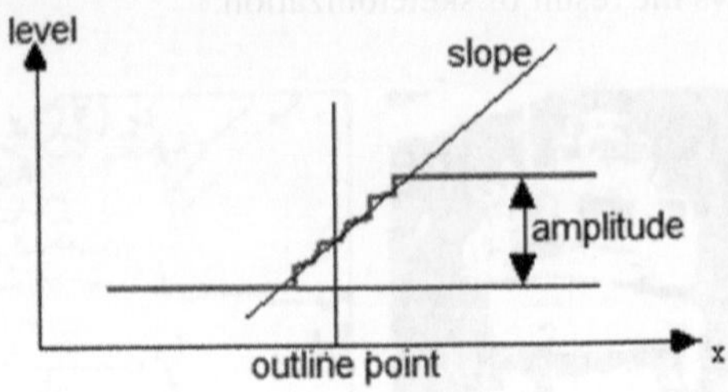

Figure 10.31. *Outline. For a color version of this figure, see www.iste.co.uk/cochard/mathematics2.zip*

To detect an isolated point, a threshold S can be used, where the detection condition being $r > S$, with r representing the absolute value of R.

To detect a straight line, four masks can be used.

Horizontal mask $\boldsymbol{A1}(hor)$ and diagonal mask (+45°) $\boldsymbol{A2}(+45°)$:

$$\boldsymbol{A1}(hor) = \begin{bmatrix} -1 & -1 & -1 \\ -2 & 2 & 2 \\ -1 & -1 & -1 \end{bmatrix} \qquad \boldsymbol{A2}(+45°) = \begin{bmatrix} -1 & -1 & 2 \\ -1 & 2 & -1 \\ 2 & -1 & -1 \end{bmatrix}$$

Vertical mask $A3(ver)$ and diagonal mask ($-45°$) $A4(+45°)$:

$$A3(ver) = \begin{bmatrix} -1 & 2 & -1 \\ -1 & 2 & -1 \\ -1 & 2 & -1 \end{bmatrix} \qquad A4(-45°) = \begin{bmatrix} 2 & -1 & -1 \\ -1 & 2 & -1 \\ -1 & -1 & 2 \end{bmatrix}$$

These masks can be applied to the image to detect various orientations of straight lines effectively.

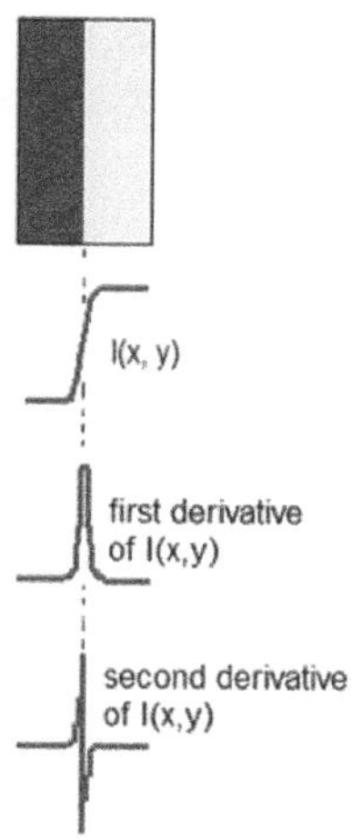

Figure 10.32. *Detecting a discontinuity. For a color version of this figure, see www.iste.co.uk/cochard/mathematics2.zip*

The detection of a straight line using a mask $\boldsymbol{A_i}$ is based on the condition $r_i > r_j$ $(j < i)$.

Detecting any form of discontinuity, however, is more complex. In general, a discontinuity is characterized by a variation in intensity levels. This can be analyzed through the behavior of the first derivative of $I(x, y)$ or the second derivative of $I(x, y)$, as shown in Figure 10.32. The first derivative highlights changes in intensity, while the second derivative can provide information about the nature of those changes, helping to identify edges and other forms of discontinuities in the image.

The outline extraction process generally consists of several key steps, which include:

1) *Outline highlighting*: this step focuses on detecting significant variations in intensity, color or other visual characteristics that define the outlines within the

image. Various techniques can be used, such as image differentiation to highlight gradients or the use of specific filters designed for outline detection.

2) *Outline reduction*: after the outlines have been detected, the next step is to reduce their thickness to a single pixel. This process ensures that the outlines are accurately represented as thin lines. Techniques such as skeletonization or thinning operations are commonly used to achieve this reduction.

3) *Outline binarization*: following the detection and reduction of outlines, binarization involves converting the outlines into a binary image, which contains only two levels of gray: black and white. This simplification aids in further processing and analysis.

4) *Outline description*: in this final step, the detected outlines are represented using a simpler, more meaningful elements. This representation facilitates the characterization and recognition of the shapes present in the image, allowing for easier interpretation and analysis.

The focus here will be on the first two stages of the outline extraction process.

10.6.1.1. *Differentiation operators*

Several methods are used to detect edges in images, with the following being the main ones:

a) Gradient method (first image derivation): for a pixel (x, y), the gradient is defined as: $A_x = I(x+1, y) - I(x, y)$ and $A_y = I(x, y+1) - I(x, y)$.

The gradient can also be characterized by its amplitude and direction:

– Amplitude: $A_0(x, y) = \left(A_x^2 + A_y^2\right)^{\frac{1}{2}}$.

– Direction: $D(x, y) = arctan\left(A_y / A_x\right)$.

To obtain the gradient, the image is convolved with specific masks derived from the above formulas. Common masks include:

– Roberts operator: $\boldsymbol{H_x} = \begin{bmatrix} -1 & 1 \end{bmatrix} \quad \boldsymbol{H_y} = \begin{bmatrix} -1 \\ 1 \end{bmatrix}$;

– Sobel operator: $\boldsymbol{H_x} = \begin{bmatrix} -1 & 0 & 1 \\ -1 & 0 & 1 \\ -1 & 0 & 1 \end{bmatrix} \quad \boldsymbol{H_y} = \begin{bmatrix} -1 & -1 & -1 \\ 0 & 0 & 0 \\ 1 & 1 & 1 \end{bmatrix}$;

– Prewitt operator: $\boldsymbol{H_x} = \begin{bmatrix} -1 & 0 & 1 \\ -2 & 0 & 2 \\ -1 & 0 & 1 \end{bmatrix} \quad \boldsymbol{H_y} = \begin{bmatrix} -1 & -2 & -1 \\ 0 & 0 & 0 \\ 1 & 2 & 1 \end{bmatrix}$.

The Prewitt and Sobel algorithms are particularly effective for edge detection while being more robust.

b) Laplacian method (second derivation of the image): the Laplacian of a function $I(\mathrm{x}, \mathrm{y})$ is defined as:

$$\Delta I = \frac{\partial^2 I}{\partial x^2} + \frac{\partial^2 I}{\partial y^2}$$

By discretization, the approximate relationship is given by:

$$\Delta I = 4I(x, y) - I(x, y+1) - I(x-1, y) - I(x+1, y) - I(x, y-1),$$

This leads to the following mask for the Laplacian:

$$\Delta = \begin{bmatrix} 0 & -1 & 0 \\ -1 & 4 & -1 \\ 0 & -1 & 0 \end{bmatrix}$$

The presence of an outline is indicated by the Laplacian crossing zero. However, the Laplacian is more commonly used to identify the region to which a pixel belongs, making it a useful tool in image analysis.

10.6.1.2. *Outline reduction*

By applying the differentiation operators, the gradient amplitude $A_0(x, y)$ and its direction can be obtained for the image (see Figure 10.32).

To accurately determine the contour or outline, it is important to select the points where $A(x, y)$ is at its maximum.

To achieve this, for each point (x, y) where the amplitude and direction of the gradient are known, the adjacent points in the direction of the gradient are identified. These points are designated as (x_1, y_1) and (x_2, y_2).

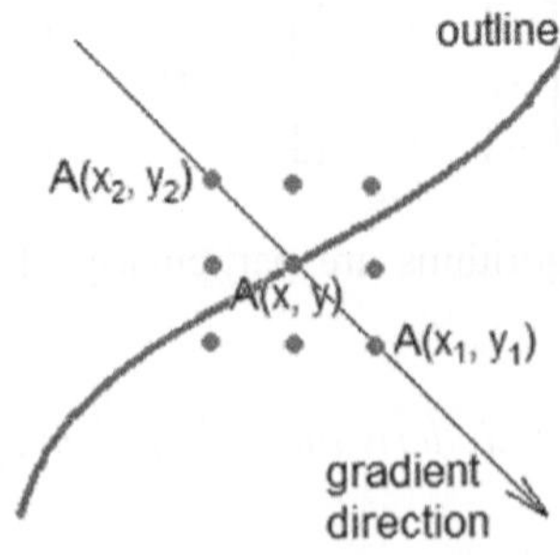

Figure 10.33. *Outline reduction. For a color version of this figure, see www.iste.co.uk/cochard/mathematics2.zip*

Next, $A_0(x, y)$ is compared to $A_0(x_1, y_1)$ and $A_0(x_2, y_2)$. If the conditions $A_0(x, y) > A_0(x_1, y_1)$ and $A_0(x, y) > A_0(x_2, y_2)$ hold true, then the point (x, y) is retained as part of the contour. If this condition is not met, the point is set to "1", indicating that it does not contribute to the outline.

10.6.2. *Regional segmentation*

The concept involves defining a region as a collection of points that share a common property, known as homogeneity. This homogeneity can be based on various criteria, such as color or grayscale values, texture or other visual characteristics.

10.6.2.1. *Separation method*

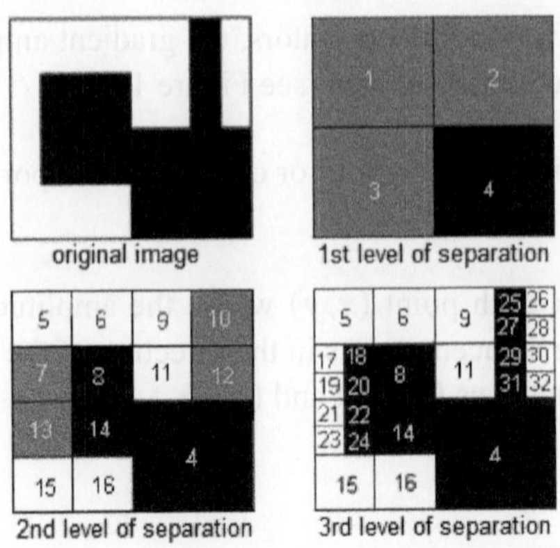

Figure 10.34. *Quad tree technique. For a color version of this figure, see www.iste.co.uk/cochard/mathematics2.zip*

The method uses the quad tree technique, which operates under the assumption that the image is square. This image is recursively divided into four quadrants, and then each quadrant is further divided into four sub-quadrants, continuing this process (see Figure 10.34). The division algorithm is straightforward and recursive, as outlined below:

```
Separation(zone)
If (criterion(zone)=TRUE)
   Classify the zone
Else
   divide the zone into four sub-zones Z1,Z2,Z3,Z4
   For (i=1 to 4)
      Apply Separation(Zi)
   EndFor
EndIf
```

The structure of the quad tree is shown in Figure 10.35.

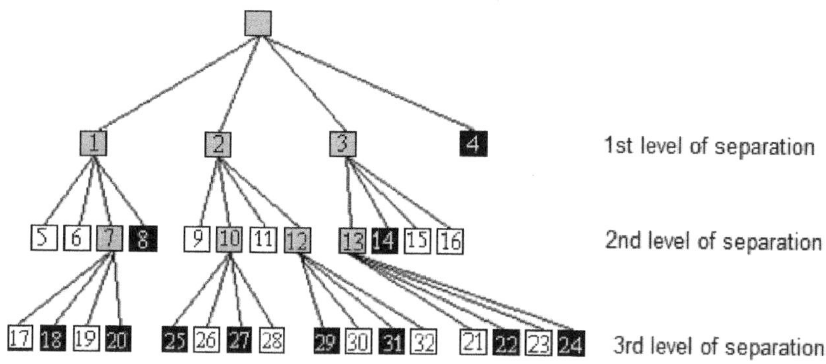

Figure 10.35. *Quad tree. For a color version of this figure, see www.iste.co.uk/cochard/mathematics2.zip*

10.6.2.2. *Fusion method*

In contrast to the previous method that involved dividing the image, fusion methods adopt an opposite principle: they allow a region to grow by incorporating neighboring regions that meet the same homogeneity criteria (region fusion). The merging rule requires that the two regions not only correspond to the same criterion but also be adjacent.

The spot coloring method consists of moving a window of three points over the image (see Figure 10.36). At each point, the following algorithm is applied, where "Color" represents a color level assigned to pixels that satisfy the same criterion:

```
For (each point (x,y))
   If (criterion(x,y) = criterion(x-1,y))
      Color(x,y) = Color(x-1,y)
   Otherwise
      If (criterion(x,y)=criterion(x,y-1))
         Color(x,y) = Colorx,y-1)
      Otherwise
         Color(x,y) = New_Color
      EndIf
   EndIf
EndFor
```

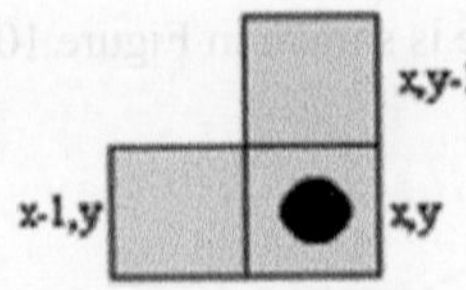

Figure 10.36. *A three-pixel window*

The local recursive method focuses on expanding a single region as much as possible before proceeding to the next. The algorithm for this method is described in what follows:

```
growth(x,y)
I(x,y) = 0 //Mark the current pixel to avoid re-examining(x,y)
// Check adjacent pixels
For (any pixel adjacent to (x,y))
   If (I(x,y)<>0 and criterion(pixel) = starting criterion)
      growth(pixel)  //Recursively grows the region
   EndIf
EndFor

// Main algorithm to initiate growth
For (each pixel (x,y))
   If (I(x,y) <> 0)
```

```
      save starting point (x,y) // Store the initial point for this
   region
      growth(x,y)               // Begin growing from this point
      increment SegmentNumber   // Move to the next segment
   EndIf
EndFor
```

11

Basics of Image Compression

CONCEPTS COVERED IN THIS CHAPTER.–

Compressing certain media is important due to the size of the corresponding files, especially for images.

Generally, compression aims to reduce redundancies within an image. The first section of this chapter describes the various forms of redundancy and introduces the general framework of the compression process.

The following section focuses on lossless compression, also known as compaction, covering the main methods: variable-length coding, bit-plane coding and predictive coding.

The third section addresses lossy compression, using predictive or transform coding. A detailed example is provided to illustrate lossy compression: the JPEG standard.

References: [GON 93, KUN 93].

The most extensive encyclopedias consist of approximately 25,000 pages of text and images. Each page can be considered as an image which, when scanned at a resolution of 300 dpi, occupies approximately 1 MB of space. Thus, the complete digitization of the encyclopedia would require 25 GB of storage space. This is a considerable volume, and if you wish to store this data on a current medium, it would encroach on the space available for other information. Therefore, it is therefore essential to "compress" the data to reduce its size.

Mathematics for Digital Science 2,
by Gérard-Michel COCHARD and Mhand HIFI.

11.1. General information

Binary-coded information can take different forms: I_1 (with n_1 bits), I_2 (with n_2 bits), I_3 (with n_3 bits), etc. Among these "equivalent" descriptions, there is one that is more compact than the others, while the other descriptions contain redundant data with respect to this one.

Let us take the descriptions I_1 and I_2 with n_1 and n_2 bits, respectively. The following characteristics can be defined:

– the compression ratio of I_1 in relation to I_2: $C = \frac{n_1}{n_2}$;

– the relative redundancy of I_1 in relation to I_2: $R = 1 - \frac{1}{C}$.

Here are a few specific cases:

– If $n_1 = n_2$, then $C = 1$ and $R = 0$. In this case, the description I_1 contains no redundant data with respect to the description I_2.

– If $n_2 \ll n_1$, then C tends towards infinity and R approaches 1. The description I_1 is strongly redundant with respect to the description I_2.

– If $n_2 \gg n_1$, then C tends towards 0 and R tends towards infinity. The description I_2 has strong redundancy with respect to the description I_1.

Three types of redundancy will be examined below: coding redundancy, interpixel redundancy and psychovisual redundancy.

11.1.1. *Coding redundancy*

Consider a grayscale image numbered from 0 to $N-1$. Let n_k denote the number of pixels with the level k, and n represent the total number of pixels. Given the large number of pixels, the probability of a pixel possessing level k is defined as:

$$p(k) = \frac{n_k}{n}, \quad k = 0, 1, 2, \dots N-1.$$

Each level k is encoded by a binary number of length $l(k)$. Therefore, the average coding length of a pixel is given by:

$$\langle l \rangle = \sum_{k=0}^{L-1} l(k)p(k)$$

The value of $\langle l \rangle$ necessarily influences the volume of the description.

EXAMPLE 11.1.– Consider eight gray levels, rated from 0 to 7, with the following probabilities: $p(0) = 0.19, p(1) = 0.25, p(2) = 0.21, p(3) = 0.16, p(4) = 0.08,$ $p(5) = 0.06, p(6) = 0.03, p(7) = 0.02.$

a) Let us perform a first coding, denoted as I_1, using a fixed-length 3-bit binary encoding (see Figure 11.1).

k	code 1	l(k)
0	000	3
1	001	3
2	010	3
3	011	3
4	100	3
5	101	3
6	110	3
7	111	3

Figure 11.1. *First coding*

b) Let us perform a second coding, denoted as I_2, which uses variable-length encoding with an average coding length of $\langle l \rangle = 2.7$ (see Figure 11.2). The compression ratio and relative redundancy of I_1 relative to I_2 are calculated as follows:

$$C = \frac{3}{2.7} \approx 1.11 \quad \text{and} \quad R = 0.099$$

Note that in the second coding, shorter codes correspond to the most frequent levels, while longer codes correspond to the rarer levels.

11.1.2. *Interpixel redundancy*

Interpixel redundancy is based on the observation that the level of a pixel can often be predicted from the levels of surrounding pixels. To "measure" this redundancy, autocorrelation coefficients are calculated between pixels at a distance of Δn. For example, along a line, the autocorrelation coefficient is calculated as:

$$\gamma(\Delta n) = \frac{A(\Delta n)}{A(0)}, \quad \text{where} \quad A(\Delta n) = \frac{1}{N - \Delta n} \sum_{x=0}^{N-1-\Delta n} I(x,y).I(x + \Delta n, y)$$

k	code 2	l(k)
0	11	2
1	01	2
2	10	2
3	001	3
4	0001	4
5	00001	5
6	000001	6
7	000000	6

Figure 11.2. *Second coding*

In this equation, n denotes the number of pixels in a line, x and y are the coordinates of a pixel and $I(x,y)$ represents the gray or color level of the pixel. In general, values of γ for low Δn values indicate a strong correlation, meaning that the information carried by a particular pixel can often be predicted by the information carried by neighboring pixels.

11.1.3. *Psychovisual redundancy*

The human eye has a specific visual perception that recognizes details carrying more information than others. It is therefore possible to eliminate unimportant details without significantly altering image quality. This operation is called quantization and is irreversible.

The IGS (Improved Gray Scale) coding method was developed to address these concerns. Suppose an original image has 256 gray levels. The image volume can be reduced by coding it with just 16 levels, using techniques developed in the chapter on image processing. However, the abrupt application of these techniques can sometimes result in unsightly or artificial effects, such as abrupt transitions between gray levels or false contours.

IGS coding offers a gentler approach. The principle is as follows: coding with 256-levels is performed on 8 bits, separated into high-order (4 bits) and low-order (4 bits) parts. For successive pixels, the 16-level (i.e. 4-bit) IGS coding is obtained by taking into account the 8-bit coding of previous pixels and retaining only the most significant bits. The precise algorithm is shown below:

```
IGS coding of a pixel of rank i:
If most significant of level (i) = 1111
   sum(i)=level(i)+0000
Else
   sum(i)= low order of sum(i-1)+level(i)
EndIf
IGS coding = high order of sum(i)
```

EXAMPLE 11.2.– Sum(0) is initialized to 0:

Pixel 1 of level 10000111: sum(1) = 10000111→ IGS coding = 1000.
Pixel 2 of level 11110100: sum(2) = 11110100→ IGS coding = 1111.
Pixel 3 of level 01101100: sum(3) = 01110000→ IGS coding = 0111.

11.1.4. *Confidence criteria*

When considering an original image with a pixel level (x, y), denoted as $I(x, y)$, and a compressed and then decompressed image with a pixel level (x, y), denoted as $I'(x, y)$, there are two types of criteria for assessing compression quality.

1) *Objective criteria*: these are based on the calculation of the mean square error, denoted as e, and the signal-to-noise ratio, denoted as r:

$$e = \sqrt{\frac{1}{MN}\sum_{x=0}^{M-1}\sum_{y=0}^{N-1}\left(I'(x,y) - I(x,y)\right)^2} \quad \text{and} \quad r = \sqrt{\frac{\sum_{x=0}^{M-1}\sum_{y=0}^{N-1} I'^2(x,y)}{e.MN}}$$

Therefore, the image dimensions are $M \times N$.

2) *Subjective criteria*: these refer to a rating scale established by a sample of users, where each value represents an assessment of quality: 1 = excellent, 2 = good, 3 = fair, 4 = borderline, 5 = poor, 6 = unusable. These subjective criteria allow for the evaluation of the perceived quality of compression according to users' opinions.

11.1.5. *Modeling image compression*

A compression system consists of an encoder at one end of the process (the source) and a decoder at the other end (the use).

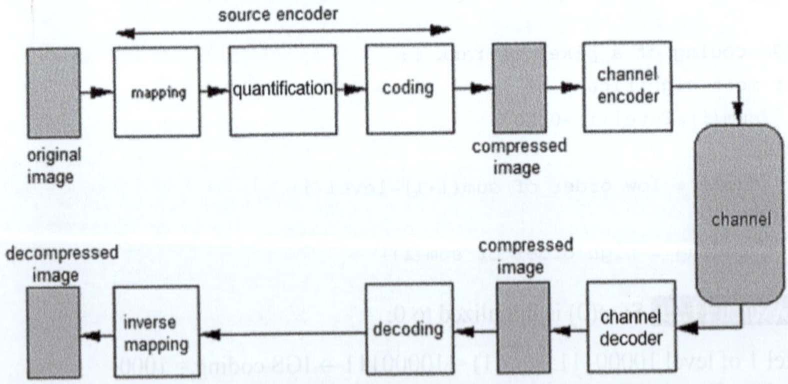

Figure 11.3. *Operation chain*

Between the encoding of the source information and the decoding for use, there is a channel that serves as the primary means of the compression operation. This channel could be a data transmission network or simply a storage device. Figure 11.3 shows the chain of operations involved in information compression and decompression.

The encoder consists of two parts, the source encoder and the channel encoder. The source encoder includes three stages:

1) *Mapping*: this stage transforms data into a format that reduces redundancy between pixels.

2) *Quantization*: this stage reduces psychovisual redundancy.

3) *Coding*: this stage precisely reduces coding-related redundancies.

Among these three operations, only quantization is irreversible, as it involves the loss of certain data.

The channel encoder is used when the transmission 262ode dis262a subject to noise. Its role is to encode data in a format that minimizes the errors introduced as the information passes through the channel, increasing immunity to noise. For data transmission channels, Hamming codes are the best-known and most widely used. These codes allow errors to be automatically detected and corrected.

At the other end of the chain, the channel decoder performs the reverse operation of the channel encoder. The source decoder then contains two steps:

1) *Decoding*: this step restores the redundancies introduced by coding.

2) *Reverse mapping*: this step restores redundancy between pixels.

Note that inverse quantization is meaningless, as the forward quantization operation is irreversible. Real compression systems do not necessarily require all the functional blocks shown in the previous figure. In fact, two categories of compression systems can be distinguished:

1) If $I'(x,y) = I(x,y)$, the compression system is ideal, meaning it allows lossless compression, sometimes referred to as compaction.

2) If $I'(x,y) \neq I(x,y)$, the compression system causes distortion, corresponding to lossy compression.

11.2. Lossless compression or compaction

Some applications, particularly in the medical field, require lossless compression. In such cases, the quantization operation is not used; only the reversible mapping and coding operations are applied.

There are several methods to achieve lossless compression, including variable-length coding, bit-plane coding and predictive coding.

11.2.1. *Variable-length coding*

The basic philosophy of this method is to reduce coding redundancy by assigning "long" codes for rare items and "short" codes for frequent items.

11.2.1.1. *Huffman coding*

Huffman coding is a member of the variable-length coding family. Its principle consists of three steps:

– Step 1: count the symbols of the information to be coded and rank them in the order of decreasing probability.

– Step 2: perform source reductions by summing the two lowest probabilities to create the probability of a "compound" symbol. The other probabilities remain unchanged, and the list is reordered in the order of decreasing probabilities.

This process continues in a dichotomous manner until only two probabilities remain.

– Step 3: coding is carried out by reversing the previous procedure.

Once the code has been created, a simple table defines the encoder and decoder.

EXAMPLE 11.3.– Figure 11.4(a) and (b) shows the three steps using a simple example: a message composed of six symbols with known probabilities of occurrence. These symbols are sorted by decreasing probability. Next, source reductions are performed, starting with the lowest probabilities. Finally, coding is performed.

symbol	prob.	step 1	step 2	step 3	step 4
m3	0.40	0.40	0.40	0.40	0.60
m1	0.30	0.30	0.30	0.30	0.40
m2	0.12	0.12	0.18	0.30	
m6	0.10	0.10	0.12		
m4	0.05	0.08			
m5	0.03				

Figure 11.4(a). *Source reduction*

symbol	prob.	coding	step 1		step 2		step 3		step 4	
m3	0.40	1	0.40	1	0.40	1	0.40	1	0.60	0
m1	0.31	00	0.31	00	0.31	00	0.31	00	0.40	1
m2	0.11	011	0.11	011	0.18	010	0.29	01		
m6	0.10	0100	0.10	0100	0.11	011				
m4	0.05	01010	0.08	0101						
m5	0.03	01011								

Figure 11.4(b). *Coding*

Note that the construction of Huffman's code can 264ode dis-intensive: 256 gray levels correspond to 254 source reductions and subsequently 254 code definitions.

The *m* most frequent symbols are coded using the Huffman method, which assigns variable-length codes based on symbol frequency. Less frequent symbols are coded using a fixed prefix (e.g. "10") followed by a binary number starting from 0. These simplifications (modifications) often reduce the complexity of the method while maintaining a level of compression without loss of information.

Other coding methods include B-code. This code consists of bits called "continuation bits" (noted as C) and information bits (noted as p), where p is a fixed value representing a binary number. For each group of p bits, there is a corresponding group of C continuation bits. For example, if $p = 2$, the letter C is followed by 2 bits. When sorting the symbols by decreasing probability, the first four symbols are coded as C00, C01, C10, C11. For the subsequent symbols, the coding expands as follows: C00C00, C00C01, C00C10, C00C11, C01C00, C01C01, etc. To obtain a binary coding, C is transformed into "0" or "1", but the bit changes with each new symbol.

EXAMPLE 11.4.– Figure 11.5 shows the resulting code for the previously processed example.

symbol	prob.	B2-code
m3	0.40	C00
m1	0.31	C01
m2	0.11	C10
m6	0.10	C11
m4	0.05	C00C00
m5	0.03	C00C01

Figure 11.5. *B2-code application*

m4m2m1→ C00C00 C10 C01

Let us choose C ="0" for the first symbol. Then, the sequence m4m2m1 becomes 000000110001.

11.2.1.2. *Block separation*

In this method, symbols are grouped into s groups. Each symbol is then encoded using its binary representation within the group, preceded by a group flag. Symbols are numbered in descending order based on their probabilities. A variation of this method encodes each symbol using the Huffman code corresponding to its group rather than its binary number.

11.2.1.3. *Arithmetic encoding*

Another widely used family of coding methods is arithmetic encoding. These encodings have the following characteristics:

1) There is no one-to-one correspondence between symbols and codes. This means that the same symbol can be represented by different code words, and vice versa.

2) A message is a sequence of source symbols, which correspond to a single code word.

3) The code word defines an interval of real numbers between 0 and 1. Each source symbol contributes to narrowing the interval, enabling the message to be identified more precisely.

4) As the number of symbols in a message increases, the interval becomes smaller and smaller, requiring more bits to describe it.

EXAMPLE 11.5.– Figure 11.6 explains the coding process.

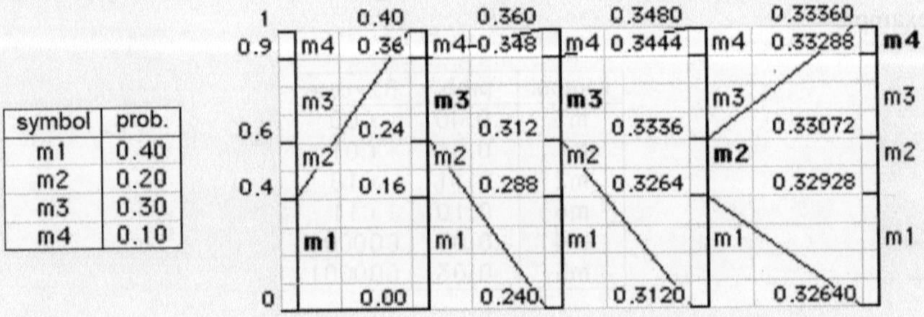

Figure 11.6. *Coding process*

The message encoding is m1m3m3m2m4.

The process does have its limitations, however, the use of an end character (here m4) limits both the message and the precision of the calculations. Specifically, any number in the interval [0.33288, 0.33360[represents the message.

11.2.2. *Bit-plane coding*

The aim here is to reduce interpixel redundancy. The principle of bit-plane coding involves decomposing a multi-level image into a series of binary images, and then compressing each binary image.

Coded on m bits, the level of a pixel is expressed as a binary number between 0 and $2^m - 1$. This level can therefore be expressed as a polynomial:

$$a_{m-1}2^{m-1} + a_{m-2}2^{m-2} + \cdots + a_1 2^1 + a_0 2^0 \quad (m \text{ levels}).$$

Each coefficient of the polynomial (0 or 1) corresponds to the presence of a black or white point in a level p plane (see Figure 11.7).

This allows the image to be decomposed into m distinct binary images. However, this method has one major drawback: even a slight change in the gray level can lead to changes in all the binary images in the decomposition. For example, the levels 127 = $(0111111)_2$ and 128 = $(10000000)_2$ are very close in terms of the gray value, but they have completely different binary representations.

This difficulty is solved by using a gray code, which for any code $a_{m-1}, a_{m-2}, \dots, a_1, a_0$ matches the code $g_{m-1}, g_{m-2}, \dots, g_1, g_0$ such that:

$$g_{m-1} = a_{m-1} \text{ and } g_i = \text{XOR}(a_I, a_{i+1}) \text{ for } 0 \leq i \leq m-2.$$

So, for 127 = $(0100000)_2$ in gray code and 128 = $(11000000)_2$ in gray code, only one bit changes.

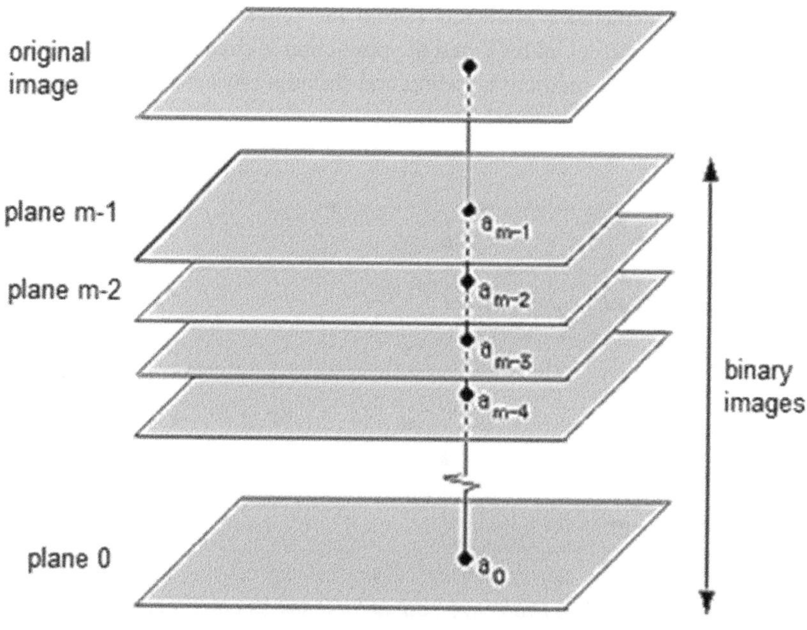

Figure 11.7. *Bit-plane coding*

The next step is to encode the binary images in a compressed form. The CAC (Constant Area Coding) method consists of dividing the binary image into pixel-sized blocks of $m \times n$ pixels. A block will be either all-black, all-white or mixed. The code 0 is assigned for the most frequent type, 10 and 11 for the other two. If the block is mixed, this code serves as a prefix, followed by the bit coding for the block's $(m \times n)$ bits.

Another popular method is RLC (run length coding), where each image line is divided into successive blocks of bits of the same level (called runs). A code is assigned to each block, often using a variable-length code.

Since these are binary images, it is sufficient to specify the value (black or white) of the first block of the line, or assume that each line begins with a "white" block of possibly zero length.

11.2.3. *Predictive coding*

Predictive coding is used to reduce interpixel redundancy: the level of the pixel to be coded is calculated ("predicted") from the previous n pixels, but only the difference between the actual level and the prediction is encoded. Figure 11.8 shows the block diagram of a predictive coding and decoding system, where the index n corresponds to the rank of a pixel.

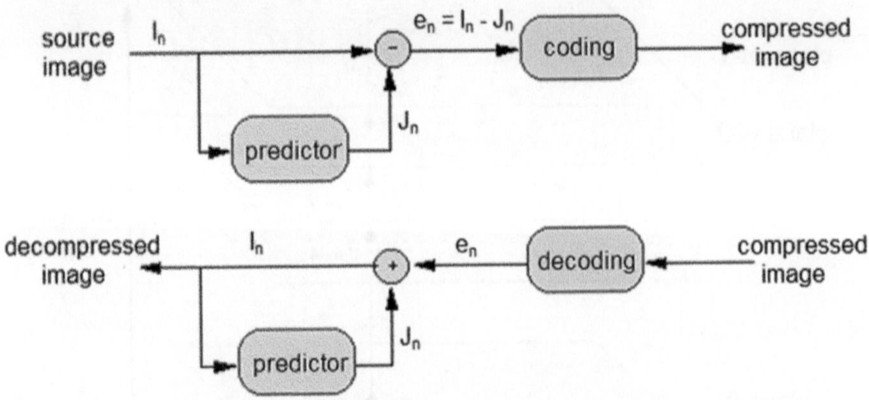

Figure 11.8. *Predictive coding for lossless compression*

The prediction $J(x, y)$ is usually given by an expression of the following form:

$$J(x,y) = \text{Int}\left(\sum_{i=1}^{m}\sum_{j=1}^{m} a_{ij} I(x-i, y-j)\right)$$

where the real coefficients a_{ij} are statistical weights assigned to the preceding pixels. A simple application of this formula is the predictive coding of a pixel from the pixel just above it:

$$J(x,y) = I(x,y-1)$$

It goes without saying that the first pixels must be absolute-coded.

11.3. Lossy compression

Compared to lossless compression methods, lossy compression methods include quantization. A distinction is made between predictive coding methods, which are extensions of the methods discussed techniques, and transform methods.

11.3.1. *Predictive coding*

Figure 11.9 shows the compression device, where the index n corresponds to the rank of a pixel. Note that quantization is applied to the difference between the actual level and the predicted level, and that this result is coded.

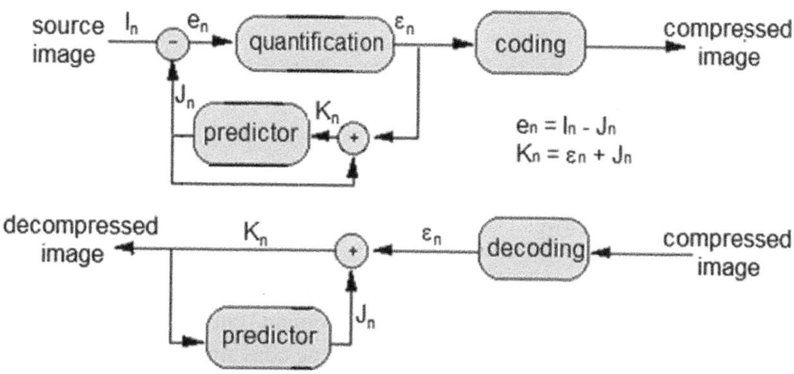

Figure 11.9. *Predictive coding for lossy compression*

An illustrative application of the process is delta modulation, defined by:

$$J_n = aK_{n-1} \quad \text{and} \quad \varepsilon_n = \begin{cases} +\xi & \text{if} \quad e_n > 0 \\ -\xi & \text{if} \quad e_n \leq 0 \end{cases}$$

If this algorithm is applied with $a = 1$ and $x = 3.5$ to the sequence of pixel levels 21, 25, 21, 24, 22, 22, the following steps occur:

– Pixel 0 : Level $I_0 = 21$; initialization at $K_0 = 21$

– Pixel 1 : Level $I_1 = 25$;

- $J_1 = K_0 = 21 \quad e_1 = I_1 - J_1 = 4 > 0 \quad e_1 = 3.5 \quad K_1 = J_1 + e_1 = 24.5$

– Pixel 2 : Level $I_2 = 21$;

– $J_2 = K_1 = 24.5 \quad e_2 = I_2 - J_2 = -3.5 < 0 \quad e_2 = -3.5 \quad K_2 = J_2 + e_2 = 21$

– Pixel 3 : Level $I_3 = 24$;

– $J_3 = K_2 = 21 \quad e_3 = I_3 - J_3 = 3 > 0 \quad e_3 = 3.5 \quad K_3 = J_3 + e_3 = 24.5$

– Pixel 4 : Level $I_4 = 22$;

– $J_4 = K_3 = 24.5 \quad e_4 = I_4 - J_{,} = -2.5 < 0 \quad e_4 = -3.5 \quad K_4 = J_4 + e_4 = 21$

– Pixel 5 : Level $I_5 = 22$;

– $J_5 = K_4 = 21 \quad e_5 = I_5 - J_5 = 1 > 0 \quad e_5 = 3.5 \quad K_5 = J_5 + e_5 = 24.5.$

The sequence 21 3.5 -3.5 3.5 -3.5 3.5 will then be sent to the encoder, reducing coding redundancy (which is certainly important here).

The predictors used above are fairly arbitrary; in reality, "optimal" predictors can be found that minimize the difference between the two e_n. This aspect will not be explored further.

11.3.2. *Transform coding*

Unlike predictive coding, which acts directly on the image geometry, transform coding converts the image from the spatial domain into a frequency domain. Specifically, the transformation applied to the image converts the set of pixels into a set of coefficients.

Figure 11.10 shows the chain of operations required to compress and decompress an image.

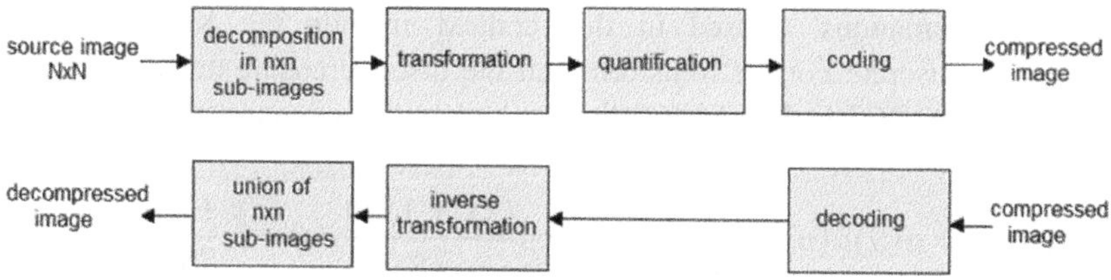

Figure 11.10. *Transform coding*

The image, assumed to be square and including $N \times N$ pixels, is divided into $(N/n)^2$ sub-images of size $n \times n$. A linear transformation is applied to each of these sub-images, providing a series of coefficients. The term $I(x,y)$ denotes a pixel in a sub-picture $n \times n$ and $J(u,v)$ represents its transform h. In this case, the following linear relationship holds:

$$I(x,y) = \sum_{u=0}^{n-1}\sum_{v=0}^{n-1} J(u,v)h(x,y,u,v) \quad \text{for } x,y = 0,1,2,\ldots,n-1$$

The set of pixels in the initial image corresponds to a square matrix $\boldsymbol{I}$ $(n \times n)$ of levels $I(x,y)$, which can be expressed as:

$$\boldsymbol{I} = \sum_{u=0}^{n-1}\sum_{v=0}^{n-1} J(u,v)\boldsymbol{H}_{uv}$$

where

$$\boldsymbol{H}_{uv} = \begin{bmatrix} h(0,0,u,v) & h(0,1,u,v) & \ldots & h(0,n-1,u,v) \\ h(1,0,u,v) & h(1,1,u,v) & \ldots & h(1,n-1,u,v) \\ \ldots & \ldots & \ldots & \ldots \\ h(n-1,0,u,v) & h(n-1,1,u,v) & \ldots & h(n-1,n-1,u,v) \end{bmatrix}$$

The initial image $\boldsymbol{I}$ can be viewed as a superposition of base images $\boldsymbol{H}_{uv}$. However, since the number of such images is generally large, it may be useful to approximate and remove those base images with minimal contributions. To achieve this, a mask $w(u,v)$ is introduced, which is equal to 0 if the coefficient $J(u,v)$ satisfies a truncation criterion, and 1 otherwise. The resulting approximation is given by:

$$I^* = \sum_{u=0}^{n-1}\sum_{v=0}^{n-1} J(u,v)w(u,v)H_{uv}$$

The transformations h used in this context include the Karhunen–Loeve transform, the discrete Fourier transform and the discrete cosine transform (DCT). Among these, the DCT is defined as follows, along with its inverse transform:

$$J(u,v) = \alpha(u)\alpha(v)\sum_{x=0}^{n-1}\sum_{y=0}^{n-1} I(x,y)\cos\frac{(2x+1)\pi u}{2n}\cos\frac{(2y+1)\pi v}{2n}$$

$$I(x,y) = \sum_{u=0}^{n-1}\sum_{v=0}^{n-1} \alpha(u)\alpha(v)J(u,v)\cos\frac{(2x+1)\pi u}{2n}\cos\frac{(2y+1)\pi v}{2n}$$

$$\text{where} \quad \alpha(u) = \sqrt{\frac{1}{n}} \quad \text{for} \quad u = 0, \quad \text{and} \quad \alpha(u) = \sqrt{\frac{2}{n}} \quad \text{for} \quad u \neq 0$$

The truncation criterion arises from two main methods: the zonal coding method, which examines the variances $\sigma^2(J(u,v))$, and the threshold coding method, which uses a coefficient magnitude threshold $J(u,v)$. A concrete example of the application of the latter method will be discussed later in the context of Joint Photographic Expert Group (JPEG) compression.

11.4. An image compression standard: JPEG

Image compression methods are standardized and follow a clearly defined and known process. Currently, the two main standards are JPEG for still images and MPEG for moving images. This discussion will focus specifically on JPEG.

The JPEG involves a sequence of three operations: the application of DCT transform, then quantization, and finally coding. The decoder performs these operations in the reverse order. Integrated circuits are available that can automatically perform the encoding and decoding processes.

To understand the JPEG compression process, consider the following example. an image is first divided into sub-images consisting of blocks of 8×8 pixels. Figure 11.11(a) shows one such sub-image $\boldsymbol{I}$, where each pixel is represented by a gray level ranging from 0 to 255. Each level is then decreased by $2^7 = 128$, producing a new image denoted as $\boldsymbol{I'}$ (see Figure 11.11(b)). Subsequently, a DCT transformation is applied to, resulting in a coefficient table $\boldsymbol{J}$ (see Figure 11.11(c)). All the images generated during this process are shown in Figure 11.11.

Quantization or truncation is performed using the threshold coding method as follows:

$$J^*(u,v) = \text{Int}\left(\frac{J(u,v)}{Z(u,v)}\right)$$

where $Z(u,v)$ is a normalization matrix. The matrix $\boldsymbol{Z}$ is shown in Figure 11.11(d).

The resulting quantized coefficients are shown in Figure 11.11(e), obtained from a zig-zag scanning of the quantized matrix J^* (see Figure 11.11(f)). This process yields the following sequence:

-25 -2 0 -3 -2 -5 2 -4 1 -4 1 1 5 0 2 0 0 -1 2 0 0 0 0 -1 -1 EOB

where EOB is the end-of-block character.

60	59	59	67	67	63	69	72
71	63	64	91	106	87	74	71
70	63	66	114	141	106	71	72
71	62	69	123	151	108	75	68
75	65	66	105	123	90	73	69
87	69	58	71	74	70	63	74
93	75	62	60	52	63	70	82
95	83	67	69	62	78	83	93

a) sub-image I

-128	-114	-111	-106	-99	-77	-41	-48
-128	-110	-106	-91	-72	-60	-19	-25
-128	-104	-93	-73	-64	-47	-24	-15
-128	-79	-64	-50	-41	-25	-7	-8
-128	-56	-36	-33	-30	-16	-28	-25
-128	-128	-128	-128	-128	-128	-128	-128
-128	-128	-128	-128	-128	-128	-128	-128
-128	-128	-128	-128	-128	-128	-128	-128

b) sub-image I'

-400	-20	-46	38	50	-8	-8	-5
6	-21	-62	8	11	-7	-7	6
-47	7	76	-26	-30	10	6	-5
-49	11	34	-15	-10	6	1	1
10	-8	-13	-3	-1	1	-5	1
-10	1	3	-4	-1	0	1	-1
-3	-2	1	0	0	-4	1	-3
-2	-1	-1	-3	-1	-1	-1	0

c) table J

16	11	10	16	24	40	51	61
12	12	14	19	26	58	60	55
14	13	16	24	40	57	69	56
14	17	22	29	51	87	80	62
18	22	37	56	68	109	103	77
24	35	55	64	81	104	113	92
49	64	78	87	103	121	120	101
72	92	95	98	112	100	103	99

d) matrix Z

-25	-2	-5	2	2	0	0	0
0	-2	-4	0	0	0	0	0
-3	1	5	-1	-1	0	0	0
-4	1	2	-1	0	0	0	0
1	0	0	0	0	0	0	0
0	0	0	0	0	0	0	0
0	0	0	0	0	0	0	0
0	0	0	0	0	0	0	0

e) table J*

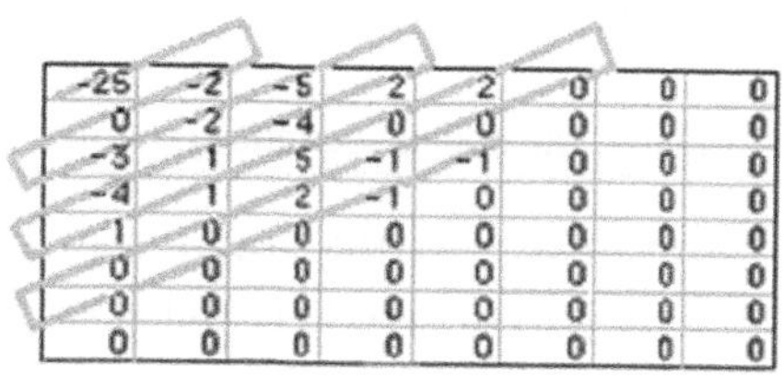

f) zig-zag path on table J*

Figure 11.11. *Initial coding process. For a color version of this figure, see www.iste.co.uk/cochard/mathematics2.zip*

The sequence of operations in the JPEG compression process is more complex, involving special coding for the most important coefficient, known as the DC coefficient, followed by a separate coding scheme for the other coefficients, referred to as AC coefficients (see Figure 11.12).

1) To calculate the difference between the DC coefficient J^* (the first coefficient) and that of the previously processed sub-image (e.g. -15), the result -10 is used to identify the appropriate category from the JPEG category table (a). In this case, the relevant category is 4. Referring to the JPEG luminance table (b), the base code is found to be 101, resulting in a total code length of 7 bits. Consequently, the 3-bit base code must be supplemented with 4 additional bits.

According to the following rule: if the difference $DC_l - DC_{l-1}$ is positive, the K least significant bits of DC_l are used (where K is the category determined earlier). Conversely, if the difference $DC_l - DC_{l-1}$ is negative, the K least significant bits of DC_l are taken, and 1 is subtracted. In this case, since the second category is involved, the calculation yields $-11 = (0101)_2$. Thus, the final DC code becomes 1010101.

2) For coding the AC coefficients, tables (a) and (c) are used, with table (c) being partially displayed. This table considers the number of "0" values preceding a non-zero AC coefficient. For each AC coefficient, its category, base code and length can be determined from the tables. The additional bits are derived based on the following rule: if the coefficient is positive, the least significant bits are taken; if the AC coefficient is negative, 1 is subtracted, and the least significant bits are taken.

For example:

– For -2, the category is 2, the base code is 01 and the length is 4, resulting in the code 0101.

– For -3, the category is 2, the base code is 111001 and the length is 8, leading to the code 11100100, etc.

Finally, the coded sequence obtained through the JPEG compression process is:

1010101 0101 11100100 0101 1000010 0110 100011 001 100011 001 001 100101 11100110 110110 0110 11110100 000 1010

To calculate the resulting compression gain, consider that the sub-image sized 8×8 was coded using 256 levels, which means it represents an uncompressed volume of:

$$8 \times 8 \times 8 = 512 \text{ bits}$$

After compression, the number of bits in the coded sub-image is 94 bits. The compression ratio can therefore be determined by the formula:

$$\text{Compression ratio} = \frac{\text{Original size}}{\text{Compressed size}} = \frac{512}{94} \approx 5.45$$

interval	category DC	category AC
0	0	N/A
-1,1	1	1
-3, -2, 2, 3	2	2
-7, ..., -4, 4, ..., 7	3	3
-15, ..., -8, 8, ..., 15	4	4
-31, ..., -16, 16, ..., 31	5	5
-63, ..., -32, 32, ..., 63	6	6
127, ..., -64, 64, ..., 127	7	7
-255, ..., -128, 128, ..., 255	8	8
-511, ..., -256, 256, ..., 511	9	9
-1023, ..., -512, 512, ..., 1023	1	A
-2047, ..., -1024, 1024, ..., 2047	B	B
-4095, ..., -2048, 2048, ..., 4095	C	C
-8191, ..., -4096, 4096, ..., 8191	D	D
-16383, ..., -8192, 8192, ..., 16383	E	E
-32767, ..., -16384, 16384, ..., 32767	F	N/A

a) categories table

category	basic code	length
0	10	3
1	11	4
2	100	5
3	0	5
4	101	7
5	110	8
6	1110	10
7	11110	12
8	111110	14
9	1111110	16
A	11111110	18
B	111111110	20

b) coefficients DC table

prec. zeros	category	basic code	length
0	0	1010 (EOBà	4
0	1	0	3
0	2	1	4
0	3	100	6
0	4	1011	8
0	5	11010	10
0	6	111000	12
0	7	1111000	14
0	8	1111110110	18
0	9	1111111110000010	25
0	A	1111111110000010	26
1	1	1100	5
1	2	111001	8
1	3	1111001	10
1	4	111110110	13
2	1	11011	6
5	1	1111010	6

c) coefficients AC (partial) table

Figure 11.12. *Coding tables*

12

Elements of Numerical Analysis

CONCEPTS COVERED IN THIS CHAPTER.–

Solving a mathematical problem is not always possible, either because no exact method of resolution is known, or because calculating the solution requires considerable resources and time. This chapter reviews some of the methods used to solve problems numerically.

It begins with solving a linear system and introducing iterative methods.

The next part explores the possibilities for numerically solving an equation $f(x) = 0$: iterative methods, secant methods, tangent methods and an overview of the numerical solution of polynomial equations.

The third section concerns numerical integration, describing various methods such as the trapezoidal, rectangular, Simpson and Poncelet methods. By applying the polynomial interpolation technique and using Lagrange polynomials, it becomes possible to calculate numerically defined integrals. This section concludes with the Monte Carlo method and point throws.

The fourth section is devoted to differential equations and their numerical solution using the Euler, parabola and Runge–Kutta algorithms, including an overview of the linked-step method.

The final section focuses on partial differential equations, which can be solved by discretization using the finite difference method to approximate first and second partial derivatives.

References: [CHA 75, DUR 60].

Mathematics for Digital Science 2,
by Gérard-Michel COCHARD and Mhand HIFI.

12.1. Numerical solution of a linear system

12.1.1. *Exact solution of a linear Cramerian system*

Let the system of n linear equations with n unknowns be solved:

$$\begin{cases} a_{11}x_1 + a_{12}x_2 + \cdots + a_{1n}x_n = b_1 \\ a_{21}x_1 + a_{22}x_2 + \cdots + a_{2n}x_n = b_2 \\ \vdots \\ a_{n1}x_1 + a_{n2}x_2 + \cdots + a_{nn}x_n = b_n \end{cases}$$

This system can be written in matrix form as $\boldsymbol{Ax} = \boldsymbol{b}$, where $\boldsymbol{A}$ is the matrix of coefficients a_{ij}, and $\boldsymbol{x}$ and $\boldsymbol{b}$ are column matrices with n rows.

If the determinant of $\boldsymbol{A}$ is non-zero, the system is a ***Cramer system***. In this case, $\boldsymbol{A}$ has an inverse $\boldsymbol{A}^{-1}$. Multiplying the previous matrix equation on the left by $\boldsymbol{A}^{-1}$, the ***unique*** solution of the system is obtained:

$$\boldsymbol{x} = \boldsymbol{A}^{-1}\boldsymbol{b}$$

which gives:

$$x_i = \sum_{j=1}^{n} (A^{-1})_{ij} b_j$$

Thus, solving the linear system reduces to matrix inversion.

The elements of the matrix A^{-1} are expressed by the following relation:

$$(\boldsymbol{A}^{-1})_{ij} = \frac{\operatorname{cof}(a_{ij})}{\det(\boldsymbol{A})}$$

where $cof(a_{ij})$ is the cofactor of element a_{ij} in the matrix $\boldsymbol{A}$, obtained by removing row i and column j, and multiplied by $(-1)^{i+j}$.

Therefore,

$$x_i = \frac{\sum_{j=1}^{n} \operatorname{cof}(a_{ij}) b_j}{\det(\boldsymbol{A})}$$

However,

$$\begin{vmatrix} a_{11} & \cdots & a_{1,i-1} & b_1 & a_{1,i+1} & \cdots & a_{1n} \\ a_{21} & \cdots & a_{2,i-1} & b_2 & a_{2,i+1} & \cdots & a_{2n} \\ \vdots & \vdots & \vdots & \vdots & \vdots & \ddots & \vdots \\ a_{n1} & \cdots & a_{n,i-1} & b_n & a_{n,i+1} & \cdots & a_{nn} \end{vmatrix} = \sum_{j=1}^{n} \operatorname{cof}(a_{ij}) b_j = \Delta_i$$

Finally, Cramer's formulas are obtained:

$$\mathrm{x_i} = \frac{\Delta_\mathrm{i}}{\mathrm{de}}$$

where Δ_i is obtained from det(**A**) by replacing the ith column with **b**.

EXAMPLE 12.1.– Solve the following system with parameters a and b (a not equal to b):

$$\begin{cases} x + ay + a^2z + a^3u = a^4 \\ x + by + b^2z + b^3u = b^4 \\ \quad y + 2az + 3a^2u = 4a^3 \\ \quad y + 2bz + 3b^2u = 4b^3 \end{cases}$$

Successively,

$$\det(A) = \begin{vmatrix} 1 & a & a^2 & a^3 \\ 1 & b & b^2 & b^3 \\ 0 & 1 & 2a & 3a^2 \\ 0 & 1 & 2b & 3b^2 \end{vmatrix} = (a-b)^4 \neq 0$$

Hence, the system is a Cramer system. The determinants for each variable are:

$$\Delta_x = \begin{vmatrix} a^4 & a & a^2 & a^3 \\ b^4 & b & b^2 & b^3 \\ 4a^3 & 1 & 2a & 3a^2 \\ 4b^3 & 1 & 2b & 3b^2 \end{vmatrix} = a^2b^2(a-b)^4$$

$$\Delta_y = \begin{vmatrix} 1 & a^4 & a^2 & a^3 \\ 1 & b^4 & b^2 & b^3 \\ 0 & 4a^3 & 2a & 3a^2 \\ 0 & 4b^3 & 2b & 3b^2 \end{vmatrix} = -2ab(a+b)(a-b)^4$$

$$\Delta_z = \begin{vmatrix} 1 & a & a^4 & a^3 \\ 1 & b & b^4 & b^3 \\ 0 & 1 & 4a^3 & 3a^2 \\ 0 & 1 & 4b^3 & 3b^2 \end{vmatrix} = (a^2 + 4ab + b^2)(a-b)^4$$

$$\Delta_u = \begin{vmatrix} 1 & a & a^2 & a^4 \\ 1 & b & b^2 & b^4 \\ 0 & 1 & 2a & 4a^3 \\ 0 & 1 & 2b & 4b^3 \end{vmatrix} = -2(a+b)(a-b)^4$$

Thus, the solutions are:

$$x = -a^2b^2, \; y = 2ab(a+b), \; z = -(a^2+4ab+b^2), \; u = 2(a+b)$$

Solving a linear Cramerian system using Cramer's formulas requires a large number of operations when there are many unknowns. For this reason, it is often preferable to use more efficient methods, such as:

– iterative methods (well-suited to computer calculations);

– direct methods.

In the following, assume that the system is Cramerian, meaning that $\det(\boldsymbol{A}) \neq 0$, ensuring a *unique* solution.

12.1.2. Principle of iterative methods

Let the matrix $\boldsymbol{A}$ be expressed as $\boldsymbol{A} = \boldsymbol{B} - \boldsymbol{C}$, where $\boldsymbol{B}$ and $\boldsymbol{C}$ are two $n \times n$ matrices satisfying a specific condition. Then,

$$\boldsymbol{Ax} = \boldsymbol{b} \Rightarrow \boldsymbol{Bx} = \boldsymbol{Cx} = \boldsymbol{b}$$

Let $\boldsymbol{x_0}$ be an arbitrary initial vector (column matrix with n rows), and calculate successive approximations as follows:

$\boldsymbol{x_1}$ such that $\boldsymbol{Bx_1} = \boldsymbol{Cx_0} + \boldsymbol{b}$

$\boldsymbol{x_2}$ such that $\boldsymbol{Bx_2} = \boldsymbol{Cx_1} + \boldsymbol{b}$

$\vdots$

$\boldsymbol{x_p}$ such that $\boldsymbol{Bx_p} = \boldsymbol{Cx_{p-1}} + \boldsymbol{b}$

This iterative procedure solves the system if the following conditions are met:

1) $\boldsymbol{B}$ is regular, ensuring that $\boldsymbol{x} = (\boldsymbol{x_1}, \boldsymbol{x_2}, \dots, \boldsymbol{x_p})$ is unique and computable.

2) $\boldsymbol{Bx_i} = \boldsymbol{z}$ is easy to solve.

3) $\boldsymbol{x}_i$ converges to the solution $\boldsymbol{x}$ as p increases.

The third condition is the most challenging. Assuming (1) and (2) are satisfied, express the iteration as:

$$\boldsymbol{B}\boldsymbol{x}_p = \boldsymbol{C}\boldsymbol{x}_{p-1} + \boldsymbol{b} \qquad \Rightarrow \quad \boldsymbol{x}_p = \boldsymbol{B}^{-1}\boldsymbol{C}\boldsymbol{x}_{p-1} + \boldsymbol{B}^{-1}\boldsymbol{b} = \boldsymbol{\Omega}\boldsymbol{x}_{p-1} + \boldsymbol{\omega}$$

where $\boldsymbol{\Omega} = \boldsymbol{B}^{-1}\boldsymbol{C}$ and $\boldsymbol{\omega} = \boldsymbol{B}^{-1}\boldsymbol{b}$.

Between two successive iterations:

$$\boldsymbol{\Delta}_p = \boldsymbol{x}_{p+1} - \boldsymbol{x}_p = \boldsymbol{\Omega}\boldsymbol{x}_p + \boldsymbol{\omega} - \boldsymbol{\Omega}\boldsymbol{x}_{p-1} - \boldsymbol{\omega} = \boldsymbol{\Omega}(\boldsymbol{x}_p p - \boldsymbol{x}_{p-1}) = \boldsymbol{\Omega}\boldsymbol{\Delta}_{p-1}$$

which, by recurrence, leads to $\boldsymbol{\Delta}_p = \boldsymbol{\Omega}^p\boldsymbol{x}_0$. Thus,

$$\boldsymbol{x}_{p+1} - \boldsymbol{x}_0 = (\boldsymbol{x}_{p+1} - \boldsymbol{x}_p) + (\boldsymbol{x}_p - \boldsymbol{x}_{p-1}) + \cdots + (\boldsymbol{x}_1 - \boldsymbol{x}_0)$$

$$= \boldsymbol{\Delta}_p + \boldsymbol{\Delta}_{p-1} + \cdots + \boldsymbol{\Delta}_0 = (\boldsymbol{\Omega}_p + \boldsymbol{\Omega}_{p-1} + \cdots + \mathbf{1})\boldsymbol{\Delta}_0$$

where $\mathbf{1}$ is the identity matrix. This brings up the convergence of the matrix series $\mathbf{1} + \boldsymbol{\Omega} + \boldsymbol{\Omega}^2 + \cdots + \boldsymbol{\Omega}^{\mathrm{p}}$. This series converges if all the eigenvalues of $\boldsymbol{\Omega}$ have a modulus less than 1, i.e. $|\lambda_i| < 1\ \forall i$, or equivalently

$$\rho(\boldsymbol{\Omega}) = \text{spectral radius of} \quad \boldsymbol{\Omega} = \max_i |\lambda_i| < 1$$

Under this condition:

$$\mathbf{1} + \boldsymbol{\Omega} + \boldsymbol{\Omega}^2 + \ldots + \boldsymbol{\Omega}^{\mathrm{p}} + \ldots = (\mathbf{1} - \boldsymbol{\Omega})^{-1}$$

In this case, $\lim_{p\to\infty}[\boldsymbol{x}_p] = \widehat{\boldsymbol{x}}$. Now, $\widehat{\boldsymbol{x}}$ must be shown to be the solution of the system, and therefore $\widehat{\boldsymbol{x}} = \boldsymbol{x}$. Indeed,

$$\widehat{\boldsymbol{x}} = \boldsymbol{x}_0 + \cdots \lim_{p\to\infty}(\mathbf{1} + \boldsymbol{\Omega} + \boldsymbol{\Omega}^2 + \cdots .. + \boldsymbol{\Omega}^{\mathrm{p}})\boldsymbol{\Delta}_0$$

$$= \boldsymbol{x}_0 + (\mathbf{1} - \boldsymbol{\Omega})^{-1}\boldsymbol{\Delta}_0$$

$$= \boldsymbol{x}_0 + (\mathbf{1} - \boldsymbol{\Omega})^{-1}(\boldsymbol{x}_1 - \boldsymbol{x}_0)$$

$$= \boldsymbol{x}_0 + (\mathbf{1} - \boldsymbol{\Omega})^{-1(}\boldsymbol{\Omega}\boldsymbol{x}_0 + \boldsymbol{\omega} - \boldsymbol{x}_0)$$

$$= \boldsymbol{x}_0 + (\mathbf{1} - \boldsymbol{\Omega})^{-1}(\boldsymbol{\Omega} - \mathbf{1})\boldsymbol{x}_0 + (\mathbf{1} - \boldsymbol{\Omega})^{-1}\boldsymbol{\omega}$$

$$= \boldsymbol{x}_0 - (\mathbf{1} - \boldsymbol{\Omega})^{-1}(\boldsymbol{\Omega} - \mathbf{1})\boldsymbol{x}_0 + (\mathbf{1} - \boldsymbol{\Omega})^{-1}\boldsymbol{\omega}$$

$$= \boldsymbol{x}_0 - \boldsymbol{x}_0 + (\mathbf{1} - \boldsymbol{B}^{-1}\boldsymbol{C})^{-1}\boldsymbol{B}^{-1}\boldsymbol{b}$$

$$= [\boldsymbol{B}(\mathbf{1} - \boldsymbol{B}^{-1}\boldsymbol{C})]^{-1}\boldsymbol{b}$$

$$= (\boldsymbol{B} - \boldsymbol{C})^{-1}\boldsymbol{b}$$

$$= \boldsymbol{A}^{-1}\boldsymbol{b}$$

proving that $\hat{\boldsymbol{x}} = \boldsymbol{x}$. Therefore, condition 3 can be expressed as follows.

3') $\rho(\boldsymbol{B}^{-1}\boldsymbol{C}) < 1$.

However, this condition is often challenging to verify directly. In practice, sufficient (though not necessary) conditions are commonly used:

– Strong condition:

$$\forall i, \quad |A_{ii}| > \sum_{j \neq i} |A_{ij}|$$

– Weak condition:

$$\forall i, \ |\boldsymbol{A}_{ii}| \geq \sum_{j \neq i} |\boldsymbol{A}_{ij}|$$

(note the symbol ≥ instead of > in the strong condition), and there exists at least one row r such that $|\boldsymbol{A}_{rr}| > \sum_{j \neq r} |A_{rj}|$.

These conditions ensure the convergence of $\boldsymbol{x}_p$ to $\boldsymbol{x}$.

12.1.3. *Diagonal iteration and Gauss–Seidel iteration*

12.1.3.1. *Diagonal iteration*

The matrix $\boldsymbol{A}$ can be decomposed into the sum of three matrices $\boldsymbol{D} + \boldsymbol{J} + \boldsymbol{K}$, where $\boldsymbol{D}$ is a diagonal matrix, $\boldsymbol{J}$ and $\boldsymbol{K}$ are strictly triangular matrices (without diagonal elements).

The decomposition $\boldsymbol{A}$ can be visualized as:

$$\boldsymbol{A} = \begin{bmatrix} \ddots & \ddots & K \\ \ddots & D & \ddots \\ J & \ddots & \ddots \end{bmatrix}$$

We can choose $\boldsymbol{B} = \boldsymbol{D}$ and $\boldsymbol{C} = -\boldsymbol{J} - \boldsymbol{K}$.

Condition 1 is easily satisfied; otherwise, we can swap the system's rows to ensure $\boldsymbol{D}$ is regular. The iteration then becomes:

$$\boldsymbol{D}\boldsymbol{x}_p = -(\boldsymbol{J} + \boldsymbol{K})\boldsymbol{x}_{p-1} + \boldsymbol{b}$$

12.1.3.2. *Gauss–Seidel iteration*

The matrix $\boldsymbol{B}$ is chosen as $\boldsymbol{D} + \boldsymbol{J}$ and $\boldsymbol{C} = -\boldsymbol{K}$. Thus, $\boldsymbol{B}$ is triangular rather than diagonal. The iteration is expressed as:

$$(\boldsymbol{D} + \boldsymbol{J})\boldsymbol{x}_p = -\boldsymbol{K}\boldsymbol{x}_{p-1} + \boldsymbol{b}$$

This method is widely used, with convergence typically faster than diagonal iteration.

12.1.4. *Direct methods*

These methods involve transforming the matrix $\boldsymbol{A}$ (and the system) into a simpler matrix, such as a triangular or diagonal matrix.

12.1.4.1. *Triangulation method (Gauss)*

The goal is to transform $\boldsymbol{A}\boldsymbol{x} = \boldsymbol{b}$ into $\boldsymbol{T}\boldsymbol{x} = \boldsymbol{b}$, where $\boldsymbol{T}$ is a triangular matrix:

$$\boldsymbol{T} = \begin{bmatrix} T_{11} & 0 & \dots & 0 \\ T_{21} & T_{22} & \cdots & 0 \\ \vdots & \vdots & \ddots & \vdots \\ T_{n1} & T_{n2} & \cdots & T_{nn} \end{bmatrix}$$

The solution to the system is then immediate:

$$x_1 = \frac{\beta_1}{T_{11}}, \quad x_2 = \frac{\beta_2 - T_{21}x_1}{T_{22}}, \quad \dots, \quad x_n = \frac{\beta_n - \sum_{j=1}^{n-1} T_{nj}x_j}{T_{nn}}$$

To achieve the desired transformation, the following rule is applied: the linear system remains unchanged when row L_p of matrix $\boldsymbol{A}$ is replaced by a linear combination of this row and the other rows:

$$L_p \longrightarrow k_p L_p + \sum_{j \neq p} k_j L_j \quad (\text{with} \quad k_p \neq 0)$$

This is known as the *Gauss pivot method.*

The process is described using a 3 × 3 matrix. The initial system is

$$\begin{bmatrix} A_{11} & A_{12} & A_{13} \\ A_{21} & A_{22} & A_{23} \\ A_{31} & A_{32} & A_{33} \end{bmatrix} \begin{bmatrix} x_1 \\ x_2 \\ x_3 \end{bmatrix} = \begin{bmatrix} b_1 \\ b_2 \\ b_3 \end{bmatrix}$$

1st step: choose the diagonal element A_{33} (assumed to be non-zero) as the pivot:

Divide row L_3 by A_{33}

$$\begin{bmatrix} A_{11} & A_{12} & A_{13} \\ A_{21} & A_{22} & A_{23} \\ A'_{31} & A'_{32} & 1 \end{bmatrix} \begin{bmatrix} x_1 \\ x_2 \\ x_3 \end{bmatrix} = \begin{bmatrix} b_1 \\ b_2 \\ b'_3 \end{bmatrix} \quad \text{with} \quad A'_{31} = \frac{A_{31}}{A_{33}}, \quad A'_{32} = \frac{A_{32}}{A_{33}}, \quad b'_3 = \frac{b_3}{A_{33}}.$$

Next, perform replacements on the other rows $L_k \rightarrow L_k - A_{lk3} L_3$:

$$\begin{bmatrix} A'_{11} & A'_{12} & 0 \\ A'_{21} & A'_{22} & 0 \\ A'_{31} & A'_{32} & 1 \end{bmatrix} \begin{bmatrix} x_1 \\ x_2 \\ x_3 \end{bmatrix} = \begin{bmatrix} b'_1 \\ b'_2 \\ b'_3 \end{bmatrix} \quad \text{with} \quad \begin{cases} A'_{k1} = A_{k1} - A_{k3} A'_{31} & \text{etc.} \\ b'_k = b_k - A_{k3} b'_3 & \text{etc.} \end{cases}$$

2nd step: choose the diagonal element A'_{22} (if non-zero) as the pivot.

Divide row L_2 by A'_{22}

$$\begin{bmatrix} A'_{11} & A'_{12} & 0 \\ A''_{21} & 1 & 0 \\ A'_{31} & A'_{32} & 1 \end{bmatrix} \begin{bmatrix} x_1 \\ x_2 \\ x_3 \end{bmatrix} = \begin{bmatrix} b'_1 \\ b''_2 \\ b'_3 \end{bmatrix}$$

Perform replacement of row $L_1 \rightarrow L_1 - A'_{k2} L_2$:

$$\begin{bmatrix} A''_{11} & 0 & 0 \\ A''_{21} & 1 & 0 \\ A'_{31} & A'_{32} & 1 \end{bmatrix} \begin{bmatrix} x_1 \\ x_2 \\ x_3 \end{bmatrix} = \begin{bmatrix} b''_1 \\ b''_2 \\ b'_3 \end{bmatrix}$$

3rd step: divide row L_1 by the pivot A''_{11} (if non-zero):

$$\begin{bmatrix} 1 & 0 & 0 \\ A''_{21} & 1 & 0 \\ A'_{31} & A'_{32} & 1 \end{bmatrix} \begin{bmatrix} x_1 \\ x_2 \\ x_3 \end{bmatrix} = \begin{bmatrix} b'''_1 \\ b''_2 \\ b'_3 \end{bmatrix}$$

The solution is:

$$x_1 = b'''_1, \quad x_2 = b''_2 - A''_{21}x_1, \qquad x_3 = b'_3 - A'_{21}x_1 - A'_{32}x_2$$

This method is applicable as long as the pivots are non-zero. If a null pivot is encountered, the offending row L_k is exchanged with another row $L_j\,(j < k)$ to change the diagonal element. The same procedure is followed if the pivot is close to zero, as division by a small pivot may lead to inaccuracies, particularly when using a computer.

12.1.4.2. *Diagonalization method*

This method is very similar to the previous one, involving the replacement of the initial matrix by a diagonal matrix: $\boldsymbol{Ax} = \boldsymbol{b} \longrightarrow \boldsymbol{Dx} = \boldsymbol{\beta}$, where

$$\boldsymbol{D} = \begin{bmatrix} D_{11} & 0 & 0 \\ 0 & \ddots & 0 \\ 0 & 0 & D_{nn} \end{bmatrix}$$

The solution is immediate:

$$x_1 = \frac{\beta_1}{D_{11}}, \qquad x_2 = \frac{\beta_2}{D_{22}}, \qquad \cdots, \; x_n = \frac{\beta_n}{D_n}.$$

To illustrate, the procedure is applied to a Cramer system of three equations with three variables (unknowns):

$$\begin{bmatrix} A_{11} & A_{12} & A_{13} \\ A_{21} & A_{22} & A_{23} \\ A_{31} & A_{32} & A_{33} \end{bmatrix} \begin{bmatrix} x_1 \\ x_2 \\ x_3 \end{bmatrix} = \begin{bmatrix} b_1 \\ b_2 \\ b_3 \end{bmatrix}$$

1st step: the element A_{11} is chosen as the pivot (if non-zero).

Divide row L_1 by A_{11}:

$$\begin{bmatrix} 1 & A'_{12} & A'_{13} \\ A_{21} & A_{22} & A_{23} \\ A_{31} & A_{32} & A_{33} \end{bmatrix} \begin{bmatrix} x_1 \\ x_2 \\ x_3 \end{bmatrix} = \begin{bmatrix} b'_1 \\ b_2 \\ b_3 \end{bmatrix}$$

Perform the replacement of rows $L_k \rightarrow L_k - A_{k1}L_1 (k \neq 1)$:

$$\begin{bmatrix} 1 & A'_{12} & A'_{13} \\ 0 & A'_{22} & A'_{23} \\ 0 & A'_{32} & A'_{33} \end{bmatrix} \begin{bmatrix} x_1 \\ x_2 \\ x_3 \end{bmatrix} = \begin{bmatrix} b'_1 \\ b'_2 \\ b'_3 \end{bmatrix}$$

2nd step: the element A'_{22} (if non-zero) is chosen as the pivot.

Divide row L_2 by A'_{22}:

$$\begin{bmatrix} 1 & A'_{12} & A'_{13} \\ 0 & 1 & A''_{23} \\ 0 & A'_{32} & A'_{33} \end{bmatrix} \begin{bmatrix} x_1 \\ x_2 \\ x_3 \end{bmatrix} = \begin{bmatrix} b'_1 \\ b''_2 \\ b'_3 \end{bmatrix}$$

Perform the replacement of rows $L_k \rightarrow L_k - A'_{k2}L_2 (k \neq 2)$:

$$\begin{bmatrix} 1 & 0 & A''_{13} \\ 0 & 1 & A''_{23} \\ 0 & 0 & A''_{33} \end{bmatrix} \begin{bmatrix} x_1 \\ x_2 \\ x_3 \end{bmatrix} = \begin{bmatrix} b''_1 \\ b''_2 \\ b''_3 \end{bmatrix}$$

3rd step: the element A''_{33} (if non-zero) is chosen as the pivot.

Divide row L_3 by A''_{33}:

$$\begin{bmatrix} 1 & 0 & A''_{13} \\ 0 & 1 & A''_{23} \\ 0 & 0 & 1 \end{bmatrix} \begin{bmatrix} x_1 \\ x_2 \\ x_3 \end{bmatrix} = \begin{bmatrix} b''_1 \\ b''_2 \\ b'''_3 \end{bmatrix}$$

Perform the replacement of rows $L_k \rightarrow L_k - A''_{k3}L_3 (k \neq 3)$:

$$\begin{bmatrix} 1 & 0 & 0 \\ 0 & 1 & 0 \\ 0 & 0 & 1 \end{bmatrix} \begin{bmatrix} x_1 \\ x_2 \\ x_3 \end{bmatrix} = \begin{bmatrix} b'''_1 \\ b'''_2 \\ b'''_3 \end{bmatrix}$$

The immediate solution is: $x_1 = b'''_1$, $x_2 = b'''_2$, and $x_3 = b'''_3$.

Again, if a pivot is zero or near-zero, row exchanges must be performed to continue the process.

NOTE 12.1.– This method can also be used to calculate the inverse of a matrix. This is because $\boldsymbol{A}$ and its inverse satisfy the equation $\boldsymbol{A}\boldsymbol{A}^{-1} = \mathbb{I}$. This process involves solving the equation $\boldsymbol{A}\boldsymbol{X} = \mathbb{I}$, or equivalently

$$\begin{bmatrix} A_{11} & A_{12} & \cdots & A_{1n} \\ A_{21} & A_{22} & \cdots & A_{2n} \\ \vdots & \vdots & \ddots & \vdots \\ A_{n1} & A_{n2} & \cdots & A_{nn} \end{bmatrix} \begin{bmatrix} X_{11} & X_{12} & \cdots & X_{1n} \\ X_{21} & X_{22} & \cdots & X_{2n} \\ \vdots & \vdots & \ddots & \vdots \\ X_{n1} & X_{n2} & \cdots & X_{nn} \end{bmatrix} = \mathbb{I} = \begin{bmatrix} 1 & 0 & \cdots & 0 \\ 0 & 1 & \cdots & 0 \\ \vdots & \vdots & \ddots & \vdots \\ 0 & 0 & \cdots & 1 \end{bmatrix}$$

This can be broken down into solving n linear systems:

$$\boldsymbol{A}\begin{bmatrix} X_{11} \\ X_{21} \\ \vdots \\ X_{n1} \end{bmatrix} = \begin{bmatrix} 1 \\ 0 \\ \vdots \\ 0 \end{bmatrix} \quad \boldsymbol{A}\begin{bmatrix} X_{12} \\ X_{22} \\ \vdots \\ X_{n2} \end{bmatrix} = \begin{bmatrix} 0 \\ 1 \\ \vdots \\ 0 \end{bmatrix} \quad \cdots \quad \boldsymbol{A}\begin{bmatrix} X_{1n} \\ X_{2n} \\ \vdots \\ X_{nn} \end{bmatrix} = \begin{bmatrix} 0 \\ 0 \\ \vdots \\ 1 \end{bmatrix}$$

Thus, solving these n linear systems provides the inverse matrix $\boldsymbol{A}^{-1}$.

12.2. Numerical solution of $f(x) = 0$

12.2.1. *Introduction*

Solving the equation $f(x) = 0$ is a common problem in mathematics. When the equation is solvable analytically, exact solutions can be obtained, if they exist. However, in many cases, an exact solution is either not possible or the equation is too complex to solve directly. In such situations, numerical methods offer a practical approach:

1) These methods do not guarantee finding *every* solution.

2) These methods provide approximate solutions.

Before applying numerical methods, it is generally advisable to first locate the roots of $f(x)$. For real values of x, the roots of the equation $f(x)$ lie within the interval]-∞, +∞. In practice, however, numerical methods are implemented using computers, which limits the search for solutions to a specific interval, typically $[-A', +A']$, where A' is the largest (and $-A'$ the smallest) value that the computer can handle, potentially eliminating some solutions.

This interval is divided into smaller subintervals of width (a, b), and the product $P = f(a)f(b)$ is computed. If $f(x)$ is continuous in the interval, the next steps can be determined based on the properties of the function:

– If $P < 0$, this indicates that $f(x)$ has an odd number of roots within the interval (a, b) (as shown in Figure 12.1).

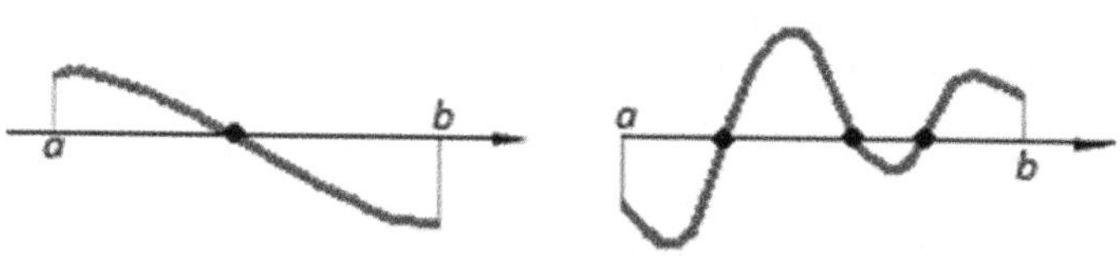

Figure 12.1. *Case P<0. For a color version of this figure, see www.iste.co.uk/cochard/mathematics2.zip*

– If $P > 0$, this indicates that $f(x)$ has either zero roots or an even number of roots within the interval (a, b) (as shown in Figure 12.2).

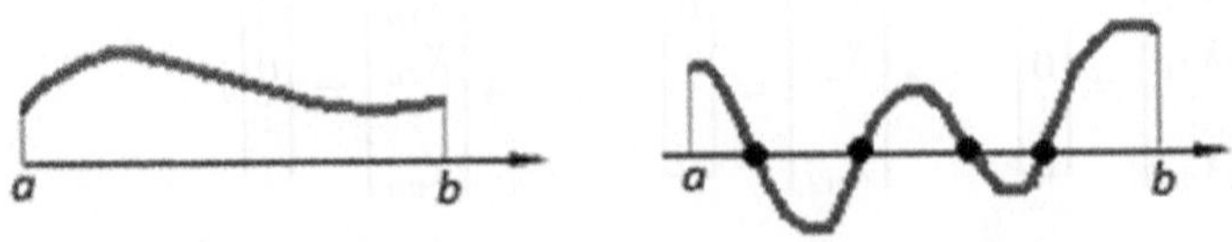

Figure 12.2. *Case P>0. For a color version of this figure, see www.iste.co.uk/cochard/mathematics2.zip*

To locate a single root, the smallest possible interval (a, b) is chosen; the function changes sign. This is important because the intermediate value theorem guarantees that a root exists within this interval if $f(a)$ and $f(b)$ have opposite signs. Once the interval is identified, the approximate value of the root can be estimated by finding the midpoint of the interval:

$$x \approx \frac{a+b}{2}$$

12.2.2. *General methods*

12.2.2.1. *Dichotomy*

The dichotomy method is a simple numerical method for solving the equation $f(x) = 0$ by repeatedly dividing an interval into two smaller intervals. This method works under the assumption that a root exists within the interval in (a, b) and that $f(x)$ is continuous on this interval.

Steps of the dichotomy method:

1) *Initial setup*

Start with an interval (a, b) where the function changes sign, i.e. $f(a) \cdot f(b) < 0$, indicating that a root exists between a and b (by the intermediate value theorem).

Compute the midpoint: $x_0 = \frac{a+b}{2}$.

The uncertainty (or error) at this step is: $\Delta x_0 = \epsilon = b - a$.

2) *First iteration*

Calculate the product $f(a) \cdot f(x0)$.

If $f(a) \cdot f(x_0) < 0$, then the root lies in the interval (a, x_0). Set the new interval as $(a_1, b_1) = (a, x_0)$.

If $f(a) \cdot f(x_0) > 0$, then the root lies in the interval (x_0, b). Set the new interval as $(a_1, b_1) = (x_0, b)$.

Calculate the new midpoint: $x_1 = \frac{a_1+b_1}{2}$.

The uncertainty at this step is halved: $\Delta x_1 = \frac{\epsilon}{2}$.

3) *Repeat the process*

Continue dividing the new interval (a_1, b_1) and recalculating the midpoint and uncertainty.

After n iterations, the uncertainty becomes: $\Delta x_n = \frac{\epsilon}{2^n}$.

This process will provide increasingly accurate approximations of the root.

EXAMPLE 12.2.– (According to G. Faccanoni). A savings account with an interest of $100t$ % has an initial deposit of A € made at the beginning of each year. At the end of each year, the total capital is updated by adding the interest generated to the existing capital. The breakdown is as follows:

1st year: the capital at the beginning of the year is A. By the end of the year, interest At is added, so the capital becomes: $A + At = A(1 + t)$.

2nd year: the capital at the beginning of the year is $A + A(1 + t)$. Interest is calculated on this amount, resulting in $A(1 + t) + A(1 + t)^2$ by year-end.

3rd year: at the beginning of the year, the capital is $A + A(1 + t) + A(1 + t)^2$. By adding the interest, the capital at the end of the year is $A(1 + t) + A(1 + t)^2 + A(1 + t)^3$.

n^{th} year: for any year n, the capital at the beginning is $A + A(1 + t) + \cdots + A(1 + t)^{n-1}$. The interest at year-end is equal to $At + At(1 + t) + \cdots + At(1 + t)^{n-1}$. By adding the interest to the current capital, the capital becomes:

$$A(1 + t) + A(1 + t)^2 + \cdots + A(1 + t)^n = A\sum_{k=1}^{n}(1 + t)^k$$

$$= A(1 + t)\frac{(1+t)^n-1}{(1+t)-1} = A\frac{(1+t)}{t}[(1 + t)^n - 1].$$

Knowing that the amount of $A = 1000$ € is paid at the beginning of each year, the goal is to find an interest rate $100t$ such that the capital at the end of the 5th year is 6000 €. Additionally, it is specified that the interest rate cannot exceed 10% (i.e. $t_{max} = 0.1$). To solve the interest rate, the following equation is set up:

$$f(t) = 1000\frac{(1+t)}{t}[(1+t)^5 - 1] - 6000 = 0.$$

Using the dichotomy algorithm (see section 12.6. Appendices) with a lower bound $a = 0$, upper bound $b = 0.1$ and a precision $\varepsilon = 10^{-12}$, an approximate interest rate of 6.14 % is obtained after 37 iterations.

12.2.2.2. *Iteration*

To solve $f(x) = 0$, the equation can be reformulated as $g(x) = x$, which is always possible. Starting from an initial arbitrary value $x_0 \in (a, b)$, the sequence is constructed as follows:

$$x_1 = g(x_0), \quad x_2 = g(x1), \quad \cdots, \quad x_p = g(x_{p-1}).$$

This construction is shown in Figure 12.3. If the sequence converges to a limit $\hat{x}$, then $\hat{x} = g(\hat{x})$, which implies that $f(\hat{x}) = 0$. Therefore, $\hat{x}$ is the root being sought.

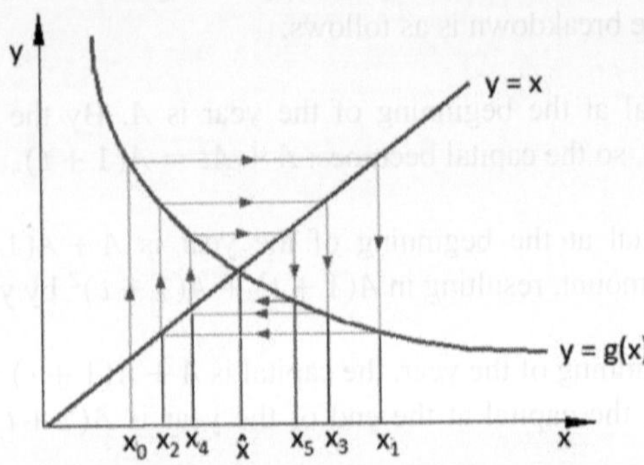

Figure 12.3. *Suite construction. For a color version of this figure, see www.iste.co.uk/cochard/mathematics2.zip*

The conditions under which the sequence is convergent need to be demonstrated. Define $u_p = x_p - x_{p-1}$. Then,

$$x_p - x_0 = (x_p - x_{p-1}) + (x_p - x_{p-2}) + \cdots + (x_1 - x_0) = u_p + u_{p-1} + \cdots + u_1$$

Or equivalently,

$$x_p = x_0 + u_1 + u_2 + \dots + u_p,$$

forming a series with the general term u_p.

Assuming $f(x)$ is differentiable on the interval (a, b), the theorem of finite increments indicates that $x_p = g(x_{p-1}) - g(x_{p-2}) = g'(x)(x_{p-1} - x_{p-2})$, where $x \in (a, b)$. Therefore,

$$u_p = g'(\xi)u_{p-1} \quad \text{and} \quad \frac{u_p}{u_{p-1}} = g'(\xi)$$

Applying d'Alembert's criterion for the convergence of the series u_p gives:

$$\left|\frac{u_p}{u_{p-1}}\right| < 1,$$

leading to $|g'(x)| < 1$, which is a sufficient condition for convergence.

EXAMPLE 12.3.– (According to J-P Calvi). Let the equation be $f(x) = x - \sin(x) - \frac{1}{4} = 0$. The function $g(x)$ is then $g(x) = \sin(x) + \frac{1}{4}$. Let us study the search for the roots of $f(x) = 0$ for $x \in \left[0, \frac{\pi}{2}\right]$. The derivative of $g(x)$ is $g'(x) = \cos(x)$, so $|g'(x)| \leq 1$. As $\frac{\pi}{2}$ is not a solution, the convergence condition is considered satisfied.

Starting from $x_0 = 0$, the solution $x = 1.17$ is obtained (see section 12.6, Appendices) after 10 iterations with a fixed precision of 10^{-3}.

12.2.2.3. *Secant method (Descartes)*

Consider the following case: $f(x)$ has a root $\hat{x}$ in (a, b); $f'(x)$ and $f''(x)$ are positive on (a, b). By plotting the successive secants as shown in Figure 12.4, a sequence $a < x_1 < x_2 < \cdots < b$ is obtained.

This increasing monotonic sequence converges to the root $\hat{x}$. The equation of the string BC is:

$$\frac{x_3 - b}{a - f(b)} = \frac{x_2 - b}{f(x_2) - f(b)}$$

Hence,

$$x_3 = b - f(b) \cdot \frac{x_2 - b}{f(x_2) - f(b)}$$

More generally:

$$x_p = b - f(b) \cdot \frac{x_{p-1} - b}{f(x_{p-1}) - f(b)}$$

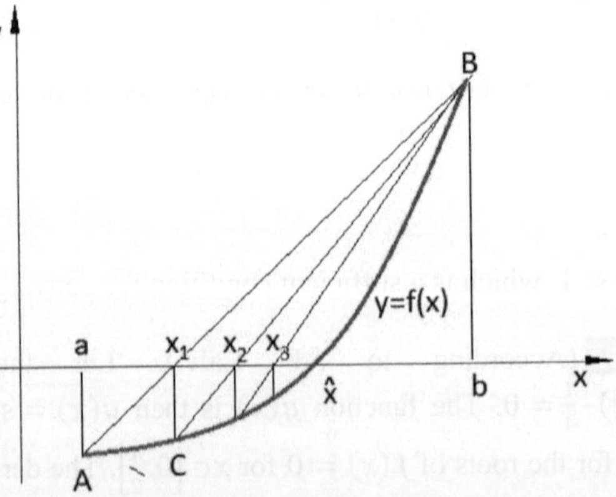

Figure 12.4. *Secant method. For a color version of this figure, see www.iste.co.uk/cochard/mathematics2.zip*

This relationship can be used to calculate, step by step, the values of x_p. It is applied until the desired precision $\varepsilon = |x_p - x_{p-1}|$ is obtained. The case studied above is only one possible scenario. There are four possible cases, as shown in Figure 12.5.

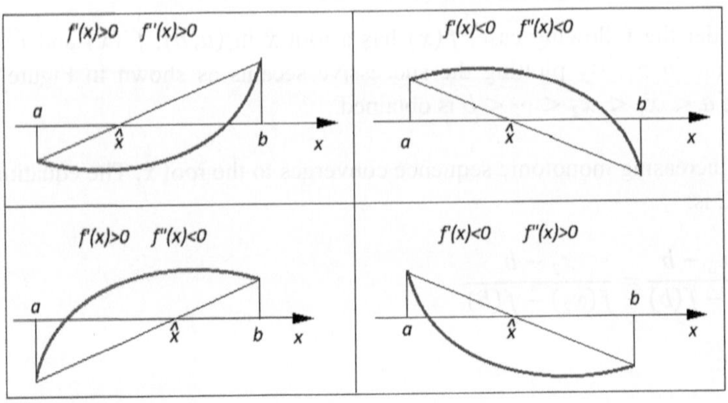

Figure 12.5. *Possible cases. For a color version of this figure, see www.iste.co.uk/cochard/mathematics2.zip*

For the curves in the first line of Figure 12.5, $\hat{x}$ is approximated by default, while for the curves in the second line, $\hat{x}$ is approximated by excess.

EXAMPLE 12.4.– Let the equation be $f(x) = 0$ with:

$$f(x) = \frac{x-4}{x+1} + x$$

Let us check if this function has a root in $[0,2]$. This is possible because $f(0) = -4$ and $f(2) = \frac{4}{3}$. Moreover, $f'(x) > 0$ and $f''(x) < 0$ in the interval under consideration. Let us take $b = 0$ (as the fixed point) and $x_0 = 2$ (as the starting point).

Let us apply the algorithm (see section 12.6, Appendices) looking for an accuracy of 10^{-6}. The root 1.236 is obtained after 15 iterations.

12.2.2.4. *Tangent method (Newton)*

Returning to the case studied in the previous section, successive tangents $T_1, T_2, T_3, \ldots$, as shown in Figure 12.6, are traced. These tangents determine the values $x_1, x_2, x_3, \ldots$, such that $b > x_1 > x_2 > \cdots > a$.

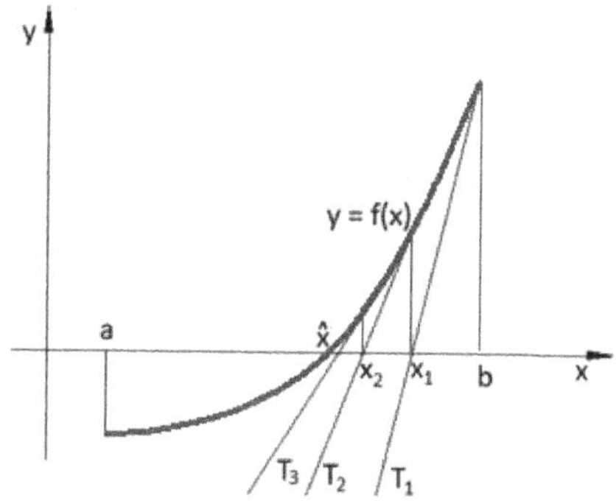

Figure 12.6. *Tangent method. For a color version of this figure, see www.iste.co.uk/cochard/mathematics2.zip*

This decreasing, lower-bounded sequence converges to the root $\hat{x}$. The equation of a tangent at the point $\left(x_{p-1}, f\left(x_{p-1}\right)\right)$ is given by:

$$f'\left(x_{p-1}\right) = \frac{y-f\left(x_{p-1}\right)}{x-x_{p-1}},$$

where (x, y) is a point on the tangent. In particular,

$$f'(x_{p-1}) = \frac{0 - f(x_{p-1})}{x_p - x_{p-1}} \quad \Rightarrow \quad x_p = x_{p-1} - \frac{f(x_{p-1})}{f'(x_{p-1})}$$

This relationship allows the successive terms of the sequence to be calculated, thus approximating $\hat{x}$.

EXAMPLE 12.5.– Apply the method to the function in example 12.3:

$$f(x) = \frac{x-4}{x+1} + x, \quad f'(x) = \frac{5}{(x+1)^2} + 1,$$

with the same precision of 10^{-6}. The first tangent is taken from the point $(0, -4)$. An identical result is obtained, but in just five iterations (see section 12.6, Appendices).

12.2.3. *Methods applicable to polynomial equations*

Assume that $f(x)$ is a polynomial of degree n with real coefficients:

$$f(x) = a_0 + a_1 x + a_2 x^2 + \dots\dots + a_n x^n$$

It is known that $f(x)$ has n roots (real or complex). To determine the real roots, the previous methods can be applied. Additionally, special methods specific to polynomial equations can be used, of which an overview is provided.

12.2.3.1. *Graeffe's method*

A polynomial $P(x) = a_0 + a_1 x + a_2 x^2 + \dots + a_n x^n$ of degree n has n real or complex roots $x_1, x_2, \dots, x_n$, which can be written as:

$$P(x) = \lambda(x - x_1)(x - x_2) \dots (x - x_n).$$

By decomposition and identification:

$$a_n = \lambda$$

$$a_{n-1} = -\lambda(x_1 + x_2 + \dots + x_n)$$

$$a_{n-2} = \lambda(x_1 x_2 + x_2 x_3 + \dots + x_{n-1} x_n)$$

$$\vdots$$

$$a_{n-k} = \lambda(-1)^k \sum x_{i_1} x_{i_2} \dots x_{i_k}$$

Suppose the roots of $P(x)$ are "very spaced" such that $|x_1| \gg |x_2| \gg \cdots \gg |x_n|$. Then,

$$-\frac{a_{n-1}}{a_n} = x_1 \left(1 + \frac{x_2}{x_1} + \frac{x_3}{x_1} + \cdots \right) \approx x_1$$

$$\frac{a_{n-2}}{a_{n-1}} = x_1 x_2 \left(1 + \frac{x_3}{x_2} + \cdots + \frac{x_{n-1} x_n}{x_1 x_2}\right) \approx x_1 x_2$$

$$\vdots$$

$$(-1)^k \frac{a_{n-k}}{a_n} = x_1 x_2 \dots x_k (1 + \cdots) \approx x_1 x_2 \dots x_k$$

The approximate values of the roots:

$$x_1 \approx -\frac{a_{n-1}}{a_n}, \quad x_2 \approx -\frac{a_{n-2}}{a_{n-1}} \quad \cdots \quad x_k \approx = -\frac{a_{n-k}}{a_{n-k-1}}$$

Consider now a polynomial $P(x) = a_n(x - x_1)(x - x_2) \cdots (x - x_n)$, where the roots are assumed to be distinct and ordered such that $|x_1| > |x_2| > \cdots > |x_n|$. Let $P_1(x)$ be the polynomial whose roots are $x_1^2, x_2^2, \dots, x_n^2$:

$$P_1(x) = a_n^2 (x - x_1^2)(x - x_2^2) \cdots (x - x_n^2)$$

If $\left|\frac{x_i}{x_{i-1}}\right| = \varepsilon < 1$, then $\frac{|x_i^2|}{|x_{i-1}^2|} = \varepsilon^2 < 1$, which means that the roots of $P_1(x)$ are "more spaced" than those of $P(x)$. By replacing x with u^2:

$$P_1(u^2) = a_n^2 (u^2 - x_1^2)(u^2 - x_2^2) \dots (u^2 - x_n^2)$$

leading the expression:

$$P_1(u^2) = a_n^2 (u - x_1)(u - x_2) \dots (u - x_n)(u + x_1)(u + x_2) \dots (u + x_n)$$

$$= (-1)^n P(u) P(-u)$$

This relationship allows for the definition of the coefficients $a_0^{(1)}, a_1^{(1)}, \dots, a_n^{(1)}$ of $P_1(x)$ as a function of the coefficients $a_0, a_1, \dots, a_n$ of $P(x)$.

Next, define $P_2(x)$, a polynomial with roots x_i^4, and coefficients $a_0^{(2)}, a_1^{(2)}, \ldots, a_n^{(2)}$. Continuing in the same manner, define $P_p(x)$, a polynomial with roots $x_i^{2^p}$ and coefficients $a_0^{(p)}, a_1^{(p)}, \ldots, a_n^{(p)}$.

If p is large enough, $P_p(x)$ is a polynomial with "very spaced" roots. Therefore, the roots of $P_p(x)$ can be approximated as:

$$x_1^{2^p} \approx -\frac{a_{n-1}^{(p)}}{a_n^{(p)}}, \quad x_2^{2^p} \approx -\frac{a_{n-2}^{(p)}}{a_{n-1}^{(p)}}, \quad \ldots, \quad x_k^{2^p} \approx -\frac{a_{n-k}^{(p)}}{a_{n-k-1}^{(p)}}$$

Thus, the approximate root values are:

$$x_k \approx \pm\left(-\frac{a_{n-k}^{(p)}}{a_{n-k-1}^{(p)}}\right)^{\frac{1}{2^p}}$$

Although this method seems simple, it has two major drawbacks:

– It can lead to the manipulation of very large parameters that exceed the computer's capacity.

– It is relatively imprecise.

12.2.3.2. *Bairstow's method*

To apply Bairstow's method, the polynomial $P(x) = a_0 + a_1x + a_2x^2 + \cdots + a_nx^n$ is divided by a trinomial $x^2 + px + q$, such that:

$$P(x) = (x^2 + px + q)Q + R$$

where $Q = b_{n-2}x^{n-2} + b_{n-1}x^{n-1} + \ldots + b_0$ is the quotient, and $R = rx + s$ is the remainder. The functions r and s depend on p and q. If p and q are chosen so that the trinomial divides $P(x)$ exactly, then $r = s = 0$.

In this case, the polynomial $P(x)$ is said to have "separated" two of its roots, as $P(x) = (x^2 + px + q)$. The remaining $n - 2$ other correspond to those of the quotient Q, and the same method can be applied to $Q(x)$.

The main problem lies in determining the pairs (p_i, q_i) at each step so that the remainders $r(p_i, q_i) = 0$ and $s(p_i, q_i) = 0$. Typically, this is done iteratively to determine the values of p_i and q_i, ensuring that the division is exact.

12.3. Numerical integration

12.3.1. *Introduction*

For a function f that is defined and continuous on an interval $[a, b]$, the integral of f can be expressed as:

$$I = \int_a^b f(x)\,dx = F(b) - F(a),$$

where F is the primitive (or antiderivative) of f. However, directly calculating the integral is often complex and may not always be possible due to the following reasons:

– $F(x)$ may not be known or expressed in terms of known functions.

– The function $f(x)$ may not be continuous.

In such cases, numerical methods become useful. These methods allow the integral to be approximated using finite values of f, which can then be computed on a computer. This approach is called the *numerical quadrature*, and it approximates the integral with a sum of weighted function values at specified points:

$$I = \int_a^b f(x)\,dx \approx \sum_{i=0}^{N} H_i f(x_i),$$

where x_i are the points (or nodes) at which the function is evaluated, and H_i are the corresponding "weights". These weights and points can be chosen in various ways depending on the numerical integration method used.

12.3.2. *Classic methods*

Assuming f is continuous, classic numerical integration methods approximate the area S under the curve $y = f(x)$, bounded by the x-axis and the vertical lines $x = a$ and $x = b$. The total area S is approximated by dividing the integral $[a, b]$ into N smaller subintervals, each of width $h = \frac{(b-a)}{N}$. The integral I is then approximated by summing the elementary areas ΔS_i, which represent the area of each subinterval. Thus, the integral is approximated as:

$$I = \int_a^b f(x)\,dx \approx S = \sum_{i=1}^{N} \Delta S_i$$

where each ΔS_i represents the contribution of a small area within the subinterval, and the sum provides the overall approximation of the integral.

12.3.2.1. *Trapezoid method*

In the trapezoid method, the area ΔS_i corresponds to the area of a trapezoid formed by the curve $y = f(x)$ and the x-axis, as shown in Figure 12.7.

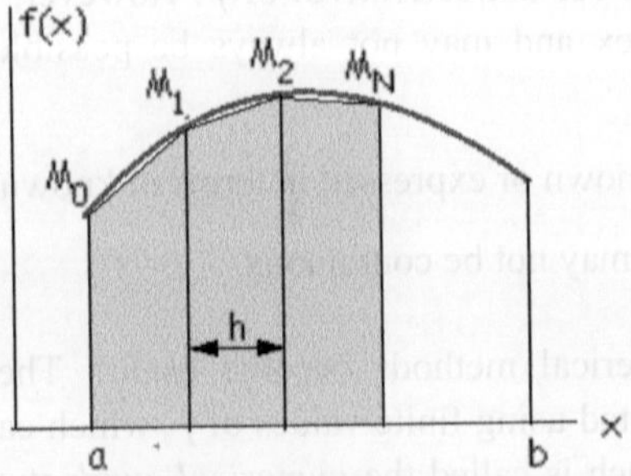

Figure 12.7. *Trapezoid method. For a color version of this figure, see www.iste.co.uk/cochard/mathematics2.zip*

The area of each trapezoid can be calculated using the formula:

$$\Delta S_i = h.\left[\frac{f(a+ih) + f(a+(i+1)h)}{2}\right]$$

where $h = \frac{b-a}{N}$ is the width of each subinterval. Summing the areas of all the trapezoids gives the approximation for the integral:

$$I \approx \frac{h}{2}[f(a) + f(a+h) + f(a+h) + f(a+2h) + \cdots + f(b)]$$

Alternatively, this can be expressed as:

$$I \approx h\left[\frac{f(a) + f(b)}{2} + \sum_{i=1}^{N-1} f(a+ih)\right]$$

Uncertainty of the approximation in the trapezoid method can be bounded by:

$$\varepsilon = \frac{(b-a)^3 \cdot M_2}{12N^2}$$

where M_2 is an upper bound of $|f''(x)|$ in the interval $[a, b]$, and N is the number of subintervals used in the approximation.

12.3.2.2. *Rectangle method*

In the rectangle method, the area ΔS_i corresponds to the area of a rectangle. The height of the rectangle can be taken at different points: $f(x_i)$, $f(x_{i-1})$, or $f\left(\frac{x_{i-1}+x_i}{2}\right)$, where $x_i = a + ih$ (see Figure 12.8).

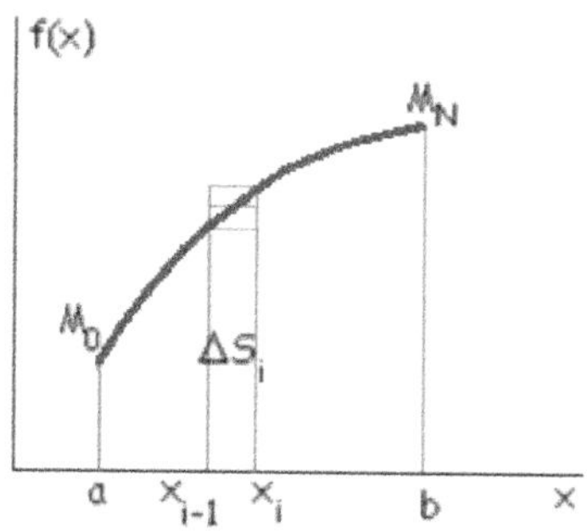

Figure 12.8. *Rectangle method. For a color version of this figure, see www.iste.co.uk/cochard/mathematics2.zip*

In the case where the height is taken at the midpoint, the formula is:

$$x_{i-1} + x_i = a + \left(i - \frac{1}{2}\right)h.$$

Thus, the integral approximation is:

$$I \approx h\sum_{i=1}^{N} f\left(a + \left(i - \frac{1}{2}\right)h\right).$$

The error uncertainty is bounded by:

$$\varepsilon = \frac{(b-a)^2 \cdot M_1}{2N},$$

where M_1 is an upper bound of $|f'(x)|$ in the interval $[a, b]$, and N is the number of subintervals.

12.3.2.3. *Simpson's method*

This frequently used method involves replacing curve arcs $M_{i-1}M_i$ by parabola arcs. It is based on the following calculation: if A, B, C are three points with abscissas $-h, 0, +h$ on a parabola of equation:

$$y = ax^2 + bx + c.$$

The area ΔS of Figure 12.9 has the value:

$$\Delta S = \int_a^b (ax^2 + bx + c)\,dx = \frac{2a}{3}h^3 + 2ch = \frac{h}{3}(2ah^2 + 6c)$$

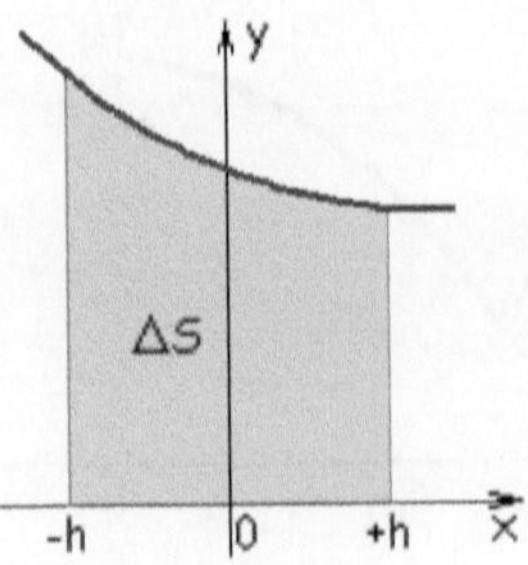

Figure 12.9. *ΔS area. For a color version of this figure, see www.iste.co.uk/cochard/mathematics2.zip*

However, $y(0) = c$ and $y(h) + y(-h) = 2ah^2 + 2c$. Therefore, the area ΔS can be expressed as:

$$\Delta S = \frac{h}{3}[y(h) + 4y(0) + y(-h)].$$

If the interval $[a, b]$ is divided into $2N$ subintervals and an interpolation parabola is used for each double interval (x_{2i-2}, x_{2i}), then the integral can be approximated as:

$$I \approx \frac{h}{3}\sum_{i=1}^{N}[f(x_{2i-2}) + 4f(x_{2i-1}) + f(x_{2i})].$$

Thus, the approximation becomes

$$I \approx \frac{h}{3}\sum_{i=1}^{N} f(a + (2i-2)h) + 4f(a + (2i-1)h) + f(a + 2ih).$$

The uncertainty error in this approximation is bounded above by:

$$\frac{(b-a)h^4}{180}\max_{(a,b)}\left|f^{(4)}(x)\right|$$

12.3.2.4. *Poncelet's method*

In this method, the curve $y = f(x)$ is replaced by a broken line joining the points with coordinates $\alpha_i = a + h(2i-1)/2, f(\alpha_i)$.

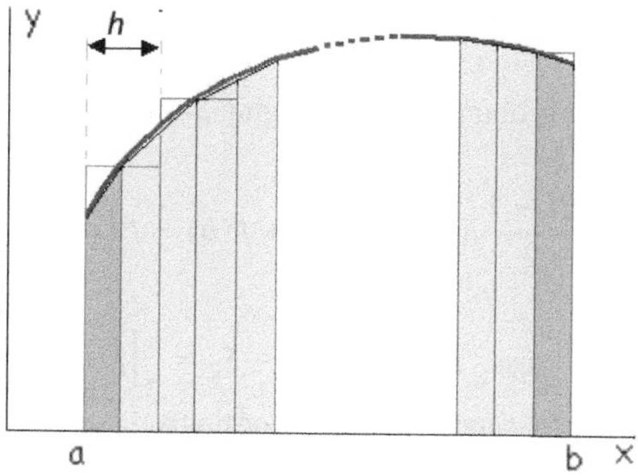

Figure 12.10. *Poncelet's method. For a color version of this figure, see www.iste.co.uk/cochard/mathematics2.zip*

The corresponding area is the sum (see Figure 12.10) of the areas of two trapezoids with base $\frac{h}{2}$:

$$\frac{h}{4}\left[f(a)+f\left(a+\frac{h}{2}\right)\right]+\frac{h}{4}\left[f(b)+f\left(b-\frac{h}{2}\right)\right]$$

and $N-1$ trapezoid areas with base h:

$$\frac{h}{2}\sum_{i=1}^{N-1}\left[f\left(a+\frac{2i-1}{2}h\right)+f\left(a+\frac{2i+1}{2}h\right)\right]$$

Thus,

$$S_1 = \frac{h}{4}\left[f(a)+f(b)+f\left(a+\frac{h}{2}\right)+f\left(b-\frac{h}{2}\right)\right]$$
$$+\frac{h}{2}\sum_{i=1}^{N-1}\left[f\left(a+\frac{2i-1}{2}h\right)+f\left(a+\frac{2i+1}{2}h\right)\right]$$

Let S_2 denote the area obtained by the rectangle method, where the height is $f((x_{i-1}+x_i)/2)$:

$$S_2 = h\sum_{i=1}^{N} f\left(a+\frac{2i-1}{2}h\right)$$

It follows that

$$S_1 = S_2 + \frac{h}{4}[f(a)+f(b) - f\left(a+\frac{h}{2}\right) - f(b-\frac{h}{2}).$$

The Poncelet formula is obtained by averaging S_1 and S_2 ($I \approx (S_1+S_2)/2$):

$$I \approx h\sum_{i=1}^{N} f\left(a+\frac{2i-1}{2}h\right) + \frac{h}{8}\left[f(a)+f(b) - f\left(a+\frac{h}{2}\right) - f\left(b-\frac{h}{2}\right)\right].$$

The uncertainty is less than $\varepsilon = S_1 - S_2 = \frac{h}{4}\left[f(a)+f(b) - f\left(a+\frac{h}{2}\right) - f\left(b-\frac{h}{2}\right)\right]$. Hence, h (or N) can be chosen accordingly.

12.3.3. *Polynomial interpolation*

It often happens that the function f is only defined for a discrete set of points $x_0, x_1, x_2, \dots, x_n, \cdots$. In this case, the analytical form of the function is unknown, and a polynomial $P(x)$ can be sought to approximate $f(x)$, taking the same values as f at the points $x_0, x_1, x_2, \dots, x_n$. A polynomial is often chosen because it is easy to integrate or differentiate. The polynomial can be written as:

$$P_n(x) = a_0 + a_1 x + \cdots + a_n x^n = a_0 + x(a_1 + x(a_2 + \cdots + a_n)) \ldots$$

This last form is convenient for computer calculations.

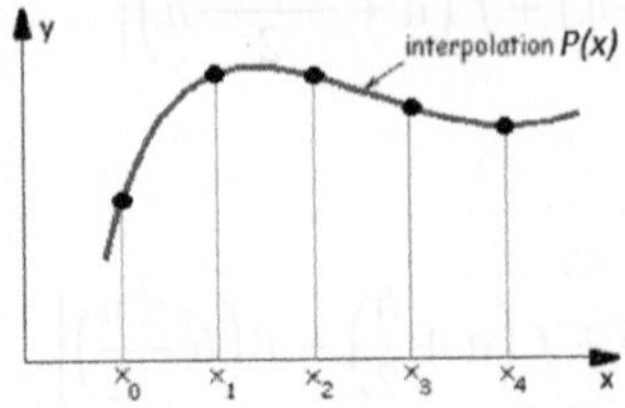

Figure 12.11. *Polynomial interpolation. For a color version of this figure, see www.iste.co.uk/cochard/mathematics2.zip*

To determine the interpolation polynomial $P_n(x)$ of a function f, the coefficients $a_0, a_1, \dots, a_n$ need to be adjusted, which requires $n+1$ relations:

$$P_n(x_i) = f(x_i), \quad i = 0,1,2,\dots,n.$$

This leads to solving a system of linear equations (the unknowns are a_i):

$$a_0 + a_1 x_i + a_2 x_i^2 + \cdots + a_n x_i^n = f(x_i), \quad i = 0,1,2,\dots,n.$$

In practice, the polynomial $P_n(x)$ is expressed in terms of base polynomials. The most commonly used base polynomials are Lagrange polynomials $L_i(x)$. Thus, the interpolation polynomial is given by: $P_n(x) = f(x_0)L_0(x) + f(x_1)L_1(x) + \cdots + f(x_n)L_n(x)$; this means that $L_i(x_j) = \delta_{ij}$, where δ_{ij} is the Kronecker delta.

From the previous relationship, the explicit expression of the Lagrange polynomial $L_i(x)$ of degree n can be reduced.

1) The polynomial $L_i(x)$ has roots $x_0, x_1, \dots, x_{i-1}, x_{i+1}, \dots, x_n$, since $L_I(x_j) = 0$ for $j \neq i$. Therefore, $L_i(x)$ has the form:

$$L_i(x) = k(x - x_0)(x - x_1) \dots (x - x_{i-1})(x - x_{i+1}) \dots (x - x_n)$$
$$= k \prod_{j \neq i} (x - x_j)$$

2) The polynomial $L_i(x)$ satisfies $L_i(x_i) = 1$, which leads to

$$k \prod_{j \neq i} (x_i - x_j) = 1$$

This fixes the constant k. As a result, the Lagrange polynomial $L_i(x)$ becomes:

$$L_i(x) = \frac{\prod_{j \neq i}(x - x_j)}{\prod_{j \neq i}(x_i - x_j)} = b_0^{(i)} + b_1^{(i)} x + \cdots + b_n^{(i)} x^n$$

EXAMPLE 12.6.– Given an unknown function $f(x)$ for which three points are known: (2, 3), (4, 5) and (6, 2). The interpolation polynomial is $P_3(x) = 3L_0(x) + 5L_1(x) + 2L_2(x)$.

The Lagrange polynomials are calculated according to their definition:

Calculation of $L_0(x)$: $L_0(x) = \frac{(x-4)(x-6)}{(2-4)(2-6)} = \frac{1}{8}(x^2 - 10x + 24)$.

Calculation of $L_1(x)$: $L_1(x) = \frac{(x-2)(x-6)}{(4-2)(4-6)} = -\frac{1}{4}(x^2 - 8x + 12)$.

Calculation of $L_2(x)$: $L_2(x) = \frac{(x-2)(x-4)}{(6-2)(6-4)} = \frac{1}{8}(x^2 - 6x + 8)$.

Forming the interpolation polynomial $P_3(x)$:

$$P_3(x) = 3L_0(x) + 5L_1(x) + 2L_2(x)$$

After substituting the values of $L_0(x)$, $L_1(x)$ and $L_2(x)$, and performing the calculations, the interpolation polynomial is:

$$P_3(x) = -\frac{5}{8}x^2 + \frac{19}{4}x - 4$$

12.3.4. *Quadrature formulas*

The classical methods described above are based on the notion of area. Alternatively, to calculate the definite integral:

$$I = \int_a^b f(x)\, dx,$$

$f(x)$ can be replaced with a function $g(x)$ such that

$$J = \int_a^b g(x)\, dx$$

which is calculable and $|I - J|$ is as small as possible. In particular, for $g(x)$ as an interpolation polynomial of f defined from points $x_0, x_1, \dots, x_n$, the following methods are distinguished:

– Newton–Côtes method for $x_0, x_1, \dots, x_n$ regularly distributed points;

– the Gauss method for points $x_0, x_1, \dots, x_n$.

12.3.4.1. *Newton–Côtes method*

Assume the points x_i are evenly spaced from a to b, with $b - a = nh$. The goal is to calculate the integral:

$$I = \int_a^b f(x)\, dx$$

Perform the change of variables $t = \frac{x-a}{h}$ and define $F(t) = f(x) = f(a + th)$. When x takes the values $x_0, x_1, \dots, x_n$, t takes the values $0, 1, \cdots, n$. The integral becomes:

$$I = h \int_0^n F(t)\, dt$$

Replace $F(t)$ with its Lagrange polynomial interpolation:

$$I \approx h \int_0^n \sum_{i=0}^{n} F(t_i) L_i(t) dt = h \sum_{i=0}^{n} F(t_i) \int_0^n L_i(t) dt = h \sum_{i=0}^{n} f(a + ht_i)\, H_i$$

where $H_i = \int_0^n L_i(t)\, dt$. This is the Newton–Côtes formula. The uncertainty error in this approximation is proportional to $h^{n+3} f^{(n+2)}(x)$ if n is even and $h^{n+2} f^{(n+1)}(x)$ if n is odd, where $x \in [a, b]$.

This formula allows us to recover previous results.

For $n = 1$, i.e. two points $x_0 = a$ and $x_1 = b$, the values of H_0 and H_1 are:

$$H_0 = -1 \int_0^1 (t-1)\, dt = \frac{1}{2} \quad \text{and} \quad H_1 = \int_0^1 t\, dt = \frac{1}{2}$$

Hence,

$$I \approx h \frac{f(a) + f(b)}{2}$$

This is the trapezoid method.

For $n = 2$, i.e. three points $x_0 = a, x_1 = \frac{a+b}{2}, x_2 = b$ and $t = 0, 1, 2$, the values of H_0, H_1 and H_2 are:

$$H_0 = \frac{1}{2}\int_0^2 (t-1)(t-2)dt = \frac{1}{3} \qquad H_1 = -\int_0^2 t(t-2)dt = \frac{4}{3} \qquad H_2 = \frac{1}{2}\int_0^2 t(t-1)dt = \frac{1}{3}$$

Thus, the approximation is:

$$I \approx \frac{h}{3}\left[f(a) + f\left(\frac{a+b}{2}\right) + f(b)\right]$$

This is Simpson's method.

12.3.4.2. *Gauss method*

Consider calculating the integral:

$$\int_{-1}^{+1} F(x)\,dx$$

The goal is to obtain points $x_0, x_1, \dots, x_n$ and coefficients $H_0, H_1, \dots, H_n$ such that:

$$\int_{-1}^{+1} F(x)\,dx \approx H_0F(x_0) + H_1F(x_1) + \dots + H_nF(x_n)$$

It is possible to choose the points x_i so that this formula is exact for polynomials of degree $2n+1$. Specifically:

1) The points x_i are chosen as the roots of the Legendre polynomial of degree n:

$$\mathfrak{L}_n(x) = \frac{1}{n!\,2^n}\frac{d^n}{dx^n}(x^2-1)^n$$

2) The coefficients H_i are given by:

$$H_i = \int_{-1}^{+1} L_i(x)\,dx$$

where $L_i(x)$ is the Lagrange interpolation polynomial.

To extend this to the general case of calculating I, perform the change of variables:

$$x = \frac{a+b}{2} + \frac{b-a}{2}t,\ dx = \frac{b-a}{2}dt,\ f(x) = f\left(\frac{a+b}{2} + \frac{b-a}{2}t\right) \equiv F(t)$$

Then,

$$I = \int_a^b f(x)\,dx = \frac{b-a}{2}\int_{-1}^{+1} F(t)\,dt \approx \frac{b-a}{2}\sum_{i=0}^{n} H_i F(t_i)$$

which leads to Gauss's formula:

$$I = \int_a^b f(x)\,dx \approx \frac{b-a}{2}\sum_{i=0}^{n} H_i f\left(\frac{a+b}{2} + \frac{b-a}{2}t_i\right)$$

12.3.5. *Monte Carlo method*

An integral defines a surface between the x-axis, the curve representing the function $f(x)$ and the parallels to the Oy-axis at points $(a, 0)$ and $(b, 0)$ (see Figure 12.12). The area of this surface, measured algebraically and denoted as A, corresponds to the integral of the function $f(x)$ between a and b:

$$\int_a^b f(x)\,dx = A$$

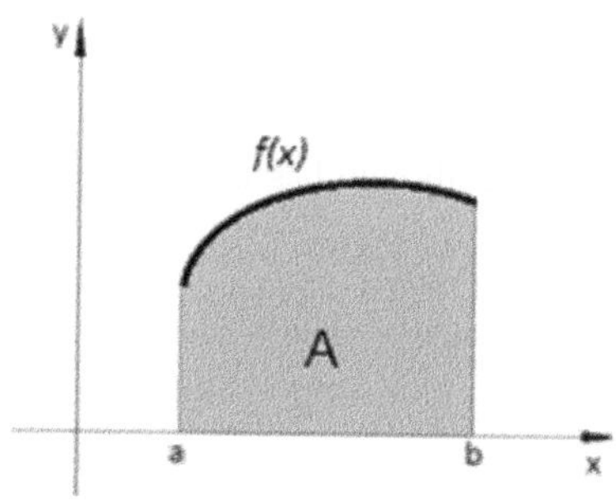

Figure 12.12. *Area A corresponds to the integral of* f(x) *between a and b. For a color version of this figure, see www.iste.co.uk/cochard/mathematics2.zip*

To approximate this area, the surface corresponding to the integral is closest in a rectangle R of known dimensions, as shown in Figure 12.13, with an area of $h(b-a)$.

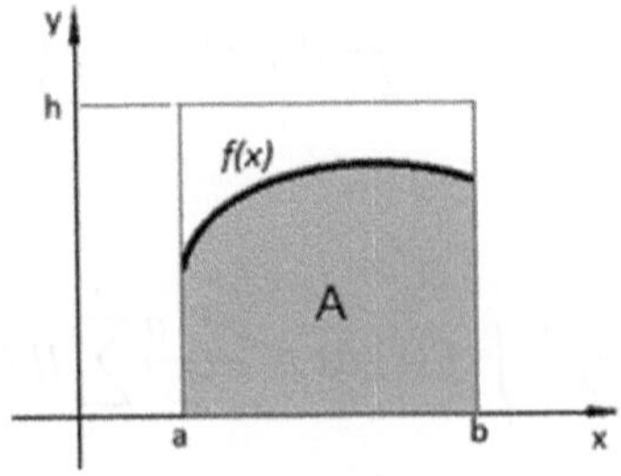

Figure 12.13. *Rectangle framing. For a color version of this figure, see www.iste.co.uk/cochard/mathematics2.zip*

A random process known as "point throwing" is then applied, where points are drawn in the rectangle R randomly but uniformly. This is done using the uniform probability distribution, which is commonly available in most programming languages. Suppose N points are thrown randomly and uniformly into the rectangle R, with n of these points falling within the area corresponding to the dark grey surface (area A). By the property of homogeneity, the following relation holds:

$$\frac{n}{N} = \frac{A}{h(b-a)} \quad \text{thus,} \quad A = \frac{n}{N} h(b-a)$$

This method is called "Monte Carlo", named after the Monte Carlo casino, due to its reliance on random sampling (or chance).

EXAMPLE 12.7.– Let the integral be:

$$I = \int_0^1 x \ln(1+x)\, dx$$

which can be calculated analytically and has the value $\frac{1}{4}$. This integral represents the area A of the dark grey surface in Figure 12.14.

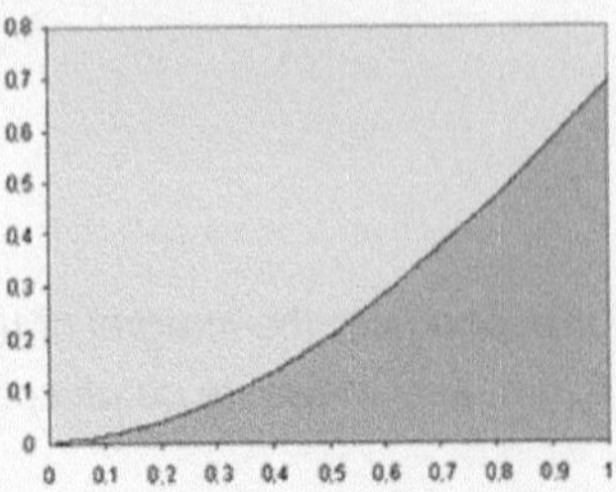

Figure 12.14. *Context for calculating the integral. For a color version of this figure, see www.iste.co.uk/cochard/mathematics2.zip*

To approximate this area, the surface enclosed in a blue rectangle with sides 1 and 0.8. Randomly, draw N points (x, y) uniformly, where $y \in [0, 0.8]$ and $x \in [0, 1]$. Count the number of points n that fall within the dark grey surface, satisfying the condition $y < x \ln(x + 1)$.

Then, the area A is given by:

$$A = \frac{0.8n}{N}.$$

Using 1000 points and a JavaScript program (see section 12.6. Appendices), an average value of $A = 0.25276$ is obtained over 10 trials.

EXAMPLE 12.8.– Consider the function:

$$f(x) = \frac{1}{\sqrt{2}} \sqrt{-(2x^2 + a^2) + a\sqrt{8x^2 + a^2}}$$

which, graphically shown in Figure 12.15, represents one-fourth $\left(\frac{1}{4}\right)$ of Bernoulli's lemniscate (the symbol for infinity). Despite the complex form of this function, it is possible to calculate the integral analytically:

$$\int_0^a f(x)\, dx = \frac{a^2}{4}$$

Thus, the area of the surface bounded by Bernoulli's lemniscate is simply a^2! To approximate this integral, A, using the Monte Carlo method, set $a = 1$, expecting a value close to $\frac{1}{4}$. To achieve this, draw the square with side a, as shown in Figure 12.15.

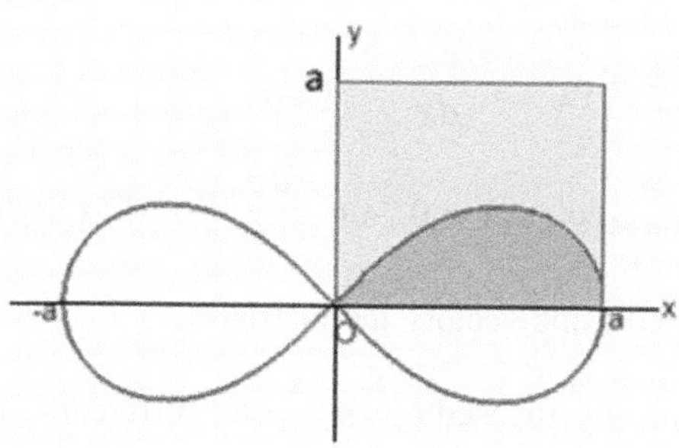

Figure 12.15. *Bernoulli lemniscate. For a color version of this figure, see www.iste.co.uk/cochard/mathematics2.zip*

Generate N random points (x, y) within the square, where $y\in[0, a]$ and $x\in[0, a]$, and count the number of points n that fall within the dark grey surface. These points must satisfy the inequality

$$(x^2 + y^2)^2 < a^2(x^2 - y^2)$$

For $a = 1$, the area approximation is $A = \frac{n}{N}$. Using a JavaScript program (see section 12.6, Appendices), perform 10 trials of 1,000 points each and take the average, yielding approximately 0.2475.

12.4. Numerical solution of differential equations

12.4.1. *Introduction*

A differential equation of order n takes the form:

$$f(x, y, y', \dots, y^{(n)}) = 0$$

where $y', y'', \dots, y^{(n)}$ are the successive derivatives of y, a function of x. The solution of such an equation is $y = g(x, C_1, C_2, \dots, C_n)$, where $C_1, C_2, \dots, C_n$ are arbitrary constants determined by initial conditions: $y(x_0) = a_0, y'(x_0) = a_1, \dots, y^{(n-1)} = a_{n-1}$

A differential equation of order n can always be reduced to a first-order differential equation. Defining:

$$v_0 = y, \quad v_1 = y', \quad v_2 = y'', \dots, v_n = y^{(n)}$$

yields the differential system:

$$\frac{dv_0}{dx} = v_1, \quad \frac{dv_1}{dx} = v_2, \cdots, \frac{dv_{n-2}}{dx} = v_{n-1}$$

$$f(x, v_0, v_1, \dots, v_{n-1}, v'_{n-1}) = 0 \quad \text{or} \quad \frac{dv_{n-1}}{dx} = F(x, v_0, v_1, \dots, v_{n-1})$$

This can be described using vectors and matrices.

So, attention is typically focused on first-order differential equations, assuming they can be expressed as:

$$\frac{dy}{dx} = f(x, y) \text{ with initial conditions } \begin{cases} y(x_0) = a_0 \\ y'(x_0) = a_1 \end{cases}$$

Numerical resolution methods calculate $y(x)$ point by point, starting from the initial point x_0:

$$\begin{array}{ccc} x_0 & \longrightarrow & y(x_0) = a_0 \\ x_1 = x_0 + h & \longrightarrow & y_1 = a_0 + \Delta_1 \\ x_2 = x_0 + 2h & \longrightarrow & y_2 = y_1 + \Delta_2 \\ \vdots & \vdots & \vdots \\ x_p = x_0 + ph & \longrightarrow & y_p = y_{p-1} + \Delta_p \end{array}$$

Algorithms differ in how increments Δ_i of y are calculated. Two main families of algorithms exist:

1) "Separate step" algorithms: the calculation of Δ_i involves only the point (x_{i-1}, y_{i-1}).

2) "Linked-step" algorithms: the calculation of Δ_i involves all preceding terms $(x_{i-1}, y_{i-1}), (x_{i-2}, y_{i-2}), \dots, (x_0, a_0)$.

The accuracy of the results depends significantly on the step h used. A smaller h reduces the theoretical error, as well as increases rounding errors (truncation errors). Thus, the choice of h results from a compromise.

12.4.2. *Separate-step algorithms*

12.4.2.1. *Euler's algorithm*

Let M_{i-1} and M_i be two successive points on the curve $y(x)$

$$x_i = x_{i-1} + h \quad \Rightarrow \quad y_i = y_{i-1} + \Delta_i$$

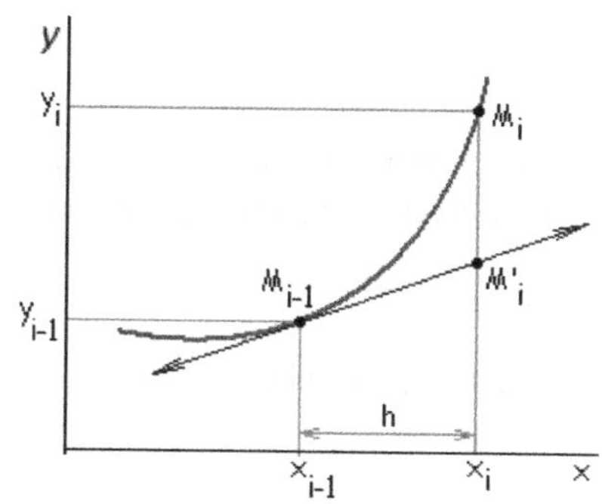

Figure 12.16. *Euler's algorithm. For a color version of this figure, see www.iste.co.uk/cochard/mathematics2.zip*

The tangent to the curve at M_{i-1} determines a point M'_i at the abscissa x_i. Euler's method involves merging the points M_i and M'_i. The tangent equation is:

$$\frac{y - y_{i-1}}{x - x_{i-1}} = y'(x_{i-1}) = f(x_{i-1}, y_{i-1})$$

so

$$y_i \approx y_{i-1} + hf(x_{i-1}, y_{i-1}), \quad \text{thus} \quad \Delta_i = hf(x_{i-1}, y_{i-1}).$$

EXAMPLE 12.9.– Consider the differential equation $y' = 1 - y$ with the initial condition $y(0) = 0$. This equation is easily solved as $y = 1 - e^{-x}$. The goal is to determine this function numerically for $x \in [0.5]$. Choosing a step size $h = 0.25$ and using a spreadsheet, the values y_i are calculated successively. The result is shown in Figure 12.17.

i	x_i	y_i	$\Delta_{i+1}=hf(x_i,y_i)$
0	0.00	0.0000	0.2500
1	0.25	0.2500	0.1875
2	0.50	0.4375	0.1406
3	0.75	0.5781	0.1055
4	1.00	0.6836	0.0791
5	1.25	0.7627	0.0593
6	1.50	0.8220	0.0445
7	1.75	0.8665	0.0334
8	2.00	0.8999	0.0250
9	2.25	0.9249	0.0188
10	2.50	0.9437	0.0141
11	2.75	0.9578	0.0106
12	3.00	0.9683	0.0079
13	3.25	0.9762	0.0059
14	3.50	0.9822	0.0045
15	3.75	0.9866	0.0033
16	4.00	0.9900	0,0025
17	4.25	0.9925	0.0019
18	4.50	0.9944	0.0014
19	4.75	0.9958	0.0011
20	5.00	0.9968	0.0008

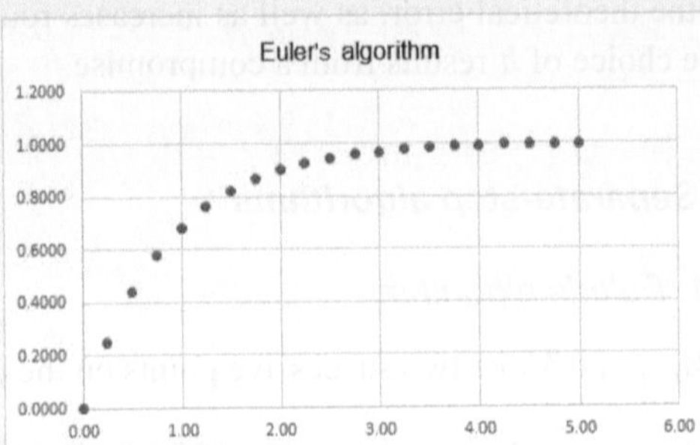

Figure 12.17. *Practical use of Euler's algorithm. For a color version of this figure, see www.iste.co.uk/cochard/mathematics2.zip*

12.4.2.2. *Parabola algorithm no. 1*

Between the points M_{i-1} and M_i, the curve $y(x)$ is approximated by the arc of a parabola.

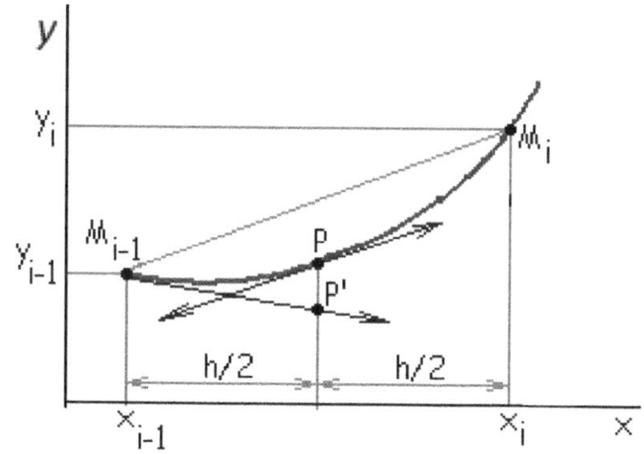

Figure 12.18. *Parabola algorithm no. 1. For a color version of this figure, see www.iste.co.uk/cochard/mathematics2.zip*

In this case, the tangent to this parabola at $x_{i-1}+\frac{h}{2}$ must be parallel to the string $M_{i-1}M_i$ (a property of the parabola). However, the corresponding point P is not directly accessible, so it is approximated by P', located on the tangent at M_{i-1}. The approximate coordinates of P are therefore:

$$x_P = x_{i-1} + \frac{h}{2}, \qquad y_P \approx y_{i-1} + \frac{h}{2} f(x_{i-1}, y_{i-1})$$

The slope of the string $M_{i-1}M_i$ is given by:

$$a = y'\left(x_{i-1} + \frac{h}{2}\right) = f\left(x_{i-1} + \frac{h}{2}, y_{i-1} + \frac{h}{2} f(x_{i-1}, y_{i-1})\right)$$

and its equation is $y - y_{i-1} = a(x - x_{i-1})$, hence:

$$y_i - y_{i-1} = hf\left(x_{i-1} + \frac{h}{2}, y_{i-1} + \frac{h}{2} f(x_{i-1}, y_{i-1})\right).$$

Thus,

$$\Delta_i = hf(x_{i-1} + \frac{h}{2}, y_{i-1} + \frac{h}{2} f(x_{i-1}, y_{i-1}))$$

EXAMPLE 12.10.– Returning to Example 12.9, apply the algorithm for Parabola no. 1. The value of Δ_i is $\Delta_i = h\left(1 + \frac{h}{2}(1 - y_{i-1})\right)$. With a step size of 0.25, the results are shown in Figure 12.19.

i	x_i	y_i	Δ_{i+1}
0	0.00	0.0000	0.2813
1	0.25	0.2813	0.2021
2	0.50	0.4834	0.1453
3	0.75	0.6287	0.1044
4	1.00	0.7331	0.0751
5	1.25	0.8082	0.0539
6	1.50	0.8621	0.0388
7	1.75	0.9009	0.0279
8	2.00	0.9288	0.0200
9	2.25	0.9488	0.0144
10	2.50	0.9632	0.0103
11	2.75	0.9736	0.0074
12	3.00	0.9810	0.0053
13	3.25	0.9863	0.0038
14	3.50	0.9902	0.0028
15	3.75	0.9929	0.0020
16	4.00	0.9949	0.0014
17	4.25	0.9964	0.0010
18	4.50	0.9974	0.0007
19	4.75	0.9981	0.0005
20	5.00	0.9986	0.0004

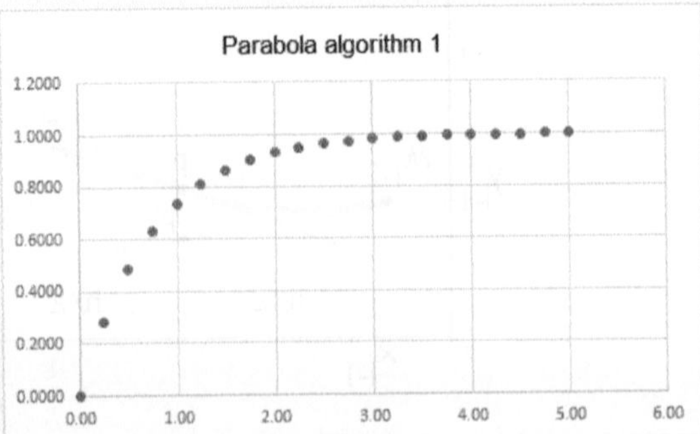

Figure 12.19. *Practical use of the parabola algorithm no. 1. For a color version of this figure, see www.iste.co.uk/cochard/mathematics2.zip*

12.4.2.3. *Parabola algorithm no. 2*

A second parabola algorithm relies on the following property: the slope of the tangent at P is half the sum of the slope of the tangent at M_{i-1} and the slope of the tangent at M_i.

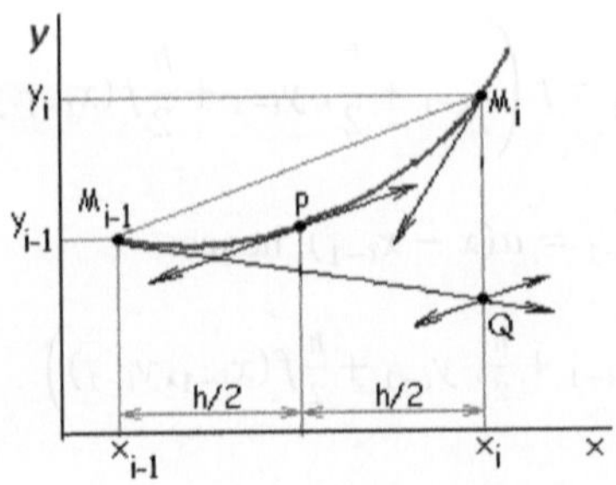

Figure 12.20. *Parabola algorithm no. 2. For a color version of this figure, see www.iste.co.uk/cochard/mathematics2.zip*

Since the latter is unknown, M_i is approximated with the point Q having coordinates:

$$x_i = x_{i-1} + h$$

$$y_i = y_{i-1} + hf(x_{i-1}, y_{i-1})$$

The corresponding slope is $f\big(x_{i-1} + h, y_{i-1} + hf(x_{i-1}, y_{i-1})\big)$. Consequently, the slope of the tangent at P (and thus of the string $M_{i-1}M_i$) is:

$$a = \frac{1}{2}\Big(f(x_{i-1}, y_{i-1}) + f\big(x_{i-1} + h,\ y_{i-1} + hf(x_{i-1}, y_{i-1})\big)\Big)$$

The string equation $M_{i-1}M_i$ is $y - y_{i-1} = a(x - x_{i-1})$. From this, it follows:

$$y_i - y_{i-1} = a(x_i - x_{i-1}).$$

Thus,

$$\Delta_i = \frac{h}{2}\Big(f(x_{i-1}, y_{i-1}) + f\big(x_{i-1} + h, y_{i-1} + hf(x_{i-1}, y_{i-1})\big)\Big)$$

EXAMPLE 12.11.– Returning to Example 12.9 and applying the algorithm for Parabola no. 2, the value of Δ_i is given by:

$$\Delta_i = \frac{h}{2}\big(2 - h(1 - y_{i-1})\big)$$

For a step size of 0.25, the results are shown in Figure 12.21.

i	x_i	y_i	Δ_{i+1}
0	0.00	0.0000	0.2188
1	0.25	0.2188	0.1709
2	0.50	0.3896	0.1335
3	0.75	0.5232	0.1043
4	1.00	0.6275	0.0815
5	1.25	0.7090	0.0637
6	1.50	0.7726	0.0497
7	1.75	0.8224	0.0389
8	2.00	0.8612	0.0304
9	2.25	0.8916	0.0237
10	2.50	0.9153	0.0185
11	2.75	0.9338	0.0145
12	3.00	0.9483	0.0113
13	3.25	0.9596	0.0088
14	3.50	0.9684	0.0069
15	3.75	0.9753	0.0054
16	4.00	0.9807	0.0042
17	4.25	0.9850	0.0033
18	4.50	0.9882	0.0026
19	4.75	0.9908	0.0020
20	5.00	0.9928	0.0016

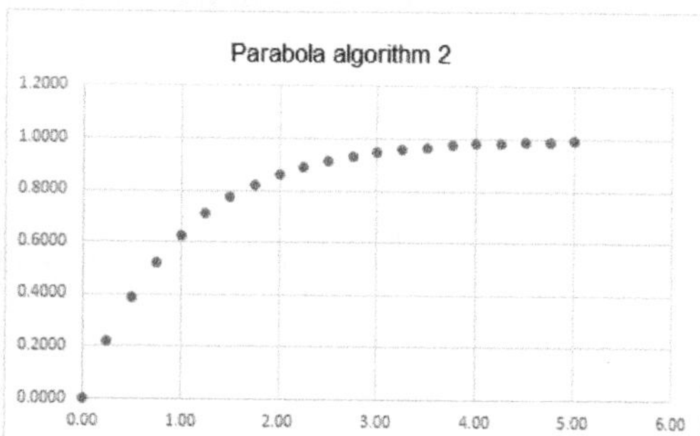

Figure 12.21. *Practical use of the parabola algorithm no. 2. For a color version of this figure, see www.iste.co.uk/cochard/mathematics2.zip*

12.4.2.4. *Runge–Kutta algorithm*

In this method, the following equation is used:

$$y_i = y_{i-1} + \sum_{n=1}^{m} R_n K_n$$

where R_n is a numerical coefficient, and K_n depends on h and $f(x, y)$:

$$K_1 = hf(x_{i-1}, y_{i-1}), \quad K_n = hf(x_{i-1} + \alpha_n h, y_{i-1} + \sum_{i=1}^{n-1} \beta_i K_i)$$

The parameters R_n, α_n, β_i, are determined to ensure that the Runge–Kutta expansion coincides with the Taylor expansion of order n:

$$y_i = y(x_{i-1}) + h\frac{dy}{dx}(x_{i-1}) + \frac{h^2}{2}\frac{d^2y}{dx^2}(x_{i-1}) + \cdots + \frac{h^n}{n!}\frac{d^ny}{dx^n}(x_{i-1}) + \varepsilon(h^n)$$

To illustrate the method, consider order 2. The Taylor expansion is:

$$y_i = y_{i-1} + h\frac{dy}{dx}(x_{i-1}) + \frac{h^2}{2}\frac{d^2y}{dx^2}(x_{i-1}) + \varepsilon(h^2)$$

This is equivalent to:

$$y_i = y_{i-1} + hf(x_{i-1}, y_{i-1}) + \frac{h^2}{2}\frac{d^2f}{dx^2}(x_{i-1}, y_{i-1}) + \varepsilon(h^2)$$

which simplifies to:

$$\begin{aligned} y_i = y_{i-1} &+ hf(x_{i-1}, y_{i-1}) \\ &+ \frac{h^2}{2}\left(\frac{\partial f}{\partial x}(x_{i-1}, y_{i-1}) + \frac{\partial f}{\partial y}(x_{i-1}, y_{i-1})\frac{dy}{dx}(x_{i-1})\right) \\ &+ \varepsilon(h^2) \end{aligned}$$

The Runge–Kutta expansion is:

$$y_i = y_{i-1} + R_1 K_1 + R_2 K_2$$
$$= y_{i-1} + R_1 h f(x_{i-1}, y_{i-1})$$
$$+ R_2 h f(x_{i-1} + \alpha_1 h, y_{i-1} + \beta_1 K_1)$$

Thus,

$$y_i = y_{i-1} + (R_1 + R_2) h f(x_{i-1}, y_{i-1}) + R_2 h^2 \alpha_1 \frac{\partial f}{\partial x}(x_{i-1}, y_{i-1})$$
$$+ R_2 h^2 \beta_1 f(x_{i-1}, y_{i-1}) \frac{\partial f}{\partial y}(x_{i-1}, y_{i-1}) + \cdots$$

By equating the two expansions:

$$R_1 + R_2 = 1, \quad R_2 \alpha_1 = \frac{1}{2}, \quad R_2 \beta_1 = \frac{1}{2}.$$

Since there are three equations for four unknowns, one of the parameters must be fixed. For instance, $\alpha_1 = 1$; hence, $R_1 = R_2 = \frac{1}{2}$, and $\beta_1 = 1$. Therefore,

$$K_1 = hf(x_{i-1}, y_{i-1}), \quad K_2 = hf(x_{i-1} + h, y_{i-1} + K_1)$$

This method is often used at order 4 with the following choice of K_n, derived through a similar calculation:

$$K_1 = hf(x_{i-1}, y_{i-1}), \quad K_2 = hf\left(x_{i-1} + \frac{h}{2}, y_{i-1} + \frac{K_1}{2}\right), \quad K_3$$
$$= hf\left(x_{i-1} + \frac{h}{2}, y_{i-1} + \frac{K_2}{2}\right),$$

and

$$K_4 = hf(x_{i-1} + h, y_{i-1} + K_3)$$

Then, the solution is:

$$y_i = y_{i-1} + \Delta_i \quad \text{with} \quad \Delta_i = \frac{1}{6}(K_1 + 2K_2 + 2K_3 + K_4)$$

EXAMPLE 12.12.– Returning to Example 12.9, the Runge–Kutta algorithm is applied. With a step size of 0.25, the results shown in Figure 12.22 are obtained.

i	x_i	y_i	K1	K2	K3	K4	Δ_{i+1}
0	0.00	0.0000	0.25	0.28	0.29	0.32	0.28
1	0.25	0.2840	0.18	0.20	0.20	0.23	0.20
2	0.50	0.4630	0.13	0.15	0.15	0.17	0.15
3	0.75	0.5973	0.10	0.11	0.11	0.13	0.11
4	1,00	0.6979	0.08	0.08	0.09	0.10	0.09
5	1.25	0.7735	0.06	0.06	0.06	0.07	0.06
6	1.50	0.8301	0.04	0.05	0.05	0.05	0.05
7	1.75	0.8726	0.03	0.04	0.04	0.04	0.04
8	2.00	0.9044	0.02	0.03	0.03	0.03	0.03
9	2.25	0.9283	0.02	0.02	0.02	0.02	0.02
10	2.50	0.9462	0.01	0.02	0.02	0.02	0.02
11	2.75	0.9597	0.01	0.01	0.01	0.01	0.01
12	3.00	0.9698	0.01	0.01	0.01	0.01	0.01
13	3.25	0.9773	0.01	0.01	0.01	0.01	0.01
14	3.50	0.9830	0.00	0.00	0.00	0.01	0.00
15	3.75	0.9872	0.00	0.00	0.00	0.00	0.00
16	4.00	0.9904	0.00	0.00	0.00	0.00	0.00
17	4.25	0.9928	0.00	0.00	0.00	0.00	0.00
18	4.50	0.9946	0.00	0.00	0.00	0.00	0.00
19	4.75	0.9960	0.00	0.00	0.00	0.00	0.00
20	5.00	0.9970	0.00	0.00	0.00	0.00	0.00

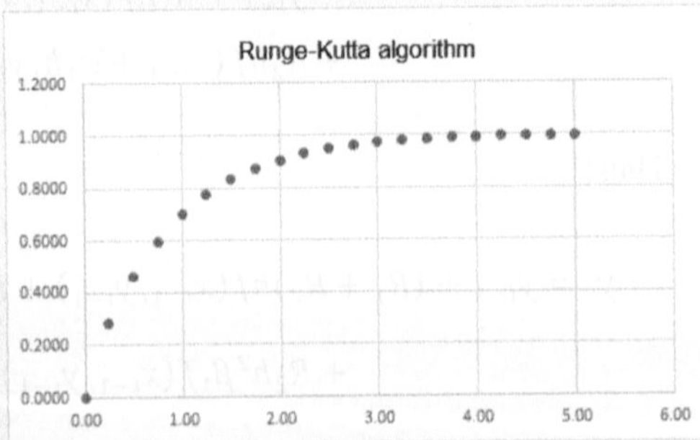

Figure 12.22. *Practical use of the Runge–Kutta algorithm. For a color version of this figure, see www.iste.co.uk/cochard/mathematics2.zip*

12.4.3. *Linked-step methods*

The preceding algorithms are split-step. To obtain a linked-step method, consider the following:

$$y_i - y_{i-1} = y(x_i) - y(x_{i-1}) = \int_{x_{i-1}}^{x_i} y'(x)\,dx = \int_{x_{i-1}}^{x_{i-1}+h} f(x, y(x))\,dx$$

Hence, $\Delta_i = \int_{x_{i-1}}^{x_{i-1}+h} f(x,y)\,dx$.

The integral can be calculated using numerical methods. There are three possible methods:

– *Prediction method*: it is assumed that the points $(x_0, y_0), (x_1, y_1)$, ..., (x_{i-1}, y_{i-1}) have already been obtained. The goal is to predict the point (x_i, y_i) from the other points. Let $F(x) = f(x, y(x))$. $F(x)$ is approximated by interpolation using Lagrange polynomials:

$$F(x) \approx P_{i-1}(x) = \sum_{k=0}^{i-1} f(x_k, y_k) L_k(x)$$

hence,

$$\Delta_i = \sum_{k=0}^{i-1} f(x_k, y_k) \int_{x_{i-1}}^{x_{i-1}+h} L_k(x)dx$$

which yields y_i.

– *Correction method*: assume $y(x_i) = y_i$ (a simple assumption). Since an extra point is available compared to the previous case, *F(x)* can be interpolated as:

$$F(x) \approx P_i(x) = \sum_{k=0}^{i} f(x_k, y_k) L_k(x)$$

hence,

$$\Delta_i = \int_{x_{i-1}}^{x_i} hL_k(x)\, dx = \sum_{k=0}^{i} f(x_k, y_k) \int_{x_{i-1}}^{x_{i-1}+h} L_k(x)\, dx = h \sum_{k=0}^{i} b_k f(x_k, y_k)$$

where the b_k are derived from the integration of Lagrange polynomials. Thus,

$$y_i = y_{i-1} + \Delta_i$$

or

$$y_i = y_{i-1} + h[b_0 f(x_0, y_0) + b_1 f(x_1, y_1) + \cdots + b_{i-1} f(x_{i-1}, y_{i-1}) + b_i f(x_i, y_i)]$$

In this equation, everything is known except what depends on y_i. This is an equation of the form $y_i = g(y_i)$, which can be solved numerically using the methods described above.

– *Prediction–correction method*: the prediction method is applied, providing an approximate value $y_i^{(0)}$ of y_i. The correction method is then used to obtain a better approximate value $y_i^{(1)} = g(y_i^{(0)})$.

12.5. Numerical solution of partial differential equations

12.5.1. *Definitions*

A partial differential equation is expressed as:

$$f\left(u, \frac{\partial u}{\partial x}, \frac{\partial u}{\partial y}, \dots, \frac{\partial^2 u}{\partial x^2}, \frac{\partial^2 u}{\partial y^2}, \dots, \frac{\partial^2 u}{\partial x \partial y}, \dots\right) = 0$$

where u is an unknown function of several variables $x, y, \cdots$.

In what follows, attention is limited to second-order linear equations of the following general form (for a function u of two variables, x and y):

$$Au_{x^2} + Bu_{xy} + Cu_{y^2} + Du_x + Eu_y + F = 0$$

where

$$u_x = \frac{\partial u}{\partial x}, \quad u_y = \frac{\partial u}{\partial y}, \quad u_{x^2} = \frac{\partial^2 u}{\partial x^2}, \quad u_{y^2} = \frac{\partial^2 u}{\partial y^2}, \quad u_{xy} = \frac{\partial^2 u}{\partial x \partial y}$$

The coefficients $A, B, \dots, F$ are functions of x and y (only functions of two variables are considered here).

In the plane, for fixed values of $A, B, \dots, F$, the equation

$$Ax^2 + Bxy + Cy^2 + Dx + Ey + F = 0$$

represents a conic curve, and its type is determined by the discriminant $B^2 - 4AC$. This leads to the following classification:

Criteria	Conical type	Equation type
$B^2 - 4AC < 0$	ellipse	elliptic equation
$B^2 - 4AC = 0$	parabola	parabolic equation
$B^2 - 4AC > 0$	hyperbola	hyperbolic equation

EXAMPLE 12.13.– Classical equations:

The Laplace equation is an elliptic equation:

$$\frac{\partial^2 u}{\partial x^2} + \frac{\partial^2 u}{\partial y^2} = 0$$

The vibrating string equation is a hyperbolic equation:

$$\frac{\partial^2 u}{\partial x^2} = \frac{1}{v^2}\frac{\partial^2 u}{\partial t^2}$$

The diffusion equation is a parabolic equation:

$$\frac{\partial^2 u}{\partial x^2} = \frac{1}{a}\frac{\partial u}{\partial t}$$

12.5.2. *Finite difference method*

12.5.2.1. *Calculation of partial derivatives*

The goal is to replace the partial derivatives of u at a point by differences in values of the function u. Consider a function $f(x)$ and seek an approximate value of $f'(x)$ at the point x_0.

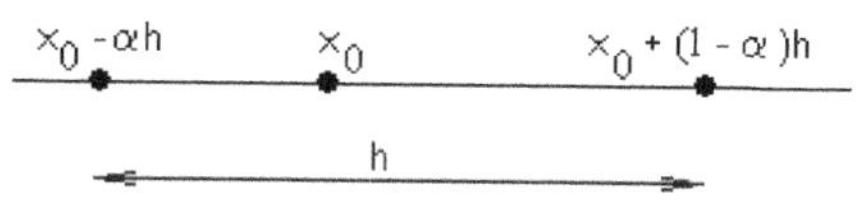

Figure 12.23. *Finding an approximate value of $f'(x)$ at point x_0*

If $x_0 - \alpha h$ and $x_0 + (1-\alpha)h$ are two points at a distance of h framing x_0 (with $\alpha \geq 0$), then:

$$f(x_0 - \alpha h) = f(x_0) - \alpha h f'(x_0) + \frac{\alpha^2}{2}h^2 f''(x_0) - \frac{\alpha^3}{6}h^3 f^{(3)}(x_0) + \cdots$$

$$f(x_0 + (1-\alpha)h) = f(x_0) + (1-\alpha)hf'(x_0) + \frac{(1-\alpha)^2}{2}h^2 f''(x_0) + \frac{(1-\alpha)^3}{6}h^3 f^{(3)}(x_0) + \cdots$$

The difference between the two expressions is:

$$f(x_0 + (1-\alpha)h) - f(x_0 - \alpha h)$$
$$= hf'(x_0) + (1-2\alpha)\frac{h^2}{2}f''(x_0)$$
$$+ [(1-\alpha)^3 + \alpha^3]\frac{h^3}{6}f^{(3)}(x_0) + \cdots$$

Thus, an approximation for $f'(x_0)$ is:

$$f'(x_0) = \frac{1}{h}[f(x_0 + (1-\alpha)h) - f(x_0 - \alpha h)] - (1-2\alpha)\frac{h}{2}f''(x_0)$$
$$- [(1-\alpha)^3 + \alpha^3]\frac{h^2}{6}f^{(3)}(x_0) + \cdots$$

Assuming $h \ll 1$, the approximation of $f'(x_0)$ simplifies to:

$$f'(x_0) = \frac{1}{h}[f(x_0 + (1-\alpha)h) - f(x_0 - \alpha h)]$$

with an error

$$|1-2\alpha|\frac{h}{2}|f''(x_0)| \quad \text{if} \quad \alpha \neq \frac{1}{2}$$

and

$$|(1-\alpha)^3 + \alpha^3|\frac{h^2}{6}|f^{(3)}(x_0)| = \frac{h^2}{24}|f^{(3)}(x_0)| \quad \text{if} \quad \alpha = \frac{1}{2}$$

The error is summarized in Figure 12.24 according to the values of α, which determine whether the difference is *forward*, *backward* or *symmetric*.

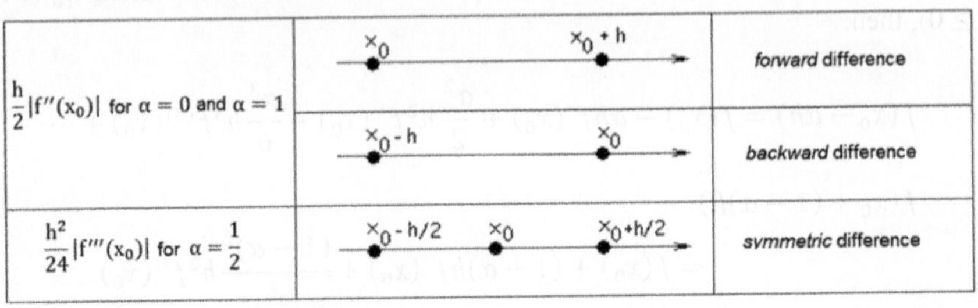

Figure 12.24. *Several types of difference*

The symmetric difference typically yields the best result and is preferred, though forward and backward differences may be simpler to use.

For the partial derivative $u_x = \frac{\partial u}{\partial x}$, with $y = y_0$ fixed, the following approximations can be used:

– Forward difference:

$$u_x \approx \frac{u(x+h,y) - u(x,y)}{h}$$

– Backward difference:

$$u_x \approx \frac{u(x,y) - u(x-h,y)}{h}$$

– Symmetric difference:

$$u_x \approx \frac{u\left(x+\frac{h}{2},y\right) - u\left(x-\frac{h}{2},y\right)}{h}$$

Similarly, for the partial derivative u_y:

– Forward difference:

$$u_y \approx \frac{u(x,y+k) - u(x,y)}{k}$$

– Backward difference:

$$u_y \approx \frac{u(x,y) - u(x,y-k)}{k}$$

– Symmetric difference:

$$u_y \approx \frac{u\left(x,y+\frac{k}{2}\right) - u\left(x,y-\frac{k}{2}\right)}{k}$$

The calculation of second partial derivatives is based on that of u_x, for example, using the symmetric difference:

$$u_{x^2} = \frac{\partial u}{\partial x} \approx \frac{u_x\left(x+\frac{h}{2},y\right) - u_x\left(x-\frac{h}{2},y\right)}{h}$$

$$= \frac{1}{h^2}[u(x+h,y) - 2u(x,y) + u(x-h,y)]$$

Similarly, for u_{y^2}:

$$u_{y^2} = \frac{\partial u}{\partial y} \approx \frac{u_y\left(x, y + \frac{k}{2}\right) - u_y\left(x, y - \frac{k}{2}\right)}{k}$$
$$= \frac{1}{k^2}\left[u(x, y + k) - 2u(x, y) + u(x, y - k)\right]$$

For the mixed partial derivative u_{xy}:

$$u_{xy} = \frac{\partial^2 u}{\partial x \partial y} \approx \frac{1}{kh}\left[u\left(x + \frac{h}{2}, y + \frac{k}{2}\right) - u\left(x + \frac{h}{2}, y - \frac{k}{2}\right) - u\left(x - \frac{h}{2}, y + \frac{k}{2}\right) + u\left(x - \frac{h}{2}, y - \frac{k}{2}\right)\right]$$

12.5.2.2. *General method*

For numerical resolution:

1) Replace the partial derivatives by the finite differences determined above.

2) Discretize the domain D of validity of the differential equation into small rectangles (mesh) of dimensions $h \times k$, at whose vertices the difference equation is applied.

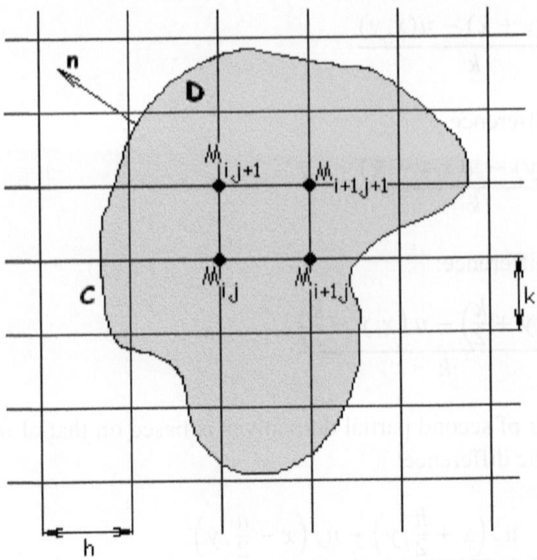

Figure 12.25. *Validity domain of the differential equation. For a color version of this figure, see www.iste.co.uk/cochard/mathematics2.zip*

In principle, there will be as many equations as there are points M_{ij}. However, these equations generally involve neighboring points, and these points must be "in" or "on" the boundary C of the domain, which limits the number of equations. This means that, in general, there will be fewer equations than unknowns.

Boundary conditions must be added to these equations (which form a linear system) corresponding to three types of problems:

– *The Dirichlet problem*: the value of $u(x, y)$ is fixed on the boundary C of the domain D: $u(x, y) = \varphi(x, y)$ for $(x, y) \in C$, where φ is a known function.

– *The Neumann problem*: the normal derivative is fixed on C:

$$\frac{du}{dn} = \psi(x, y), \text{ for } (x, y) \in C,$$

where ψ is a known function.

– *The Cauchy problem*: this combines the two previous cases.

12.5.3. *Resolution examples*

EXAMPLE 12.14.– Laplace equation.

The Laplace equation is of the elliptic type

$$\frac{\partial^2 u}{\partial x^2} + \frac{\partial^2 u}{\partial y^2} = 0$$

Using finite differences, the equation becomes:

$$\frac{1}{h^2}\{u(x+h, y) - 2u(x, y) + u(x-h, y)\} + \frac{1}{k^2}\{u(x, y+k) - 2u(x, y) + u(x, y-k)\} = 0$$

Assuming the domain D is rectangular with dimensions that are multiples of h, the domain D can be divided into squares of sides h. For each point M_{ij}, the equation becomes:

$$u_{i+1,j} - 2u_{i,j} + u_{i-1,j} + u_{i,j+1} - 2u_{i,j} + u_{i,j-1} = 0$$

This simplifies to:

$$u_{i+1,j} + u_{i-1,j} + u_{i,j+1} + u_{i,j-1} - 4u_{i,j} = 0$$

This equation involve the values of $u(x,y)$ at the points M_{ij} and its neighboring $M_{i,j-1}, M_{i,j+1}, M_{i-1,j}$ and $M_{i+1,j}$.

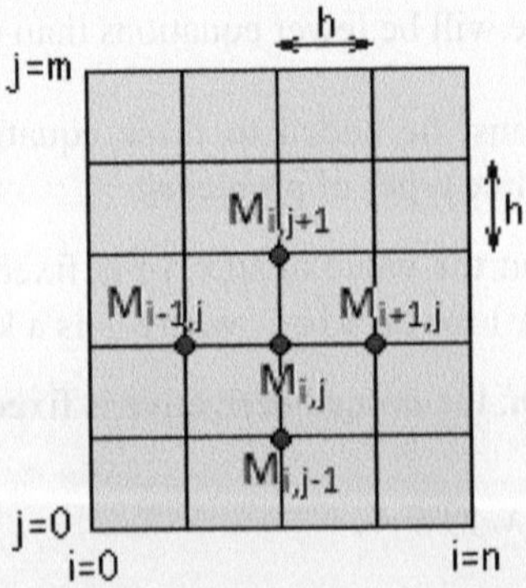

Figure 12.26. *Validity domain D. For a color version of this figure, see www.iste.co.uk/cochard/mathematics2.zip*

The equation is valid only for the interior of the domain D. Therefore, there are $(n-1)(m-1)$ equations of this type (one for each M_{ij} inside the domain) and $(n+1)(m+1)$ values u_{ij} that need to be determined. In the Dirichlet problem, the values of u on the boundary are given: $u_{ij} = \varphi_{ij}$ for $(i = 0 \text{ or } n, j = 0 \text{ or } n)$, resulting in $2(n+m)$ equations. In the end, a system of $(n+1)(m+1)$ linear equations with $(n+1)(m+1)$ unknowns are obtained.

This system can be solved iteratively using the following method. The system is written in the form:

$$u_{11} = \frac{1}{4}[u_{21} + u_{12} + \varphi_{01} + \varphi_{10}]$$

$$u_{12} = \frac{1}{4}[u_{22} + u_{13} + u_{11} + \varphi_{02}]$$

$$\ldots$$

$$u_{ij} = \frac{1}{4}[u_{i+1,j} + u_{i-1,j} + u_{i,j-1} + u_{i,j+1}]$$

$$\ldots$$

$$u_{n-1,m-1} = \frac{1}{4}[\varphi_{n,m-1} + u_{n-2,m-1} + \varphi_{n-1,m} + u_{n-1,m-2}]$$

By posing:

$$\boldsymbol{u} = \begin{bmatrix} u_{11} \\ u_{12} \\ \vdots \\ u_{n-1,m-1} \end{bmatrix} \quad \text{and} \quad \boldsymbol{A} = \begin{bmatrix} 0 & 1 & 0 & \dots & 1 & 0 & \dots \\ 1 & 0 & 1 & \dots & 0 & 1 & \dots \\ \vdots & \vdots & \vdots & \ddots & \vdots & \vdots & \ddots \\ \dots & \dots & \dots & \dots & \dots & \dots & \dots \end{bmatrix}$$

where $\frac{1}{4}\boldsymbol{A}$ is the matrix applied to the u_{ij} of the second member, and

$$\boldsymbol{\varphi} = \begin{bmatrix} \varphi_{01} + \varphi_{10} \\ \varphi_{02} \\ \vdots \\ 0 \\ \vdots \\ \varphi_{n,m-1} + \varphi_{n-1,m} \end{bmatrix}$$

is the vector of boundary terms, the system is

$$\boldsymbol{u} = \frac{1}{4}\boldsymbol{A}\boldsymbol{u} + \boldsymbol{\varphi},$$

Or equivalently

$$\boldsymbol{B}\boldsymbol{u} = (\boldsymbol{A} - 4\boldsymbol{I})\boldsymbol{u} = -\boldsymbol{\varphi}$$

The solution process starts with an initial value $u^{(0)}$:

$$\boldsymbol{u}^{(1)} = \frac{1}{4}\boldsymbol{A}\boldsymbol{u}^{(0)} + \boldsymbol{\varphi}$$

$$\boldsymbol{u}^{(2)} = \frac{1}{4}\boldsymbol{A}\boldsymbol{u}^{(1)} + \boldsymbol{\varphi}$$

...

$$\boldsymbol{u}^{(p)} = \frac{1}{4}\boldsymbol{A}\boldsymbol{u}^{(p-1)} + \boldsymbol{\varphi}$$

Convergence is ensured because the "weak" sufficient condition is verified: $|B_{ii}| \geq \sum_{j \neq i} |B_{ij}|$, and there exists an index r such that $|B_{rr}| > \sum_{j \neq r} |B_{rj}|$.

The diagonal terms of $\boldsymbol{B}$ are -4.

In each row where no boundary term is involved, the sum of the non-diagonal terms is 4, so $|B_{ii}| \geq \sum_{j\neq i} |B_{ij}|$.

In a row where a boundary term is involved (e.g. the first row), the following condition holds:

$$|B_{11}| = 4 > \sum_{j\neq 1} |B_{1j}| = 2.$$

EXAMPLE 12.15.– Vibrating equation for vibrating strings.

This is a hyperbolic equation, known as the string equation:

$$\frac{\partial^2 u}{\partial x^2} = \frac{1}{v^2}\frac{\partial^2 u}{\partial t^2}$$

The function $u(x,t)$ represents the amplitude of the string at point x and at time t. Assuming the string is fixed at $x = 0$ and $x = l$, the domain D is a rectangular strip, with $x \in [0.1]$ and $t \geq 0$.

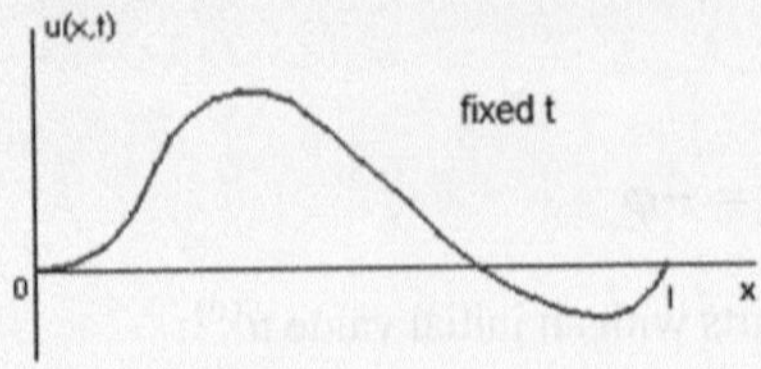

Figure 12.27. *Vibrating string at fixed t. For a color version of this figure, see www.iste.co.uk/cochard/mathematics2.zip*

The boundary conditions are as follows:

– At the ends of the string: $u(0,t) = u(l,t) = 0$.

– Initial position at $t = 0 : u(x,0) = f(x)$.

– Initial velocity at $t = 0: \frac{\partial u}{\partial t}(x,0) = g(x)$ with $g(0) = g(l) = 0$.

Next, a mesh of the domain D is created using rectangles of dimensions $h \times k$ (see Figure 12.28).

Using finite differences, the equation is written as:

$$\lambda^2[u(x+h,t) - 2u(x,t) + u(x-h,t)] = u(x,t+k)$$
$$-2u(x,t)u(x,t-k)$$

or equivalently:

$$u_{i,j+1} = \lambda^2 u_{i+1,j} + \lambda^2 u_{i-1,j} + 2(1-\lambda^2)u_{i,j} - u_{i,j-1}$$

with $\lambda = \frac{kv}{h}$. Thus, for each point M_{ij} inside the domain:

$$\lambda^2 u_{i+1,j} + \lambda^2 u_{i-1,j} - u_{i,j+1} - u_{i,j-1} + 2(1-\lambda^2)u_{i,j} = 0$$

This equation determines the amplitude of the string at time $j+1$ as a function of the amplitudes at the previous times j and $j-1$.

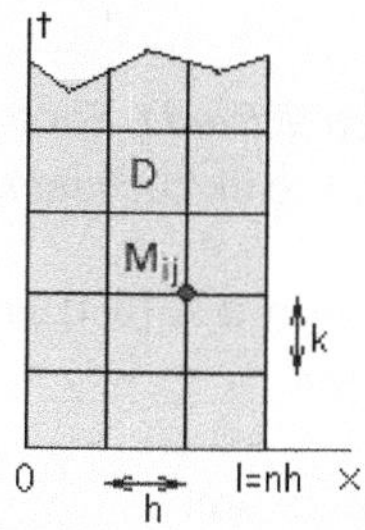

Figure 12.28. *Scope of validity. For a color version of this figure, see www.iste.co.uk/cochard/mathematics2.zip*

The solution is obtained step by step:

At $t = 0, j = 0$, the initial condition is $u_{i0} = 0$ for $i = 0$ or $i = n$; otherwise, $u_{i0} = \mathrm{f}(\mathrm{x_i})$.

At $= k, j = 1$, the initial velocity $\frac{\partial u}{\partial t}(x,0) = g(x)$ is approximated by $\frac{u_{i1}-u_{i0}}{k} = g_i$, so $u_{i1} = u_{i0} + kg_i$.

At $t = 2k\ j = 2$, the equation becomes:

$$u_{i2} = \lambda^2 u_{i+1,1} + \lambda^2 u_{i-1,1} + 2(1-\lambda^2)u_{i1} - u_{i0}$$

And this process continues for subsequent time steps.

EXAMPLE 12.16.– Diffusion equation.

The diffusion equation, also known as the heat equation, is of the parabolic type with $a > 0$:

$$\frac{\partial^2 u}{\partial x^2} = \frac{1}{a}\frac{\partial u}{\partial t}$$

The function $u(x,t)$ represents, for example, the instantaneous temperatureat time t at a section located at x of a homogeneous straight bar whose ends are kept at constant temperature θ and θ' ($\theta > \theta'$, for example).

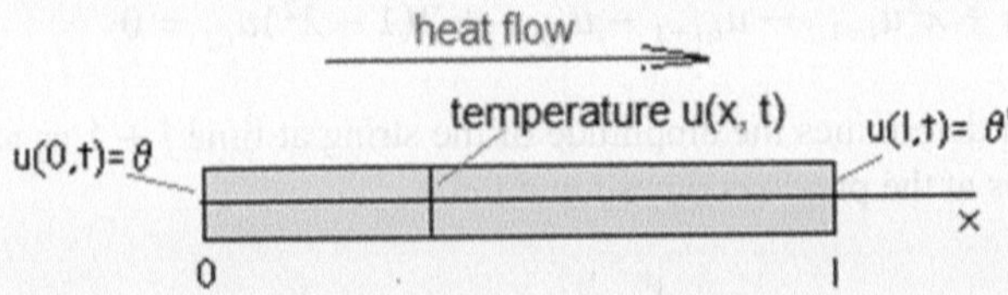

Figure 12.29. *Situation at fixed t. For a color version of this figure, see www.iste.co.uk/cochard/mathematics2.zip*

The domain D is a rectangular strip $x \in [0,1]$ and $t \geq 0$. It is decomposed into small rectangles of dimensions $h \times k$.

Using finite differences, the equations are:

$$\frac{\partial^2 u}{\partial x^2} = \frac{1}{h^2}[u(x+h,t) - 2u(x,t) + u(x-h,t) \text{ (symmetric case)}$$

$$\frac{\partial u}{\partial t} = \frac{1}{k}[u(x,t+k) - u(x,t) \text{ (forward case)}$$

At each point M_{ij} inside domain D, the equation becomes:

$$a\left[u_{i+1,j} - 2u_{i,j} + u_{i-1,j}\right] = u_{i,j+1} - u_{i,j}$$

with $a = \frac{ak}{h^2}$, leading to:

$$u_{i,j+1} = au_{i+1,j} + (1-2a)u_{i,j} + au_{i-1,j}$$

This means that the function u at time $t + k$ is expressed as a function of u at time t. Therefore, the solution can be computed step by step.

At $t = 0,\ j = 0$, the initial condition is:

$$u_{i0} = \begin{cases} \theta & \text{if } \ i = 0 \\ \theta' & \text{if } \ i = n \end{cases}$$

At $t = k, j = 1$, the boundary condition is:

$$u_{i1} = \begin{cases} \theta & \text{if } i = 0 \\ \alpha u_{i+1,0} + (1-2\alpha)u_{i1} + \alpha u_{i-1,1} & \text{if } i \neq 0 \text{ or } i \neq n \\ \theta' & \text{if } i = n \end{cases}$$

And this process continues for subsequent steps.

12.6. Appendices

12.6.1. *Dichotomy method*

JavaScript

```
function f1(x)
{
    f=1000*(1.+x)*(Math.pow(1.+x,5)-1.)/x-6000.;
    return f;
}
dichotomy function(a,b,epsilon)
{
    x= (a+b)/2;
    N=0;
    while(b - a > epsilon)
    {
        if(f1(a)*f1(x) < 0) b=x;
        else a = x;
        x=(a+b)/2;
        N=N+1;
    };
    document.getElementById('rate').value =100*x;
    document.getElementById('iterations').value =N;
}
```

HTML

```
<table border="1" cellpadding="5" cellspacing="0" align="center" style="font-size:14px;">
  <tr>
    <td colspan="2" align="center">
      <input type="button" value="calculate" onClick="dichotomy(0.,0.1,0.000000000001);" style="cursor:pointer;">
    </td>
  </tr>
  <tr>
    <td>solution : </td>
    <td align="center">
      <input type="text" id="rate" value="" style="text-align:right;">
    </td>
  </tr>
  <tr>
    <td>iterations : </td>
    <td align="center">
      <input type="text" id="iterations" value="" style="text-align:right;">
    </td>
  </tr>
</table>
```

12.6.2. *Iterative method*

JavaScript

```
function g(x)
{
 G=Math.sin(x)+0.25;
 return G;
}
iteration         function(x0,
epsilon)
{
 k=0;
 x=Array();
 x[0]=x0;
 x[1]=g(x[0]);
 while(Math.abs(x[k+1]-x[k])
 >epsilon)
 {
   k=k+1;
   x[k+1]=g(x[k]);
 }
 document.getElementById
 ('solution').value =x[k+1];
 document.getElementById
 ('iterations2').value =k;
}
```

HTML

```
<table border="1" cellpadding="5"
cellspacing="0" align="center"
style="font-size:14px;">
  <tr>
    <td colspan="2" align="center">
    <input type="button" value=
    "calculate" style=
    "cursor:pointer;" onClick=
    "iteration(0,0.001);">
    </td>
  </tr>
  <tr>
    <td>solution : </td>
    <td align="center">
      <input type="text"
      id="solution" value=""
      style="text-align:right;">
    </td>
  </tr>
  <tr>
    <td>iterations : </td>
    <td align="center">
     <input type="text"
     id="iterations2" value=""
     style="text-align:right;">
    </td>
  </tr>
</table>
```

12.6.3. *Secant method*

JavaScript

```
function f2(x)
{
 f=(x-4)/(x+1)+x;
```

HTML

```
<table border="1" cellpadding="5"
cellspacing="0" align="center"
style= "font-size:14px;">
```

```javascript
  return f;
}
secant
function(fixed,mobile,epsilon)
{
  x=Array();
  k=0;
  x[0]=mobile;
  do
  {
     k=k+1;
     x[k]=fixed-
     f2(fixed)*(x[k-1]-
     fixed)/(f2(x[k-1])-
     f2(fixed);
     delta=Math.abs(x[k]-x[k-
     1]);
  }while (delta> epsilon);
  document.getElementById
  ('solution3').value=x[k];
  document.getElementById
  ('iterations3').value=k;
}
```

```html
  <tr>
    <td colspan="2" align=
     "center">
      <input type="button"
       value="calculate"
       onClick="secant(0,2,
      0.000001);"
       style="cursor:pointer;">
    </td>
  </tr>
  <tr>
    <td>solution : </td>
    <td align="center">
       <input type="text"
        id="solution3" value=""
        style="text-align:right;">
    </td>
  </tr>
  <tr>
    <td>iterations:</td>
    <td align="center">
       <input type="text"
        id="iterations3" value=""
        style="text-align:right;">
    </td>
  </tr>
</table>
```

12.6.4. *Tangent method*

JavaScript

```javascript
function f2p(x)
{
  d=5/Math.pow(x+1,2) + 1..;
  return d;
}
tangent function
(fixed,epsilon)
{
```

HTML

```html
<table border="1" cellpadding="5"
cellspacing="0" align="center"
style="font-size:14px;">
  <tr>
     <td colspan="2"
     align="center">
        <input type="button"
        value="calculate"
```

``
x=Array();
x[0]=fixed;
k=0;
do
{
  k=k+1;
  x[k]=x[k-1]-f2(x[k-
  1])/f2p(x[k-1]);
  delta=Math.abs(x[k] -
  x[k-1]);
}while(delta>epsilon);
document.getElementById('sol
ution4').value=x[k];
document.getElementById('ite
rations4').value=k;
}
``
	``
	      onClick="tangent(0,0.0000
	      01);"
	      style="cursor:pointer;">
	    </td>
	  </tr>
	  <tr>
	    <td>solution : </td>
	    <td align="center">
	      <input type="text"
	      id="solution4" value=""
	      style="text-
	      align:right;">
	    </td>
	  </tr>
	  <tr>
	    <td>iterations:</td>
	    <td align="center">
	      <input type="text"
	      id="iterations4" value=""
	      style="text-
	      align:right;">
	    </td>
	  </tr>
	</table>
	``

12.6.5. *Monte Carlo method, Example 12.7*

JavaScript	HTML

``
unifo function(a,b)
{
  var z=Math.random() ;
  alf=(b-a)*z+a ;
  return alf ;
}
montecarlo2 function(N)
{
  var canvas =
  document.getElementById("canvas2"
  );
``
	``
	<table border="0"
	width="100%">
	  <tr>
	    <td>
	      <form
	      name="simulation3">
	        <table border="0"
	        cellpadding="5"
	        align="center">
	          <tr>
	            <td>Number of
	``

```
var ctx =
canvas.getContext("2d");
N=parseInt(N);
ctx.globalCompositeOperation =
"destination-over";
var image = new Image();
image.onload = function() {
   var
   pattern=ctx.createPattern(imag
   e,'no-repeat');
    ctx.rect(0,0,365,276);
   ctx.fillStyle=pattern;
   ctx.fill();
}
image.src =
'images_S5/fig13.png';
ctx.beginPath();
ctx.fillStyle="#FF0000";
ctx.font = "16pt Arial";
ctx.textAlign = 'center';
ctx.textBaseline = 'middle';
pN=0;
for(i=1;i<N+1;i++)
{
   x=unifo(0,1);
   y=unifo(0,0.8);
   e=x*Math.log(1+x);
   if(y < e)
   {
     pN=pN+1;
   }
   xcan=parseInt(Math.round(325*x
   +30,0));
   ycan=parseInt(Math.round(-
   296.25*y+246,0));
   ctx.fillRect(xcan-1,ycan-
   1,2,2);
}
ctx.closePath();
s=parseFloat(pN*0.8/N);
return s;
```

```
        points : </td>
        <td><input
        type="text"
        name="nbpoints"
        value=""></td>
      </tr>
      <tr>
        <td><span
        class="Style5">
        Estimating the
        value of <span
        class="Style1">I
        </span> :
        </span></td>
        <td><input
        type="text"
        name="I" value=""
        disabled="disabled
        "></td>
      </tr>
  </table>
  <br>
  <div align="center">
    <input type="button"
    name="simulate"
    class="validate"
    value="Simulate"
    onClick="I.value=
    montecarlo2(nbpoints.
    value);">
       
      
    <input type="button"
    name="reset"
    class="valider"
    value="Reset"
    onClick="I.value='';
    nbpoints.value='';
    erase_canvas('canvas2
    ');">
  </div>
```

}
erase_canvas function(id)
{
var canvas =
document.getElementById(id);
var ctx =
canvas.getContext("2d");
ctx.clearRect(0,0,canvas.width,
canvas.height);
}
	</form>
	</td>
	<td>
	<div align="center">
	<canvas id="canvas2"
	width="365"
	height="276"
	style="border:1px solid
	#000000;background-
	color:#EEEEEE;">
	</canvas>
	</div>
	</td>
	</tr>
	</table>

12.6.6. *Monte Carlo method, Example 12.8*

JavaScript	HTML

unifo function(a,b)
{
var z=Math.random() ;
alf=(b-a)*z+a ;
return alf ;
}
montecarlo3 function(N)
{
var canvas =
document.getElementById("canvas
3");
var ctx =
canvas.getContext("2d");
N=parseInt(N);a = 60;
ctx.globalCompositeOperation =
"destination-over";
var image = new Image();
image.onload = function() {
var
pattern=ctx.createPattern(imag
	<table border="0" width="100%">
	<tr>
	<td>
	<form name="simulation3">
	<table border="0"
	cellpadding="5"
	align="center">
	<tr>
	<td>Number of points :
	</td>
	<td><input type="text"
	name="nbp"
	value=""></td>
	</tr>
	<tr>
	<td>Estimation of the
	value of α :
	</td>
	<td><input type="text"
	name="A" value=""

```
    e,'no-repeat');
    ctx.rect(0,0,300,300);
    ctx.fillStyle=pattern;
    ctx.fill();
  }
  image.src =
  'images_S5/fig15.png';
  ctx.beginPath();
  ctx.fillStyle="#0000FF";
  pN=0;
  for(i=1;i<N+1;i++)
  {
   x=unifo(0,a);
   y=unifo(0,a);
   if((x*x+y*y)*(x*x+y*y) <
   a*a*(x*x-y*y)) pN=pN+1;
   xcan=Math.round(37.5+3.75*x,0)
   ;
   ycan=Math.round(-
   3.75*y+262.5,0);
   ctx.fillRect(xcan-1,ycan-
   1,2,2);
  }
  ctx.closePath();
  GA=parseFloat(pN/N);
  return GA;
}
erase_canvas function(id)
{
  var canvas =
  document.getElementById(id);
  var ctx =
  canvas.getContext("2d");
  ctx.clearRect(0,0,canvas.width,
  canvas.height);
}
```

```
          disabled="disabled"></t
          d>
        </tr>
      </table>
      <br>
      <div align="center">
        <input type="button"
        name="simulate"
        class="valider"
        value="Simulate"
        onClick="A.value=
        montecarlo3(nbp.value);">
            &
        nbsp;
        <input type="button"
        name="reset"
        class="validate"
        value="Reset"
        onClick="A.value='';
        nbp.value='';
        erase_canvas('canvas3');"
        >
      </div>
    </form>
  </td>
  <td>
    <div align="center"><canvas
    id="canvas3" width="300"
    height="300"
    style="border:1px solid
    #000000;background-
    color:#EEEEEE;"></canvas></d
    iv></td>
  </tr>
</table>
```

References

[BEL 98] BELLANGER M., *Traitement numérique du signal. Théorie et pratique*, Dunod, Paris, 1998.

[BER 14] BERGOUNIOUX M., *Mathématiques pour le traitement du signal*, Dunod, Paris, 2014.

[BRO 82] BRONSON R., *Operations Research*, Schaum/McGraw Hill, New York, 1982.

[CHA 75] CHARET G., *Cours d'analyse numérique*, SEDES-Informatique, 1975.

[CHO 17] CHOUINARD J.Y., Systèmes de communications, GEL-3006 (82878), Course notes, available at: http://wcours.gel.ulaval.ca/2017/a/GEL3006/default/5notes/GEL-3006_2017_09_introduction.pdf, 2017.

[COS 89] COSTER M., CHERMANT J.L., *Précis d'analyse d'image*, Presses du CNRS, Paris, 1989.

[COT 97] COTTET F., *Traitement des signaux et acquisition de données*, Dunod, Paris, 1997.

[COU 92] COUWENBERGH J.P., *L'indispensable pour maitriser la couleur*, Marabout, Paris, 1992.

[DUR 60] DURAND E., *Solutions numériques des équations algébriques*, Masson, Paris, 1960.

[FOL 91] FOLEY J.D., VAN DAM A., FEINER S.K., HUGHES J.F., *Computer Graphics, Principles and Practice*, Addison Wesley, New York, 1991.

[FON 83] FONTOLLIET P.G., *Systèmes de télécommunications. Bases de transmission*, Dunod, Paris, 1983.

[GAR 84] GARDANS Y., LUCAS M., *Techniques graphiques interactives et CAO*, Hermes, Paris, 1984.

[GON 93] GONZALES R.C., WOODS R.E., *Digital Image Processing*, Addison Wesley, New York, 1993.

[HAY 99] HAYES M.H., *Digital Signal Processing*, Schaum/McGraw Hill, New York, 1999.

[HEA 97] HEARN D., BAKER M.P., *Computer Graphics*, Prentice Hall, Upper Saddle River, 1997.

[HOL 94] HOLSINGER E., *Le multimédia*, Dunod, Paris, 1994.

[KUN 93] KUNT M., GRANDLUND G., KOCHER M., *Traitement numérique des images*, Presses Polytechniques et Universitaires Romandes, Lausanne, 1993.

[LEL 00] LELANDAIS S., Initiation au traitement numérique de l'image, Leaflet, Université Paris VIII, 2000.

[LEP 91] LEPECQ C.H., RIMOUX N., *Les fichiers graphiques sur ordinateur*, Armand Colin, Paris, 1991.

[LIP 83] LIPSCHUTZ S., *Mathématiques pour informaticiens*, Schaum-McGraw Hill, Paris, 1983.

[MAC 87] MACCHI C., GUIBERT J.-F., *Téléinformatique*, Dunod, Paris, 1987.

[MUL 12a] MULLER J.-P., jipehemphyap, Radiocommunications, YouTube video, available at: https://www.youtube.com/watch?v=QwRtf4qSOyk, 2012.

[MUL 12b] MULLER J.-P., jipehemphyap, Modulation d'amplitude, YouTube video, available at: https://www.youtube.com/watch?v=Q1HUnpl9zE4, 2012.

[MUL 12c] MULLER J.-P., jipehemphyap, Modulation de fréquence, YouTube video, available at: https://www.youtube.com/watch?v=izmpoR1d4X4&t=509s, 2012.

[PAP 77] PAPOULIS A., *Signal Analysis*, McGraw Hill, New York, 1977.

[PLA 87] PLASTOCK R.A., KALLEY G., *Infographie*, Schaum/McGraw Hill, Paris, 1987.

[POL 76] POLY J., POULAIN P., *Initiation à l'informatique*, Dunod/Bordas, Paris, 1976.

[RIM 93] RIMMER S., *Gérer vos fichiers graphiques*, Addison Wesley, Paris, 1993.

[STA 07] STALLING W., *Data and Computer Communications*, Prentice Hall, New York, 2007.

[THO 14] THONKY GUIDES, QR code tutorial, available at: https://www.thonky.com/qr-code-tutorial/, 2014.

[TIS 08] TISSERAND E., PAUTEX J.-F., SCHWEITZER P., *Analyse et traitement des signaux. Méthodes et applications au son et à l'image*, Dunod, Paris, 2008.

[TOU 87] TOUMAZET J.-J., *Traitement de l'image sur micro-ordinateur*, SYBEX, Paris, 1987.

[VEL 19] VELU J., *Méthodes mathématiques pour l'informatique*, Dunod, Paris, 2019.

List of Authors

Gérard-Michel COCHARD
Eco-PRocédés Optimisation et Aide à la Décision (EPROAD, EA 4669)
Université de Picardie Jules Verne
Amiens
France

Mhand HIFI
Eco-PRocédés Optimisation et Aide à la Décision (EPROAD, EA 4669)
Université de Picardie Jules Verne
Amiens
France

Index